Springer Geology

The book series Springer Geology comprises a broad portfolio of scientific books, aiming at researchers, students, and everyone interested in geology. The series includes peer-reviewed monographs, edited volumes, textbooks, and conference proceedings. It covers the entire research area of geology including, but not limited to, economic geology, mineral resources, historical geology, quantitative geology, structural geology, geomorphology, paleontology, and sedimentology.

More information about this series at http://www.springer.com/series/10172

Oleg V. Petrov · Shuwen Dong
Editors

Tectonics of Asia (Northern, Central and Eastern Asia)

Springer

Editors
Oleg V. Petrov
Russian Geological Research Institute
Saint Petersburg, Russia

Shuwen Dong
SinoProbe Center
Chinese Academy of Geological Sciences
Beijing, China

School of Geoscience and Engineering
Nanjing University
Nanjing, China

ISSN 2197-9545 ISSN 2197-9553 (electronic)
Springer Geology
ISBN 978-3-030-62000-4 ISBN 978-3-030-62001-1 (eBook)
https://doi.org/10.1007/978-3-030-62001-1

This Springer imprint is published by the registered company Springer Nature Switzerland AG
The registered company address is: Gewerbestrasse 11, 6330 Cham, Switzerland

Contents

Tectonic Model of Northern, Central and Eastern Asia

O. V. Petrov, S. P. Shokalsky, A. F. Morozov, I. I. Pospelov, S. N. Kashubin, and Dong Shuwen

Abstract The chapter focuses on the Tectonic map of Northern, Central and Eastern Asia at 1: 2.5 M scale, which reflects the current state of knowledge on tectonics of this vast area and which was compiled in the framework of the International project "3D Geological Structures and Metallogeny of Northern, Central and Eastern Asia" initiated in 2002 by the geological surveys of five countries: Russia, China, Mongolia, Kazakhstan and the Republic of Korea. The charter outlines a brief history of the tectonic map compilation, principles of legend compilation and its description, describes the scheme of tectonic zoning of Northern, Central and Eastern Asia (Central Asian fold belt and adjacent tectonic structures).

1 History of Compilation of the Tectonic Map of Northern-Central-Eastern Asia and Adjacent Areas at Scale 1:2,500,000

New digital *Tectonic Map of Northern-Central-Eastern Asia and Adjacent Areas at scale 1:2,500,000* (Fig. 1) was compiled by A. P. Karpinsky Russian Geological Research Institute (CGMW Subcommission for Northern Eurasia) and Geological Institute, Russian Academy of Sciences, (CGMW Subcommission for Tectonic Maps) in the framework of the International project "3D Geological Structures and Metallogeny of Northern, Central and Eastern Asia". In this project the Russian party (years 2007–2013) was represented by the Federal Agency on Mineral Resources

O. V. Petrov (✉) · S. P. Shokalsky · I. I. Pospelov · S. N. Kashubin
Russian Geological Research Institute (VSEGEI), 74 Sredny Prospect, St. Petersburg 199106, Russia
e-mail: vsgdir@vsegei.ru

A. F. Morozov
Federal Agency On Mineral Resources (Rosnedra), 4/6 Gruzinskaya, Moscow 125993, Russia

D. Shuwen
PI of SinoProbe Center, Nanjing University, Chinese Academy of Geological Sciences, No. 26 Baiwanzhuang street, Beijing 100037, China

O. V. Petrov and S. Dong (eds.), *Tectonics of Asia (Northern, Central and Eastern Asia)*, Springer Geology, https://doi.org/10.1007/978-3-030-62001-1_1

Fig. 1 Tectonic map of Northern—Central—Eastern Asia and Adjacent Areas at scale 1:2,500,000 (2014)

of the Ministry of Natural Resources and Environment, VSEGEI, GIN RAS and geological institutes of the Siberian Branch of the Russian Academy of Sciences. The geological surveys of China, Mongolia, Kazakhstan, and the Republic of Korea were the foreign participants. Compiled maps of the Atlas cover the territory of the Asian part of Russia (including Urals and Pre-Uralian region), Kazakhstan, China, Mongolia, Korean Peninsula, and Central Asian republics (Kyrgyzstan, Uzbekistan, Tajikistan, and Turkmenistan).

History of creation of the *Tectonic map of Northern-Central-Eastern Asia and Adjacent Areas at scale 1:2,500,000* originates from 2001; it has become the second tectonic map compiled in a result of multipartite scientific cooperation between Russia, China, Mongolia, Kazakhstan, and the Republic of Korea.

In July 2001 in St. Petersburg, VSEGEI, General Director of VSEGEI Dr. O. Petrov and the then President of the Chinese Academy of Geological Sciences Dr. Zhang Yanying signed the bilateral agreement about cooperative works on compilation of the geological maps of different contents.

The **first** workshop meeting took place in Beijing in October 2002. During it, the representatives of four countries (China, Russia, Mongolia, and Kazakhstan) discussed the organizational problems. There have been determined: the title of project *Atlas of Geological Maps of Central Asia and Adjacent Areas*, map scales—1:2,500,000, map territories and boundaries, content of topographic base map and others. Very important decision was accepted about the Atlas structure. The Atlas must consist of four maps: geological, tectonic, metallogenic and mineragenic fuel-energy resources ones. Also the responsible executors have been determined for maps of the Atlas. The Russian party presented by the Federal Agency on Mineral Resources of the Ministry of Natural Resources of the Russian Federation (Rosnedra) and A. P. Karpinsky Russian Geological Research Institute (VSEGEI) would be responsible for tectonic and metallogenic maps, and the Chinese party (Chinese Academy of Geological Sciences—CAGS), for two another maps.

At the **second** workshop meeting in October 2003 (St. Petersburg, VSEGEI) already the particular problems of the map compilation were discussed. In that time, the intimate scientific cooperation between VSEGEI and Geological Institute RAS on the tectonic map creation started. Geological Institute RAS continued to work under the project of the Commission for the Geological Map of the World (CGMW) *International Tectonic map of Asia, 1:5,000,000—ITMA-5000*, which was realized

Fig. 2 October 2002, Beijing. Singing of the agreement on the starting of the International (China, Russia, Kazakhstan, Mongolia) project "*Atlas of Geological maps of Central Asia and Adjacent Areas at scale 1:2,500,000*"

under the leadership of CGMW Subcommission for Tectonic maps. To that time, the Subcommission for Tectonic maps had worked out and approved the legend for ITMA-5000, and collected numerous author's drafts for continental (mainland) part of Asia.

This ITMA-5000 legend has been accepted as a basis for the future *Tectonic map of Central Asia and Adjacent Areas*. This meeting has made a decision to increase legend for the continental part of Asia in accordance with the map scale and also to exclude a part of legend concerning sea and oceanic water areas. Some regional drafts for tectonic map were transferred from the Geological Institute RAS to VSEGEI for compiling the first general draft of the map.

At the **third** workshop meeting in May 2004 (Ulan Bator, Mongolia) (Fig. 2) project participants made a decision about the first demonstration of four maps of the Atlas at the forthcoming International Geological Congress. Here, in Ulan Bator, geologists from the Republic of Korea (Korea Institute of Geoscienc and Mineral Resources—KIGAM) joined this international project.

For the first time the draft tectonic map was shown at the 32nd Session of the International Geological Congress in August 2004 in Florence (Italy) at the Exhibition GEOEXPO-2004 at booths of the Federal Agency on Mineral Resources, Ministry of Natural Resources of the Russian Federation. Tectonic map with different degree of detail showed the majority of fold belts and areas, whereas the structures of numerous sedimentary basins and platform covers had to be worked out in greater detail.

Renewal of membership of CGMW Subcommission for Northern Eurasia had a decisive importance in further work under tectonic map: at the CGMW Bureau Meeting and General Assembly, Dr. O. Petrov was elected CGMW Vice-President for Northern Eurasia and Dr. S. Shokalsky, Secretary General of this Subcommission. From this moment, teamwork of VSEGEI and Geological Institute RAS in compilation of the tectonic map has been carried out not only in the framework of the five-country international project, but also under the aegis of UNESCO Commission for the Geological map of the World (CGMW). After the **forth** workshop meeting (April 2005, Borovoe, Kazakhstan) executive editors of the tectonic map began to receive new separated regional drafts from the responsible authors of another countries, and also the proposals for legend improvement and corrections for map demonstrated earlier at the Geological Congress.

Thus, to the end of 2005, almost whole territory of the tectonic map was covered by regional map drafts. The rest "white stains" of the map mainly concerned to the sedimentary basins—Junggar, Tarim, Turpan-Hami, Qaidam, Huabei, Songliao, number of the Russian Far East depressions, sedimentary platform covers of North China and Yangtze cratons. But the West Siberian Basin and Turan epi-Variscan platform claimed special attention because on the first map draft only schematic structure of the basements was shown. Participants from Korean Institute of Geosciences and Mineral Resources (KIGAM) promised to create the tectonic map draft for the Korean Peninsula in June 2006 before starting of the **fifth** workshop meeting in Daejeon (Republic of Korea). The work out on the tectonic map, as on other maps of the Atlas, accompanied by cooperative geological field investigations which took place not only in time of the workshop meetings but also in specially organized field trips for decision of problems of geological and geochronological correlations. In June 2005, project participants visited the key regions of the Yangtze Platform where studied in detail features of the Precambrian sedimentary successions and then the Mesoproterozoic-Neoproterozoic sedimentary cover of North China craton near Beijing (Jixian section). In July 2006, project members could compare the Precambrian successions of both Chinese cratons with the sections of Pre-Uralian part of East European platform (Bashkirian anticlinorium, Republic of Bashkortostan). Before Bashkirian field trip in June 2006, participants visited the outcrops of the Yangtze Platform basement in the southern part of the Korean Peninsula and studied its structure. Later, such comparison has been used in revision of both geological and tectonic maps of Atlas (see Chap. 3).

At the end of 2006, the tectonic map began to look like a united whole map (see Fig. 3). The above mentioned territories of sedimentary basin and covers of platforms were covered by map drafts, their structures and relationships with surrounding fold belts and areas were shown. Map was supplemented by draft for the Korean Peninsula. Also the structural features of Tarim and Ordos sedimentary covers, Junggar Basin, depressions of Northeast China and Russian Far East were shown in detail.

Workshop meeting of the project participants in December 2006 in Shanghai became a significant moment in compilation of the *Tectonic map of Northern-Central-Eastern Asia and Adjacent Areas*. The progress in realization of the project *Atlas of Geological Maps of Central Asia and Adjacent Areas* and compilation of

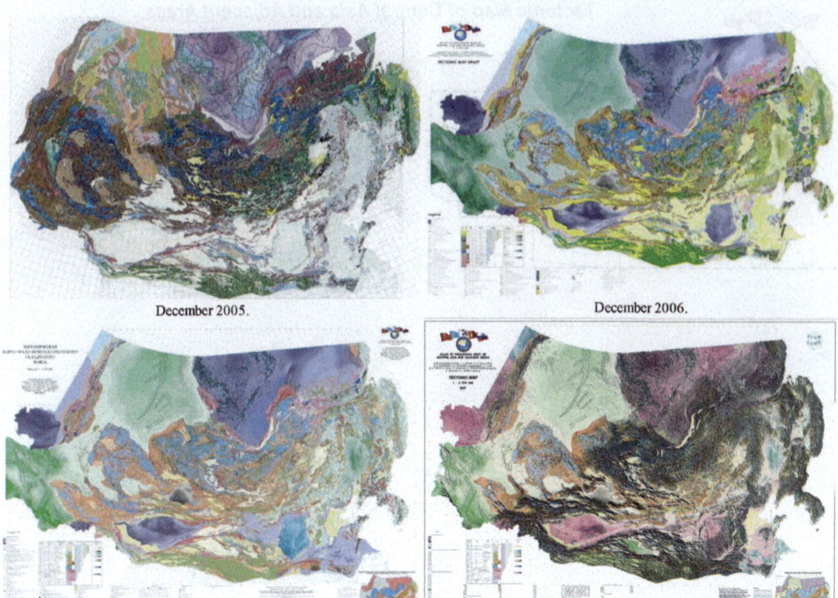

Fig. 3 The Tectonic Map of Central Asia and Adjacent Areas Combined with Shuttle Radar Topography Map (2007)

all its four maps was clearly seen. Therefore, by the initiative of Chinese colleagues, it was suggested to continue of the ongoing project with increasing the territory for future maps and with an accent to study the deep tectonics and metallogenic features connected with it. So in Shanghai the next project "3D Geological Structures and Metallogeny of Northern, Central and Eastern Asia" was substantiated. Practically new project gradually "grew up" from the previous one without waiting for its completion.

To the beginning of February 2007, the tectonic map was prepared for demonstration in Russian (see Fig. 4), as it had to be beforehand seen by numerous Russian scientists and geologists who could make the suggestions for improvement of scientific content, changes into some tectonic structures. Such version of the *Tectonic map of Central Asia and Adjacent Areas* was demonstrated at the annual Russian Tectonic Conference held in Moscow State University, where it got approval and received a high appraisal from Russian tectonists.

In July 2007 at the workshop meeting in Ulan Bator (Mongolia), the *Tectonic map of Central Asia and Adjacent Areas* was highly appreciated by the representatives of all counties-participants and recommended to final editing and preparing for printing to its demonstration as a complete set map of the Atlas at the International Geological Congress in Oslo. To the beginning of December, all low-thick Cenozoic sedimentary covers, under which the tectonic structure was known, were

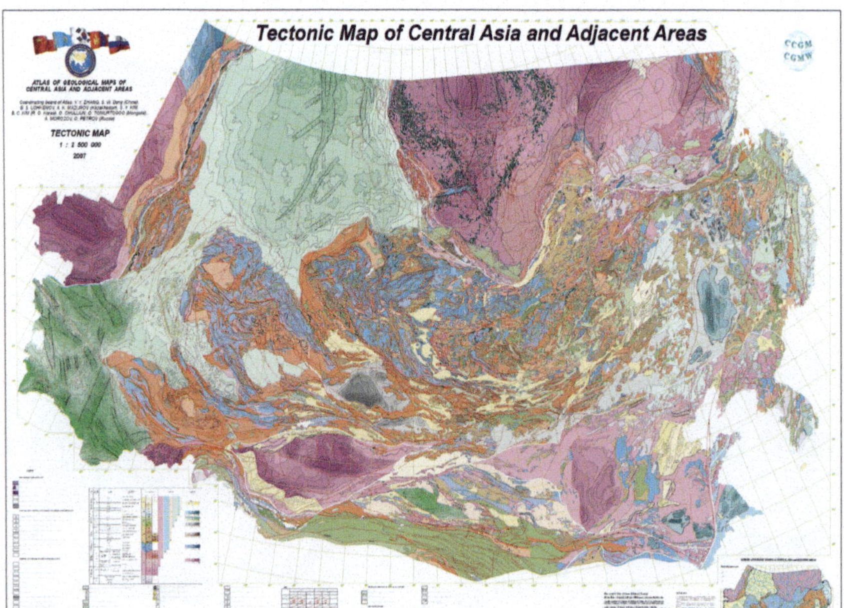

Fig. 4 *Tectonic map of Central Asia and Adjacent Areas at scale 1:2,500,000.* The map was published in 2008 before the 33rd International Geological Congress (August 2008, Oslo, Norway)

deleted. Carried out age analysis of the ancient craton crystalline basements, young platforms and sedimentary basins fold basements allowed to give a classification by the age of basements and lowermost rocks of sedimentary covers. The structural characteristics of the Qaidam Basin were added onto map.

As such appearance the tectonic map (see Fig. 3) was represented at the **sixth** workshop meeting in December 2007 in Sanya (China, Hainan Island). The main result of that meeting was generally recognized that the works under the Atlas's maps were completed and countries responsible for the individual maps (China—Geological and Mineragenic Fuel-Energy Resources, and Russia—Geological and Tectonic ones) could set out to their final appearance and published printing.

Published *Tectonic map of Central Asia and Adjacent Areas* (Fig. 4) (Tectonic Map… 2008) was demonstrated at the exhibition GEOEXPO-2008 during the 33rd session of the International Geological Congress in August 2008 in Oslo (Norway) together with the Metallogenic map at booths of the Federal Agency on Mineral Resources of the Ministry of Natural Resources of the Russian Federation and VSEGEI. Two another maps—Geological and Mineragenic Fuel-Energy Resources—were placed at the neighbouring booths of the Chinese Academy of Geological Sciences. Therefore, the exhibition's visitors could receive the full impression about the composition and scientific content of the *Atlas of Geological Maps of Central Asia and Adjacent Areas at scale 1:2,500,000*. Atlas had a huge success at the Congress. By the results of the completed project, a special session of the Congress "Tectonics and Metallogeny of Central Asia and Adjacent Areas" took place under the leadership of the heads of the Russian and Chinese author's groups—Dr. O.Petrov and Prof. Dong Shuwen.

After termination of the International Geological Congress, participants of the new project started (besides special investigations on deep structure of the Earth's crust, geodynamic features of magmatic events etc.) to compile the map's complete set (Geological, Tectonic, Metallogenic and Mineragenic Fuel-energy Resources ones) for a more large territory as compared with the previous project. Framework of the new maps circumscribed with the whole Asian part of Russia including the Arctic islands and southern part of China. The **seventh** workshop meeting in June 2009 (St. Petersburg, VSEGEI) decided to narrow the maps only to the continental (i.e. mainland) part of such a huge territory. To that time the 'nucleus" of the tectonic map was been already published, and the legend was approved and shown that it was the best version for the tectonic map at such scale, which would be used as a scientific basis for the metallogenic map. Therefore, the main task for the tectonic map compilers was to draw up the drafts for the northern part of Asian Russia and southern part of China according to the legend used for the *Tectonic map of Central Asia and Adjacent Areas at scale 1:2,500,000*.

The **eighth** workshop meeting of five countiesparticipants of the project took place in June 2010 again in Daejeon (Republic of Korea). During it, two separated drafts of the tectonic map were considered: one for the northern part of Asian Russia, another for the southern half of China. The legend of the tectonic map was supplemented with by symbols for volcanic and intrusive formations with special types of mineralization (proceeding from the task of compiling the metallogenic map including unique ore deposits).

At the **ninth** workshop meeting in Ulan Bator (Mongolia, August 2011) the *Tectonic map of Northern-Central-Eastern Asia and Adjacent Areas* was represented for review to all participants of the international project (Fig. 5). Map practically fully covered the territory of the project with exception of the Arctic islands, but showed the tectonic composition of the Russian shelves (Arctic and Far Eastern) and also of the adjacent territories of the China's neighbouring countries. At this workshop meeting participants decided to expose the new *Atlas of Geological maps of Northern, Central and Eastern Asia* at the forthcoming in 2012 International Geological Congress, as the *Atlas of Geological Maps of Central Asia and Adjacent Areas* which was shown in Oslo (33rd IGC).

In new version of the *Tectonic map of Northern, Central and Eastern Asia* (see Fig. 5) the data on tectonics of the Arctic islands were added. Tectonic map was limited mainly by coastal line, excepting some regions of the Arctic shallow shelf, and the main tectonic disjunctive lines were shown on the map. These faults are necessary for understanding the seafloor structures of some water areas (Sea of Okhotsk, Yellow Sea etc.) and adjacent structures of the land.

This variant of the *Tectonic map of Northern-Central-Eastern Asia and Adjacent Areas* was demonstrated at the Exhibition GEOEXPO during the 34th International Geological Congress (August, 2012, Australia, Brisbane) together with the Metallogenic map and also with Geological and Mineragenic FuelEnergy Resources maps at the neighbouring booths of the Chinese Academy of Geological Sciences.

As in 2008, the new complete set of maps aroused interest of the world-wide scientific geological community.

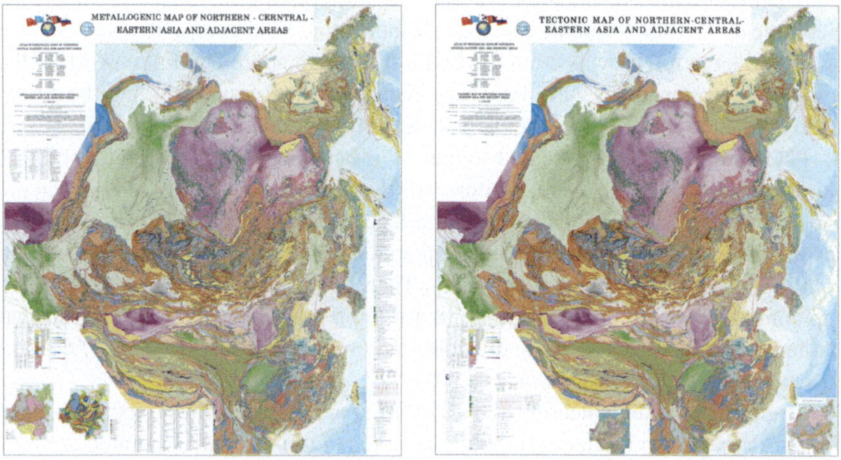

Fig. 5 Stages of compilation of the *Tectonic map of Northern, Central and Eastern Asia at scale 1:2,500,000* (2012 and 2014)

The latest corrected version of the tectonic map was inspected and reviewed at the jubilee **tenth** workshop meeting (10th anniversary of the International map compiling cooperation) in Beijing (November 2012), and in 2014 after final design the *Tectonic map of Northern-Central-Eastern Asia and Adjacent Areas at scale 1:2,500,000* (see Fig. 1 and 5b) was put for publication to VSEGEI Publishing House.

2 Principles and Legend of the Tectonic Map of Northern-Central-Eastern Asia and Adjacent Areas

For magmatic rock do not carry the tectonic information. At present, i.e. in the beginning of the XXI century, it is rather difficult to make general conclusions from a huge volume of geological information without tectonic cartography as a base for understanding all processes which took place during the Earth's history. Geological map as the "official document" allows us to "judge" only about the composition and age of outcropped rocks on the Earth's surface. But actually modern tectonic map is essentially a 3-D model of the Earth's crust structure reflected on the "2-D" paper (or virtual) model. As the tectonic map shows structures of different regions and independent tectonic elements (and also of continents, oceans and the Earth *en masse*) it must be volumetric. Previously, the map's volumetric content was achieved by a system of conventional symbols and by methods of representation of basic structural elements on the map. For example, the system of the color-gradation and isopachs reflected the structure of the platform basements and sedimentary basins.

However, changes in geotectonic conceptions during the twentieth century connected with different degree of studies of crustal structural and petrologic composition did not contribute to the unification of the principles for compiling tectonic maps. During the last century, the principles and legends for tectonic maps were changed not once. The changes in tectonic conceptions compelled map's authors to devise each time new legend putting an emphasis on those principles which, in the author's opinion, were determinative to that moment. In all probability, it was necessary to collect the huge potential of geological knowledge so that long-term search for all-acceptable ideas should be transformed into a stable conception which could be reproduced on tectonic map.

Until the 70s of the twentieth century, a fundamental feature for compiling middleand small-scale tectonic maps was the time of main folding. Fold areas and zones of different ages composing continents or major structures of the Earth were shown. But such an approach to the compilation of tectonic maps did not show processes of material reorganization (or reworking) of the Earth's crust and the evolution of the Earth as a whole. So, it were structural (but not petrologic-formational) researches, who mostly influenced the map compilation. Therefore, on the multi-sheeted and multi-colored *Tectonic map of the USSR* (editions of 1953 and 1956; Chief compiler Academician N. S. Shatsky), first of all, different-aged platforms and fold areas with reflection of their internal structure were shown. These maps gave new and quite important representations about time and spatial patterns of the processes of tectonic evolution in large region of the world and also were very useful in forecasting different types of mineral deposits. At the same time, the principle of differentiation of folding epochs (as a basis for tectonic map compilation) did not allow deep understanding the essence of tectonic evolution and objective and unique summary on such problems as horizontal displacement of crustal blocks, patterns of thrust-folded formation, relationship of orogenic belt formation and magmatism but it rather reflected the fixism concept as dominating for geologists.

Critical moment in tectonic cartography occurred in late 60s–early 70s of the twentieth century as a result of the development of a theory about the oceanic crust of the geological past and its transformation into continental one (Peive 1969).

A radical change towards mobilism ideas took place due to the disclosure of absolutely new geotectonic facts. The relationship between ophiolite series of continents and the crust of recent oceans was identified. This discovery caused another interpretation of "eugeosyncline areas" paleotectonics and gave the possibility to expose the fragments of the past oceanic structures in the structure of continents. Revelation of the geological essence for the oceanic crust layers and upper mantle in the recent oceans made it possible to define concretely the process of continental crust formation instead of oceanic one. A doctrine appeared concerning oceanic, transitional ("geosynclines" proper) and continental stages of crustal evolution that gave a possibility to distinguish such above mentioned stages not only for the recent tectonic plan of the Earth but for its geological past as well. And this, in turn, engenders a new approach to the tectonic zoning of the continents—depending on the time of continental-type crust formation, and also of granite-metamorphic layers of different ages that compose it. Among these granite-metamorphic layers, not only crustal and mantle substrata are determinant, but also the rock assemblages appropriate to different stages of crust formation.

It became possible to recognize in spaces and time rock complexes appropriate to different stages of transformation of the oceanic crust into continental one. Substantiation of zones with different time of formation of the continental crust gave an opportunity to determine spatial appropriateness of this process in a huge area such as, for example, Eurasia.

Thus, in 70s of the twentieth century, tectonic cartography approached the creation of "4-D" tectonic models, taken into account the factor of time and evolution processes in the earth crust. The concept of creating a new tectonic map of Northern Eurasia was first substantiated and a system of legend's symbols for different complex-indicators of individual stages of the Earth's crust and lithosphere as a whole was elaborated (Peive et al. 1976). In the course of work over the Legend for Tectonic map, it was decided that the Legend could be a basis in compiling the metallogenic and prognosis maps, fuel-and-energy maps, deep structure maps, etc.

The *Tectonic map of Northern Eurasia* (1979) shows areas with continental crust that formed in the different geological time were located both in autochthonous and allochthonous positions, and recent areas with non-completed formation of the continental crust and oceans. Dry land corresponds to the areas with continental crust and zones of ocean-continent transition to areas with non-completed granite-metamorphic crust layer. These theoretical conclusions opened new ways to the compilation of general and regional tectonic maps. New ideas were used while compiling the *Tectonic map of the Urals, scale 1:1 M* (1977) and its Explanatory Note, the *Tectonic map of Northeast Asia, scale 1:2.5 M* (1992) and the *Tectonic map of Kazakhstan, scale 1:2.5 M*.

These ideas and principles of tectonic cartography were used by the Subcommission for Tectonic maps (President V. E. Khain, Secretary General Yu. G. Leonov) of the Commission for Geological map of the World (CGMW). The *International Tectonic map of the World, scale 1:15 M* (1982) and the *International Tectonic map of Europe, scale 1:5 M* (3rd edition, (1996)) have received the worldwide acceptance. Both Soviet and Russian experience was used in the compilation of smalland middle-scale tectonic maps of continents, countries and global tectonic structures.

However, further evolution of the concepts concerning the structure and transformation of the earth crust, and to a greater degree of the whole lithosphere, caused not so the revision of the idea of reworking the oceanic crust to continental one, as the development and addition of these concepts. It appears that not everywhere the paleooceanic crust within the continents is transformed to continental one. Sometimes, this process was interrupted at the stage of accretion of an island arc, marginal seas and deep-trench complexes. In this case, the newformed accretionary crust had no geological and geophysical features of the real continental crust as, for an example, in ancient cratons. The structure of the Alazeya Upland (centre of the Kolyma foldthrust accretion-collision area, Northeast Russia) is a typical example of such “non-finished” process. Here the total thickness of the crust is low, and granitemetamorphic layer is almost absent.

On the other hand, once formed continental crust can grow as a result of tectonic processes. These processes concern the fragments of the ancient continental crust situated in more widespread younger fold zones. Large areas of fold belts with blocks of ancient continental crust occur in the Altay-Sayan fold system, Kazakhstan mosaic structure, Ural linear fold-thrust belt (particularly in the South Urals).

These blocks were the parts of huge continental masses until the structures with oceanic crust have come to existence, which later reorganized into younger (sometimes considerably younger) continental crust. Correspondingly, these blocks have all features of the most ancient continental crust (metamorphic and folded complexes, volcano-plutonic associations, shelf sedimentary cover etc.). Of course, the ancient crust in process of formation of younger framing crust in the fold belts was exposed to considerable thermal-magmatic and structural reworking. These processes are reflected in its thickness, composition and (it is very important) in metallogenic specification of ancient blocks. In fact, this is a fragment of ancient crust with “new properties”.

Thus, in the final crust structure, the fragments of different ages and types are found closely as to the structural position and the degree of magmatic and metamorphic reworking. But they can differ in the thickness and structural position, because in the process of fold belt formation the blocks (paleo-microcontinents) underwent the horizontal movements and now usually form allochthonous or para-allochthonous thrust or overthrust complexes in the "new crust".

Recently, researches have shown that the continental crust and the transitional crust, and even the oceanic crust are polygenous, i.e. during evolution they permanently changed their geological and geophysical properties, i.e. they repeatedly structurally, magmatically and metamorphically changed. Once coming to existence in spreading zones, the oceanic crust continues to change features even before its reorganization (for example, in subduction zones) in zones of transform faults, above "hot spots" etc. And these ***varieties of polygenous crust compose the integrated Earth's layer with certain features different from the underlying and overlying layers***. This layer is known as the ***consolidated crust***.

Modern stage of compilation of tectonic maps started in the mid-90s of the last century. Such notions as consolidated crust, its evolution, geological rock assemblages (in Russian—rock formations) of the different geodynamic environments have become the determinative condition in analysis of tectonic structure of the Earth's crust. Age of consolidated crust, its principal geological and geophysical difference from upper mantle and sedimentary covers, horizontal accretion with transformation of the crust, its structural and material reworking during the evolution process are the fundamental principles in compiling maps of new generation. The modern progress in geosciences with using of Hi-Tech methods for study of material composition of the rocks, absolute age, and their paleomagnetic properties are very helpful for compiling present-day tectonic maps which take into account practically all information collected by the present. Digital tectonic maps, which compiling is based upon modern geological maps, become the objects of continuing monitoring. In turn, tectonic maps are a basis for the compilation the various geological maps—from metallogenic maps to maps of natural geoecological environment and water resources.

The compilation the modern tectonic maps and tectonic bases for metallogenic maps and special-purpose maps of applied economic significance is based upon the modern concept of the ***consolidated crust***.

Consolidated Earth's crust is the Earth's layer with a basement and a roof, which underwent (partially or completely) folding, metamorphism and granitization, distinguished on composition, structure and physical parameters from overlapping (plate sedimentary cover) and underlying (upper mantle rocks) formations of lithosphere. Earth crust has doublelayered structure: (a) consolidated folded basement of various ages and, (b) plate sedimentary cover (sometimes with pre-plate complexes) (Leonov 2008). According to paleotectonic aspect we can offer the following formulation of the consolidated crust. ***Consolidated earth crust*** is a complex of rocks, rock associations and formations (sedimentary, metamorphic and magmatic), intensively deformed, arisen during evolution of fold-thrusting areas, belts and structural-formational zones during indivisible completed geotectonic (geodynamic) cycle and having approximately homogeneous reological properties in general structure of the Earth's crust.

Consolidated crust, as above mentioned, can pass through several stages of structural-material transformations. And such processes can take place in the present time and be related to geotectonic activity both in blocks of the consolidated crust and in their peripheries. So, for example, southeast part of the Siberian craton with Paleoproterozoic crust continues to grow upward in the region of the Pacific volcanic belt. But the most ancient craton of the Earth—North China (=Sino-Korean)—at the present time has three main fragments in its structure, which differ in time of final (to the present time) consolidation. They are (from west to east): Alashan (two stages of reworking and epi-Variscan sedimentary cover), Ordos (with extant fragment of pre-Mesoproterozoic consolidation) and North China (Sino-Korean) (with intensive Jurassic—Cretaceous Yanshanian intra-plate granitic magmatism).

The authors believe that the concept of the analysis of condition and evolution of the Earth's consolidated crust without its composition (continental, transitional or oceanic) in the shows in the best way the evolution processes both of separate structures and the Earth's crust as a whole. At the same time this concept allows geotectonic zoning at different levels—from global (for example, cratons and fold collision and accretion belts) to regional structures, including magmatic, metamorphic and others rock assemblages, rock complexes formed in the integrated geodynamic environment. As a consequence, they have not only similar geological characteristics, but also similar geochemical and metallogenic specifications that allow using the tectonic maps as a basis for further creation of various metallogenic maps. For example, the legend for the Tectonic map of Northern, Central and Eastern Asia (2008–2014), similar to the *Tectonic map of Central Asia and Adjacent Areas, scale 1:2.5 M* (2008), is composed following the principle of definition of consolidation stages and similarity of rock assemblages of similar geodynamic environments.

It is worthy to note that tectonic maps based on the concept of consolidated crust are the immediate continuation of the maps which reflected the transformation of the oceanic crust into continental one. The definition of the stages and complexes-indicators of consolidation gives the possibility to imprint not only the presence of the crust block of a such an age but also it suggests the whole historical-tectonic process of the crust's changing and growth vertically and laterally. And the time of consolidation is the most important criterion for delimitation of blocks with continental crust.

In process of the map creation we proceeded from the fact that the consolidated crust occurs in fold belts of different ages and in blocks (i.e. microcontinents) of continental margins. Accordingly, the tectonic map must reflect (a) time of consolidation, (b) major sedimentary and volcanogenic rock assemblages (rock formations) with pre-consolidated evolution of the crust, (c) ophiolite complexes in their recent structural position, (d) intrusive magmatism, and first of all granitoid magmatism of different geodynamic environments (supra-subduction, collision, post-collision etc.), (e) prevalence of regional metamorphism or separated metamorphic facies, (f) molasse deposits. Consolidated crust reworking (structurally, thermalmagmatic) once or many times during later stages is shown separately by special symbols.

Proceeding from the definition of consolidated crust we must recognize that for areas with continental crust double-layer structure is settled: folded basement represented by consolidated crust of different ages and sedimentary (plate) cover (sometimes with pre-plate complexes). According to the conception of the consolidated crust, the "proto-platform" (or "pre-plate") complexes are the lowermost elements of sedimentary (plate) cover. By analogy, the consolidated crust cannot include deformed cover's sedimentary complexes such as those of the Verkhoyansk fold-thrust system located on the thin consolidated crust of the Siberian craton, folded complexes of Pre-Uralian fore-deep depression, Pre-Kopet Dagh depression, Songpan-Garze fold belt of NE Tibetan Plateau and others.

The age of lowermost horizons and their thickness is the main principle of designation for sedimentary cover (plate complex). But between the age of consolidated crust (basement) and age of the base of sedimentary cover (plate complex) the considerable time interval might be. In some cases the age of "pre-plate" complex is the time of the start of its accumulation (but only if it is widespread). Among intraplate complexes, the intra-plate magmatism is shown—volcanic and plutonic rocks of the different volumes): fold and fault deformations of different kinematics and morphology; salt tectonics; consedimentation morpho-structures; buried deformations, zones of superimposed fold-thrust deformations related to the evolution of the adjacent fold belts (see above). Particularly shown are molasse (superimposed on the "plate" complex, or sedimentary cover of platforms) basins of fore-deep and intra-mountain depressions, which have been formed or being formed near fold belts with intensive orogenic processes.

2.1 Legend

Legend for the *Tectonic Map of Northern-Central-Eastern Asia and Adjacent Areas* practically is the broadened legend to the *Tectonic map of Central Asia and Adjacent Areas published* in 2008 as the map of the *Atlas of Geological maps of Central Asia and Adjacent Areas at scale 1:2,500,000* (Tectonic Map… 2008).

Fundamental principle of the legend to the *Tectonic Map of Northern-Central-Eastern Asia and Adjacent Areas* appeared in the legend to the 3-rd edition of the *International Tectonic Map of Europe* (International Tectonic Map… 1996). In it, the legend for the first time was presented tabular which has shown the sequence of tectonic events in the geological history of the European continent and the main stages of formations of the sedimentary covers and sedimentary basins on ancient (Archean-Proterozoic) and young (Paleozoic–Mesozoic) crystalline and folded basements. The periods of ophiolite and ophiolite structure originations, stages of continental riftogenesis, fold deformations in the structures of fordeep and intra-mountainous depressions and basins etc. corresponded to this sequence of tectonic events. That legend was created for tectonic map at scale 1:5,000,000 and reflected not only the main stages of origination and deformation of the earth's crust (continental, transitional or oceanic) but also the principal rock assemblages (sedimentary, volcanogenic, and intrusive) of different paleodynamic settings.

The same principle of composition was built into legend of the started in 2000 *International Tectonic map of Asia at scale 1:5,000,000* as the project of the Commission for the Geological Map of the World (CGMW). For this legend, the consecutive row of tectonic events was increased, their geochronological correlations, tectonic events exclusive for South and East Asia (Indosinian and Yanshanian tectonic ones) were added.

Legend to the *International Tectonic map of Asia at scale 1:5,000,000* was directly used in the *Tectonic map of Central Asia and Adjacent Areas at scale 1:2,500,000.* Without changing the principles of legend creation, the map's executive editors and editors-in-chief had to supplement it considerably with some new parts and symbols pursuant to the following conditions.

1. Considerable changing of the scale gave in detail the image data in the map, tectonic bodies with different rock assemblages. The legend for the following blocks was added:

    - Rock assemblages of paleooceanic crust,
    - Volcanic and volcano-sedimentary rock assemblages of continental and transitional crust,
    - Sedimentary rock assemblages of continental and transitional crust,
    - Metamorphic rock assemblages,
    - Mafic, ultramafic and alkaline rock assemblages,
    - Granite assemblages.

2. There appeared the possibility to show in the map the tectonic nappes and thrusts formed by both rocks of basement and sedimentary cover, or only by sedimentary cover. Great importance was attached to ophiolitic allochtons.
3. Considerable extension of legend is explained by the destination of tectonic map in the Atlas. It reflected not only the composition of consolidated crust and sedimentary layer but was also planned as a base map for the metallogenic map. Therefore, it was very important to show in the tectonic map a great diversity of rock associations with different mineragenic orientation. This makes clear not only the wide spectrum of volcanic, volcano-sedimentary, intrusive rocks (including different granitoids). The authors were compelled to include into legend a large group of non-scale symbols (first of all, for intrusive complexes). In contrast to non-scale symbols for glaucophane schist (blueschist) and eclogite, the non-scale symbols

Legend to the *Tectonic map of Northern-Central-Eastern Asia and Adjacent Areas at scale 1:2,500,000* has underdone the insignificant changes in connection with broadening of map territory and including into map the new structural elements which did not come into previous tectonic map of 2008.

Informational block of legend (Fig. 6) includes the sequence of tectonic events combined into tectonic cycles: Pre-Riphean, Riphean-Cadomian, Caledonian, Variscan, Kimmerian, Alpine-Himalayan. Pre-Caledonian part of tectonic events shows the correlation of Russian and Chinese geochronological subdivisions and corresponding tectonic events. This scheme appeared as a result of cooperative investigations of the authors of the Atlas in the Pre-Uralian part of the East-European Platform in Russia and North China (Sino-Korean) and Yangtze platforms in China in 2005–2007. Therefore, for convenience of the tectonic map users, the authors

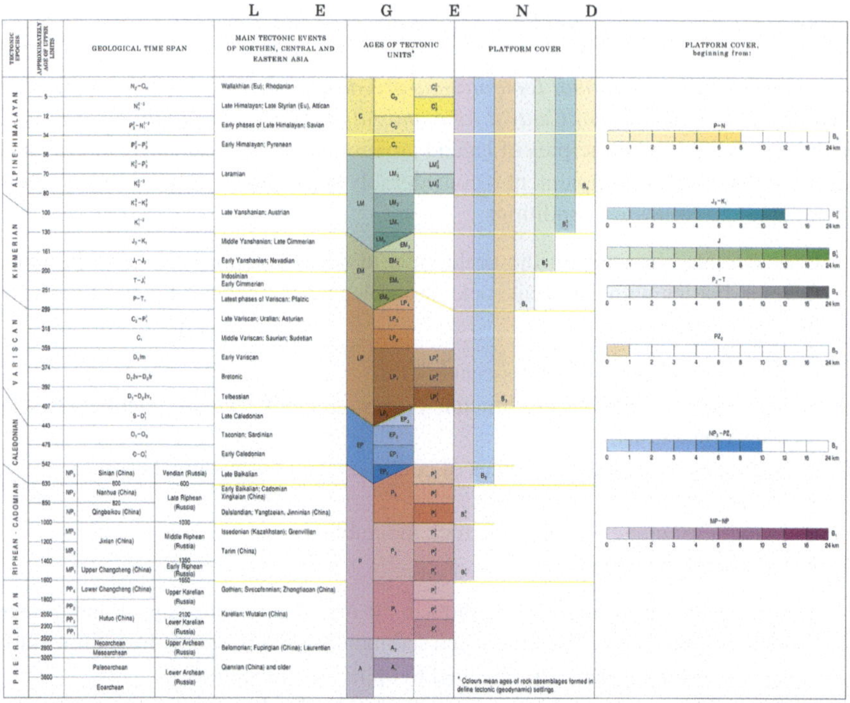

Fig. 6 Informational block of the Legend to the *Tectonic map of Northern-Central-Eastern Asia*

have kept the traditional Russian and Chinese stratigraphic (geochronological) subdivisions and names of important tectonic events.

Tectonic events in informational block correspond to the following reorganizations of the Earth's crust: age of cratonization (mainly in the Archaean-Proterozoic), age of crust consolidation (metamorphism, magmatism, and deformations), time of fold-thrust deformations (both in fold-thrust belts, and in sedimentary covers), and time of remobilization and destruction of the crust (continental riftogenesis, thermal-magmatic and structural reworking).

Column "Ages of tectonic units" includes the color spectrum for all tectonic bodies reflected in the tectonic map and indexes for their designation.

Column "Platform cover" shows the diversity of sedimentary covers of the ancient and young platforms (and sedimentary basins) in dependence from the age of the lowermost rocks in sedimentary succession. By this indicator (time of the beginning of sedimentation) all sedimentary covers are divided into the following groups:

- Riphean—Early Vendian (for North China Platform this sedimentary complex starts earlier that the Early Riphean with accumulation of Changcheng Fm. with the age of basal layers of 1800 Ma, while in the Bashkir anticlinorium in the Cis-Urals part of the East European Platform, since 1600 Ma);

- Late Vendian—Early Paleozoic. These sedimentary complexes started to form during interval of the Late Vendian (in China from Sinian)—Cambrian;
- Late Paleozoic—beginning of sedimentary covers formation in the Early—Middle Devonian, corresponds to origination of the epi-Caledonian platforms.
- Late Permian—Triassic. In that time the largest epi-Variscan platforms started to form; among them Turan young platform, West Siberian sedimentary basin and others.
- Jurassic. Origination of such sedimentary basins as Songliao, Huabei in China.
- Late Jurassic—Early Cretaceous.
- Paleogene—Neogene. Numerous intra-mountainous and fore-deep depressions are mostly situated in South Eurasia and along the Pacific active continental margins; their origination depended on the active Cenozoic tectonics. Sedimentary covers of this age continue to accumulate along the Arctic Ocean coast. Block ***Rock assemblages of paleooceanic crust*** (Fig. 7) includes the standard complex of symbols for designation both preserved successions of paleooceanic crust (ophiolite with preserved sequence) and dismembered ophiolite (ultramafic protrusions etc.), the outcrops of which occur more frequent than ophiolitic successions. Symbol "ophiolitic mélange" (serpertinite mélange) is more widespread because namely mélanges accompany the majority of collisional and accretionary structures formed at the places of paleooceanic basin and marginal seas. The symbol "Basalt-carbonate cover of intraoceanic seamounts..." is used much rarely because these paleostructures very rarely remain in the fold-thrust belts and areas

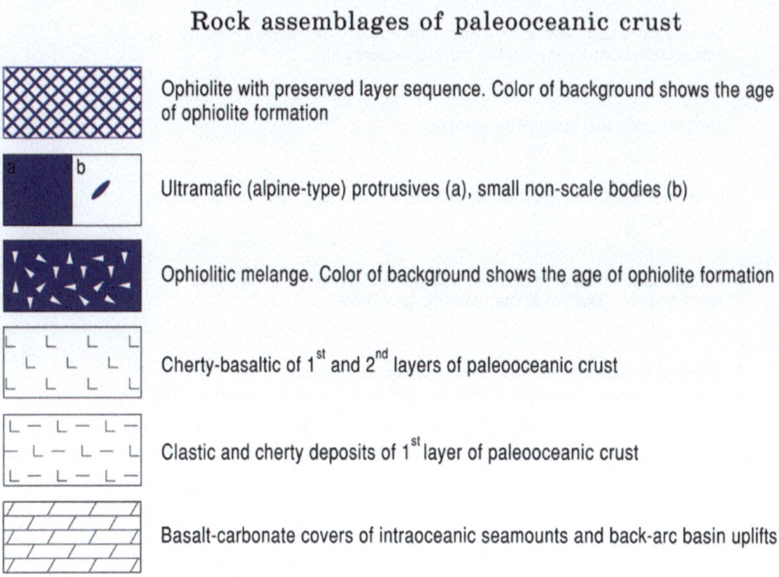

Fig. 7 Rock assemblages of paleooceanic crust

(the examples are known in the Mongol-Okhotsk belt of the Central Mongolia and in some ophiolite sutures of Kunlun Mts. and Tibetan Plateau).

Volcanic and volcano-sedimentary rock assemblages of continental and transitional crust (Fig. 8). This block mostly includes the symbols for widespread volcanic associations; among them there are recent complexes and paleoanalogues: rock assemblages of ensimatic and ensialic island arcs, active Andeantype continental margins, and also assemblages of marginal seas and intra-arc basins. Special place in the legend belongs to volcanic complexes associated with destruction of the crust: rift and riftogenous structures, "hot spots", basaltic floods etc.

Volcanic and volcano-sedimentary rock assemblages of continental and transitional crust

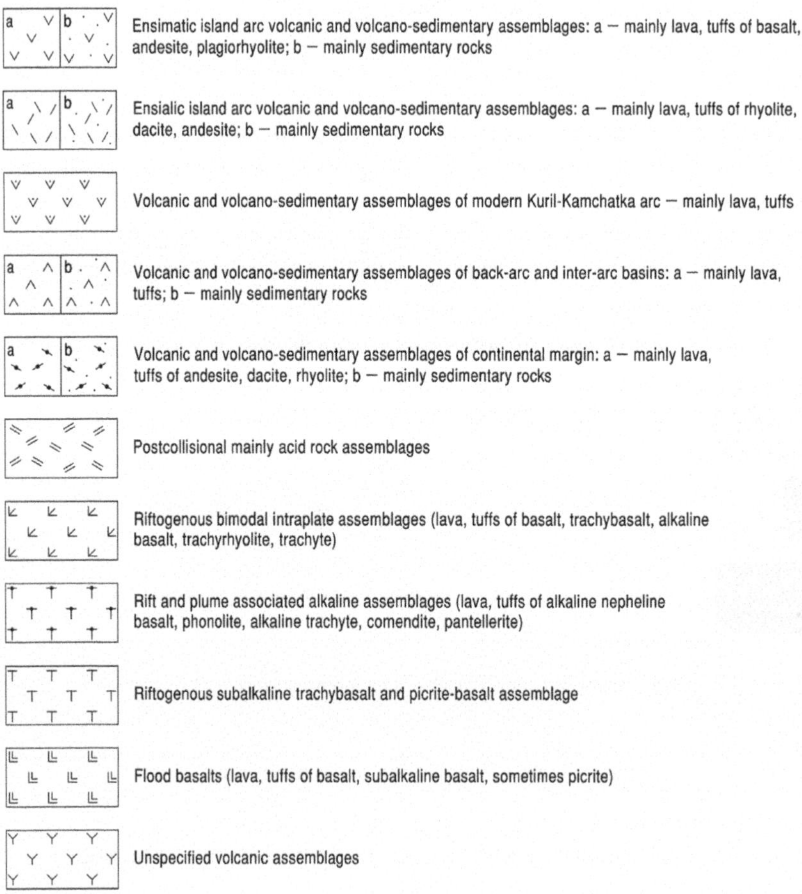

Fig. 8 Volcanic and volcano-sedimentary rock assemblages of continental and transitional crust

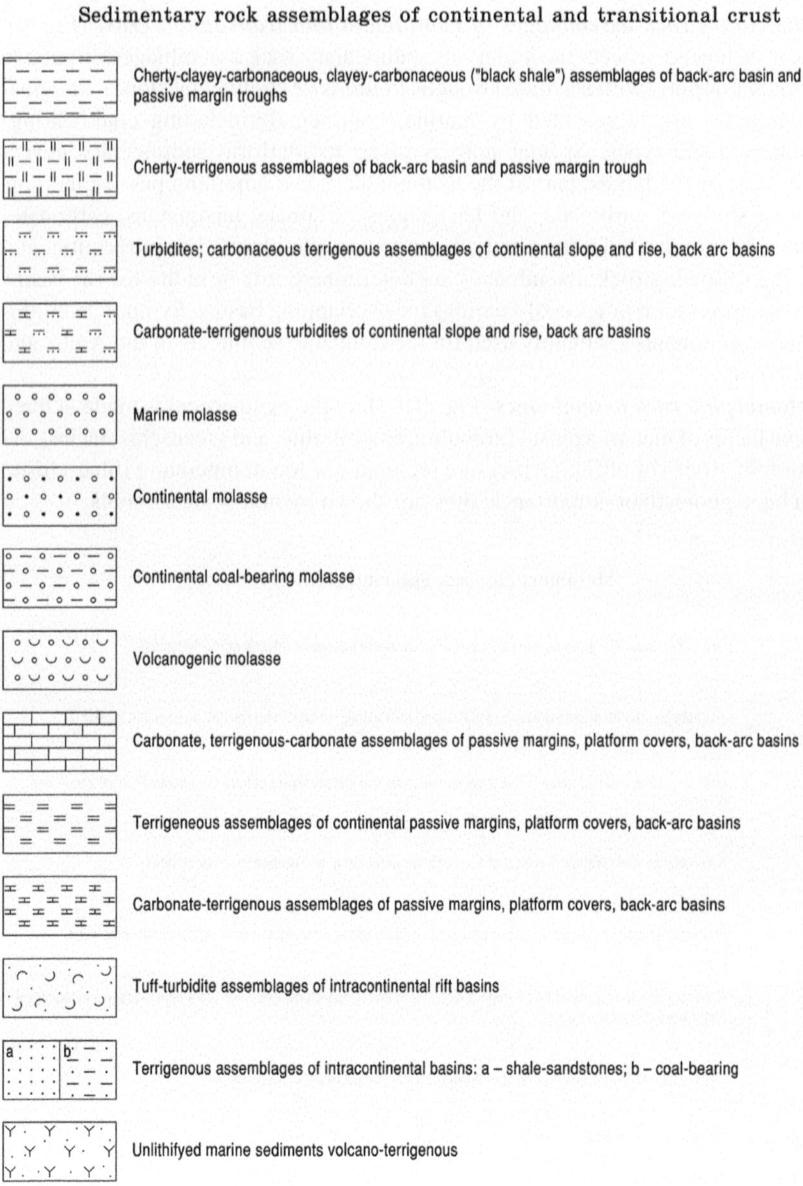

Fig. 9 Sedimentary rock assemblages of continental and transitional crust

Sedimentary rock assemblages of continental and transitional crust (Fig. 9). This part of legend reflects the variety of sedimentary rock assemblages of passive continental margins (with separated troughs in limits of continental slope), marginal seas. Molasses are represented by marine, continental (including coal-bearing) and volcanogenic types. Special note is given to platform sedimentary covers (cratons occupy the largest part of the tectonic map) and adjoining passive margins (including shelves): carbonate, and terrigenous-carbonate, terrigenous, carbonate-turbidite (calcarenite, calcilutite and others) rock assemblages. For intracontinental basins the following rock assemblages are determined: tuff-turbidite for rift basins and terrigenous (sometimes coal-bearing) for overlapping basins. Symbol "nonlithified marine sediments" is mainly used for the Cenozoic sediments of the Arctic and Pacific coasts.

Metamorphic rock assemblages (Fig. 10). Here the main accent is made at three principal facies of metamorphism (granulite, amphibolite, and greenschist facies). As metamorphic rocks of ultrahigh pressure (eclogite) or low-temperature (blueschists) have a huge geotectonic importance, they are shown by non-scale symbols.

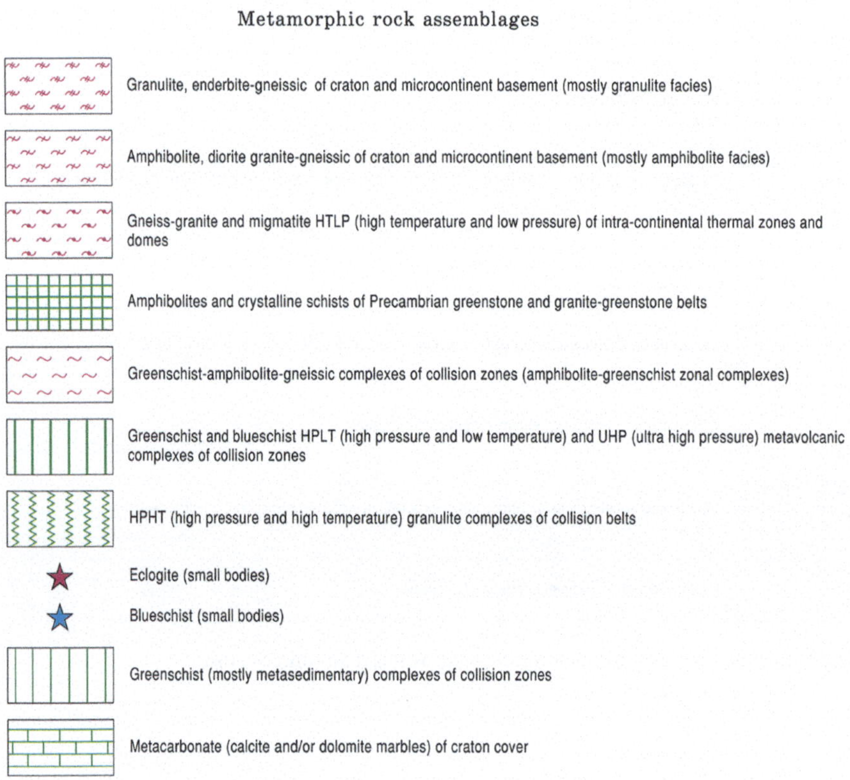

Fig. 10 Metamorphic rock assemblages

Separate symbol group shows the metamorphic degree in collisional belts and sutures. Symbol of the Precambrian greenstone and granite-greenstone belts intends for basements of cratons.

Mafic, ultramafic and alkaline rock assemblages (Fig. 11). This group of symbols is created especially for reflecting mineragenic orientation of mafic, ultramafic, and alkaline rocks, some of them have widespread distribution (for example, Platinum-bearing belt of the Urals, see Chap. 4), or peridotite-gabbro differentiated massifs

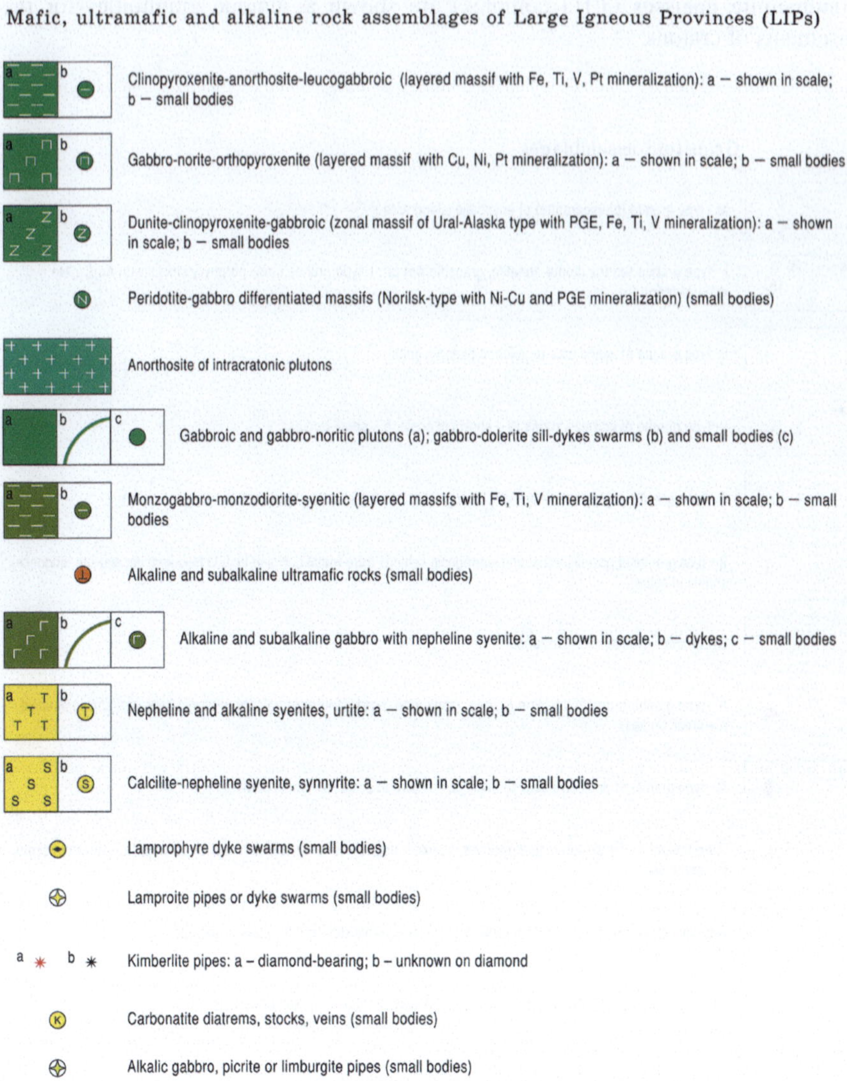

Fig. 11 Mafic, ultramafic, and alkaline rock assemblages

like Norilsk one. Also the anorthosite plutons and kimberlite pipes are shown in the map for the cratons.

Granitoid assemblages (Fig. 12). This group of symbols unites granitoids of all most important paleodynamic settings of their formation. We determined among them: plagiogranites of ensimatic island arcs, granites of ensialic island arcs and volcano-plutonic belts, wide variety of granites of collisional belts, syntectonic granites of shear zones. Also granites with specific mineralization (scale and non-scale symbols) are presented in the legend. Rapakivi, grey "gneisses" and tonalite-trondhjemite gneisses (TTG complex) are shown as unique granitoides for the basements of cratons.

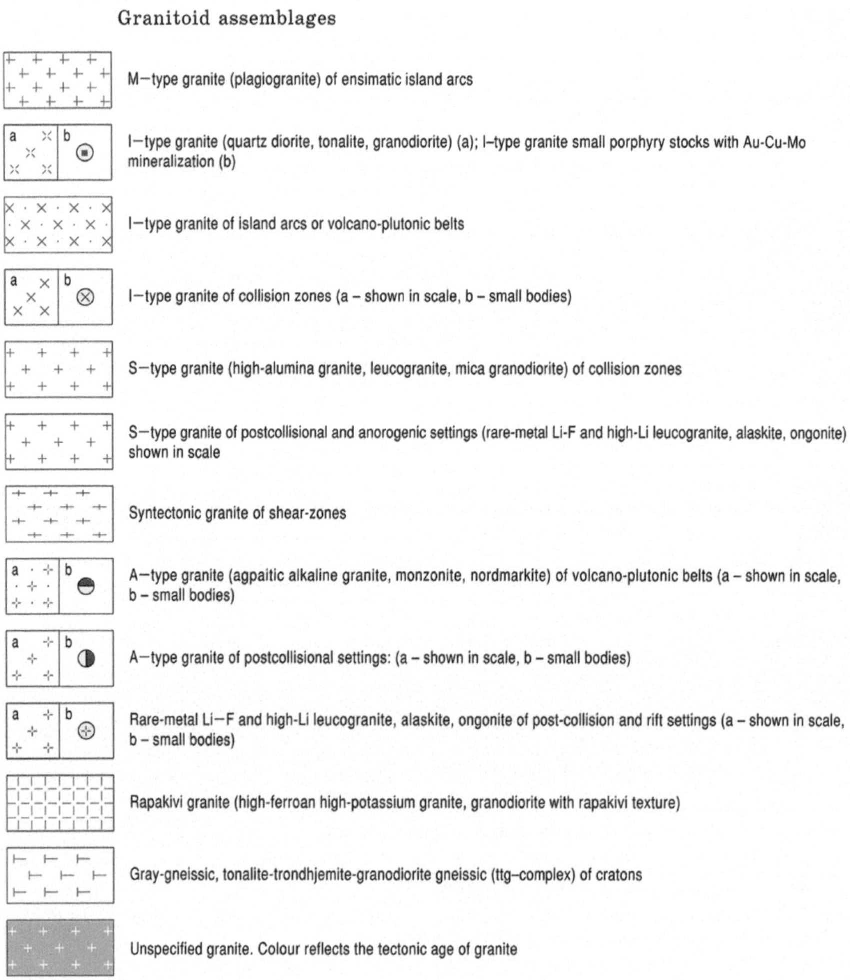

Fig. 12 Granitoid assemblages

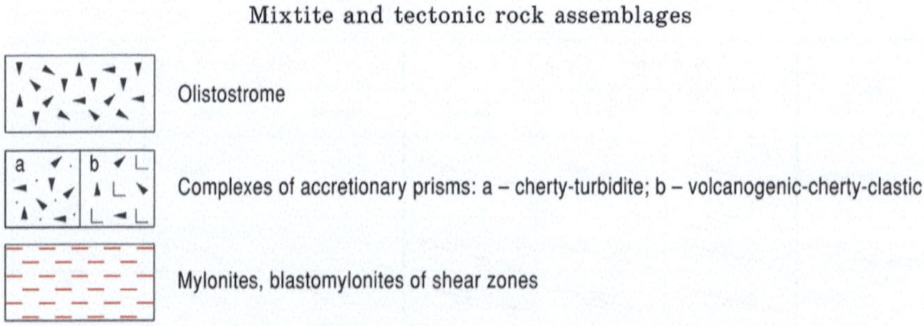

Fig. 13 Mixtite and tectonic rock assemblages

Mixtite and tectonic rock assemblages (Fig. 13). Complexes of the accretionary prisms are very widespread in the fold-thrust belts and sutures, especially in the Phanerozoic ones. Olistostromes are typical of rock assemblages underlying tectonic allochtons, overthusts, nappes. Olistostromes are very often presented in the fore-deep basin complexes and are the elements of continental molasses.

Mylonites and blastomylonites in some outcrops have enough width to be shown as lengthy shear zones (Irtysh shear zone, several zones of the Anabar Shield).

Structural and metamorphic reworking of rock assemblages (Fig. 14). Structural and metamorphic (or thermal-magmatic) reworking of ancient rock assemblages results in progressive (with intensive magmatism) or regressive (without magmatic events) changes of older complexes. In the map, the time of reworking is shown by the color of the strongest changing (with recombination and origination of new metamorphic mineral association), or the time of the latest reorganization (in case of retrograde metamorphism).

Zones of superimposed structural reworking involve substantially the pre-mountainous parts of the fore-deep depressions and basins. The best examples are represented by the folded sedimentary cover of Tajik block, pre-Kopet Dagh depression, Kalpin fold-thrust system in the north of Tarim Basin and numerous others.

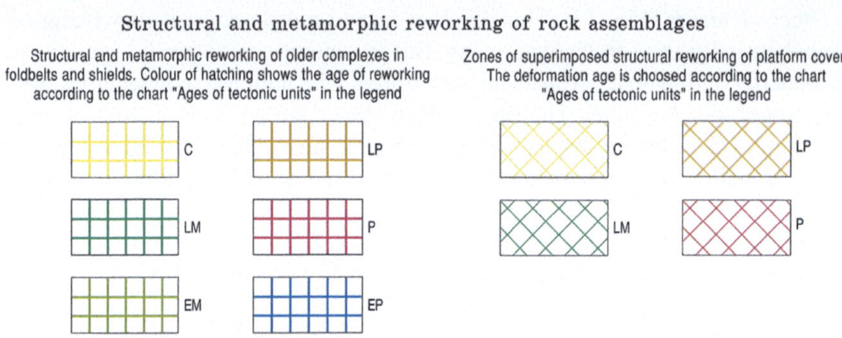

Fig. 14 Structural and metamorphic reworking of rock assemblages

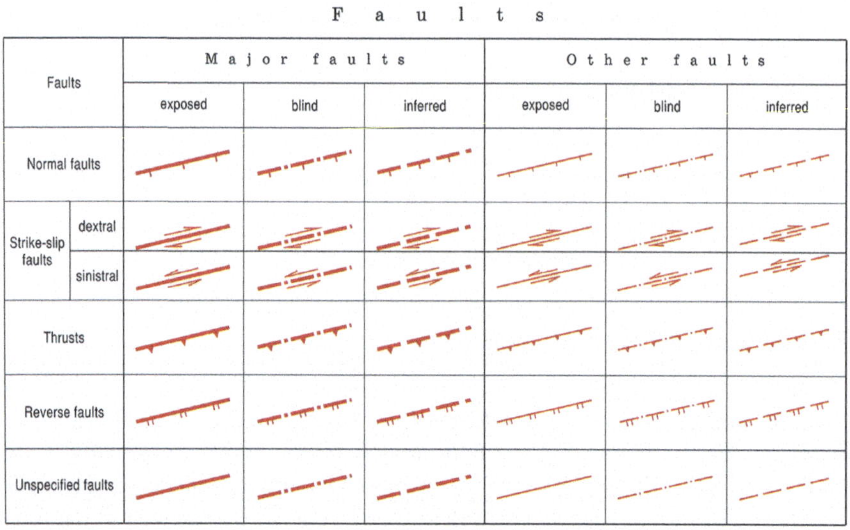

Fig. 15 Faults

Faults (Fig. 15). This block of legend includes practically all types of faults (with the exception of tectonic nappes) which are needed to explain the structural features of both consolidated crust and deformed sedimentary covers, basin sedimentary complexes.

Nappe boundaries (Fig. 16). Among nappe boundaries, three main types are distinguished: involving only sedimentary cover (thin-skinned tectonics), involving both fragments of the basement and sedimentary cover rocks (thick-skinned tectonics), ophiolitic allochtons. In the map, the most famous ophiolitic allochtons such as West Uralian are shown (see Chap. 4): Sarmara, Kraka (southern part of the Urals), and Polar Uralian allochtonous maficultramafic massifs.

Mesozoic complexes of cordillera-type metamorphic cores and elements (see Fig. 16). Here, the typical faults as listric and detachment displacements are shown. For the Mesozoic metamorphic cores authors have included the special scale symbol.

Other elements (Fig. 17). This block of legend mostly consists of widespread symbols usually used in tectonic maps. But it contains such specific symbols as "Boundaries of fold regional elements for the areas with removed sedimentary cover" and "Boundary of craton". The first of them shows the position of front of overthrusting (or detachment displacements) along the areas between platform and its cover (or sedimentary basin), on the one hand, and younger fold-thrust belt, on another one.

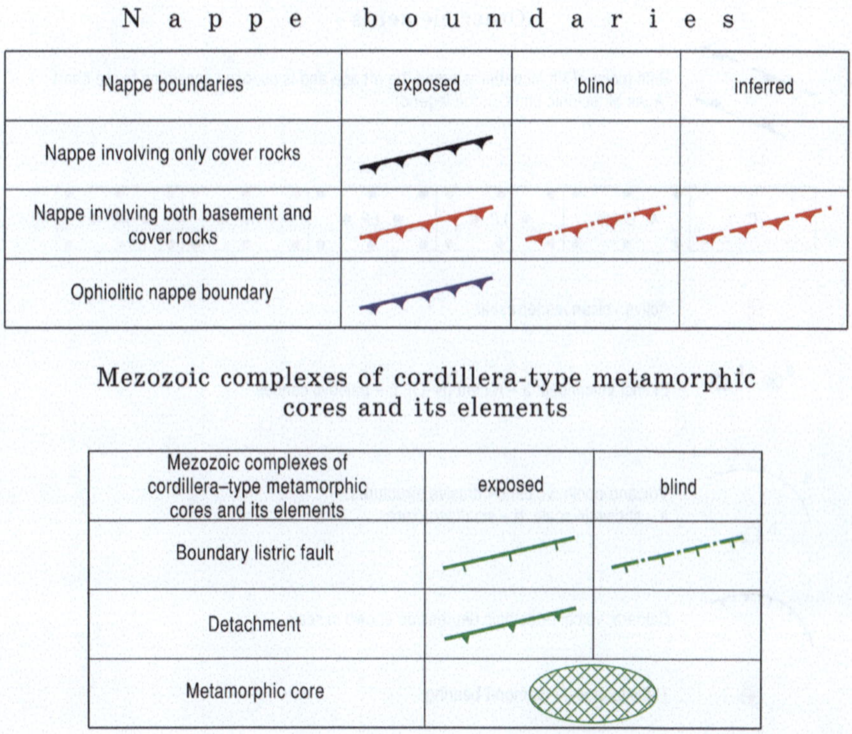

Fig. 16 Nappe boundaries. Mesozoic complexes of cordillera-type metamorphic cores and elements

3 Tectonic Zoning of Nothern, Central and Eastern Asia (Central Asian Fold Belt and Adjacent Tectonic Structures)

3.1 Tectonic Zoning

Being a part of the *Tectonic map of Northern-Central-Eastern Asia and Adjacent Area*, the territory of Northern, Central and Eastern Asia, by the tectonic essence, represents a compound ensemble of tectonic structures of different age and origin. This ensemble includes ancient cratons (Siberian, East European, Sino-Korean, or North China), massifs and blocks with thick continental crust of different age (Cathasian blocks, Yangtze Platform, Paleo-Gondwanaland blocks of Tibet and Pamirs, Himalaya, Tarim–Tajik Platform) and blocks of unidentified origin (basement of the Junggar Basin, Qaidam, Hami-Turpan, Fergana depressions etc.).

Central position of the map belongs to the Central Asian fold belt (in Russian part—to the Ural-Mongolian mobile belt), originated as a result of long-time (more than 400 Ma) evolution at the place of the Paleoasian Ocean as a part of the

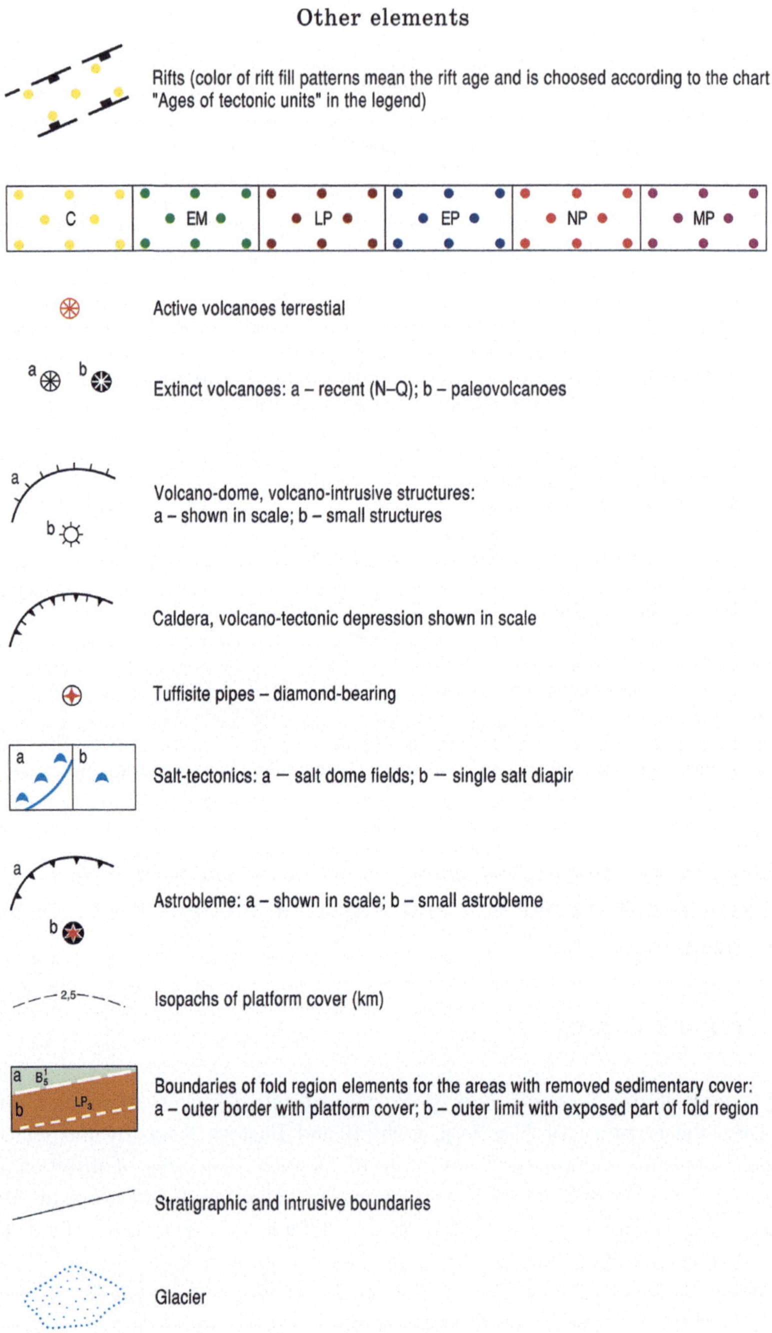

Fig. 17 Other elements

Paleopacific Ocean (Fig. 18). During evolution of the Paleoasian Ocean, accretion processes prevailed over others when isolated cratons (paleocontinents) "accreted" by paleooceanic, paleo-island arc and others complexes, joined to the paleocontinent margins by the processes of subduction or obduction. Simultaneously, separated blocks with continental type of crust (microcontinents) of different origination were accreted—Paleo-Siberian, Cathasian, Laurentian and others.

Ultima analysi, originated structure at the place of the Paleoasian Ocean acquired the features of a mosaic structure which is characteristic of accretionary type (or ***Pacific style***) of the Earth's crust evolution. In this Explanatory Note (Chap. 15), the tectonic evolution of the Central Asian mobile belt and origination of its mosaic structure is described in detail. Linearity of the accretionary complexes of the Urals

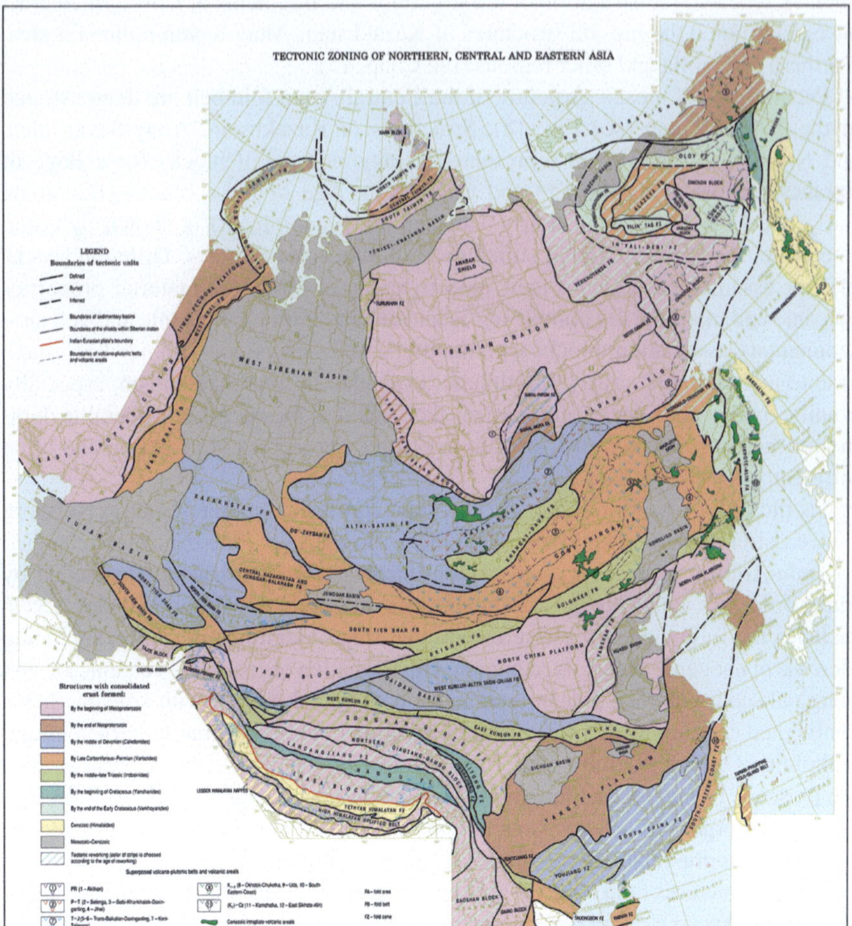

Fig. 18 Map of Tectonic zoning of Northern, Central and Eastern Asia (inset map to the *Tectonic map of Northern-Central-Eastern Asia and Adjacent Areas, 1:2,500,000)*

is conditioned, first of all, by the structural homogeneity of the continental margin of the East European craton, to which paleooceanic, paleo-island arc complexes and isolated microcontinents (such as East Uralian block) have been accreted. The process of accretion and obduction of the Uralian different complexes onto the East-European continental margin is described in Chap. 4. In tectonic plan, accretionary and obduction complexes of the Urals are similar to those of the Eastern Appalachians on the structurally homogenous margin of the North American craton. It is quite interesting because Appalachian structures have been formed along the Iapetus (*syn.* Yapetus) boundary which was a direct prolongation of the Paleoasian Ocean.

Repeated processes of accretion and destruction in the internal parts of the Paleoasian Ocean (in the modern structure—in the central part of the Central Asian fold belt), presence of numerous continental blocks with different sizes (microcontinents), repeated redistribution of the structures and fragments of new-formed crust have determined the mosaic structures of Kazakhstan, Altay-Sayan mountain area, Northern Mongolia and other regions (see Chap. 15).

Peculiarities of mosaic structure of the Central Asian fold belt are demonstrated in the Tectonic map (see Fig. 18). Structures of Kazakhstan, Altay-Sayan area, and Northern Mongolia are represented as agglomerate of blocks (or collage of terranes), differing by the age and origination. Age of these blocks (Baikalian, Early and Late Caledonian, Early Variscan are predominant) is shown by color. Therefore, the largest regional structures look like color mosaics. Different specks of rock assemblages explain their paleodynamic or structural-material properties (oceanic and marginal sea complexes, island arc and intra-arc volcanic and volcano-sedimentary association; blocks with mature, granite-gneiss, or with newly-formed, granite-metamorphic, crust). Multiphase, expanded in time magmatism, especially granitic, in a greater degree emphasizes the structural mosaicity, considerable duration of the consolidated crust formation, complex processes of accretion and often differently directed collision.

Pacific style of evolution when the new supercontinents emerged (such as Northern Eurasia after closure of the Paleoasian Ocean in the Late Paleozoic) was changed by ***Indo-Atlantic style***. Processes of non-linear geodynamics were superseded by linear processes which are the constituent parts of standard "Wilson's Cycle". Cycle includes the following successive evolutional stages: continental rifting; spreading increase of neogenic ocean; its reduction with formation of ensimatic volcanic arc and marginal sea (or of the Andean-type active continental margin with volcanic belt); continental collision with a widespread formation of collision granite intrusions and batholiths. One of the most important features of the Indo-Atlantic structures is the presence of narrow and lengthy ophiolite zones and sutures. They include, as a rule, metamorphosed paleooceanic and intra-oceanic (island arc and marginal sea) complexes, supra-subduction ophiolites and ophiolitic mélanges.

Linear structures (and sutures) are typical, first of all, for the Tethyan domain of the Central Asia, i.e. from the Paleotethys ophiolite zone of the Northern Pamirs—Kunlun—Qinling—Dabeishan in the north and southward in the Pamirs—Tibetan—Himalayan part of the Tectonic map (see Fig. 18). These narrow and lengthy (on tectonic map more than 3000 km) zones are controlled by outcrops of ophiolitic (serpentinite) mélange and neighboring belts of supra-subduction and collision granitoid intrusions. The seismo-active strike-slips are situated in the recent structures along zones and sutures, as the most rheologically unstable structures.

Mongol-Okhotsk linear belt is one of the best representatives of changing of the Pacific accretionary style of crustal tectonic evolution by the Indo-Atlantic collision one. It seems to us that this tendency is the general regularity of the Earth's evolution when supercontinents combined due to accretion processes start to break-up with formation of paleooceans of the Atlantic type. Origination of the Mongol-Okhotsk paleoocean (the same of the South Anyui Paleoocean) took place as the result of rifting of heterogeneous consolidated crust including both the ancient blocks with thick continental crust and relatively thin and weakly consolidated accretionary systems of paleooceanic complexes (see Chaps. 8 and 12). This rifting and later collision in conditions of heterogeneous crust determined the peculiarity of collision and post-collision magmatism of the Mongol-Okhotsk belt, specificity of geochemistry and, as consequence, variety of different ore deposits.

Carrying out analysis of the tectonic composition of the Central Asian fold belt we separate in the map the following main structures:

- cratons,
- fold belts,
- sedimentary basins.

Cratons in the Russian part of the map are represented by the largest continental blocks: East European and Siberian platforms. The most ancient continental crust has a more widespread distribution than their shield and platform parts. For example, the largest part of Siberian craton is covered by complexes of peri-continental marginal basins (in Mesozoides of Northeast Russia) or by the Mesozoic volcanogenic rocks of the Pacific (Okhotsk-Chukotka) volcanic belt.

Formation of the continental crust in both cratons had finished at the end of the Paleoproterozoic and, since the Riphean (Mesoproterozoic) the platform regime was settled. Siberian Craton in the Riphean (since Mesoproterozoic) was a very stable structure and the Riphean (Mesoproterozoic) sediments are distributed practically on the whole territory of the platform. In the East European Platform, the Riphean (beginning from the Mesoproterozoic) is represented by aulacogen's piftogeneous facies; cover sediments, beginning from the Vendian (Ediacaran), are widespread everywhere. We regard both platforms as structures with consolidated crust formed to the beginning of the Mesoproterozoic.

Age of the crystalline basement and time of the crust consolidation for North China (Sino-Korean) craton and Yangtze Platform are described in Chaps. 13 and 14 of this Explanatory Note.

Fold belts of different age in Northern, Central and Eastern Asia include in their structures both neogenic consolidated crust which was formed in the process of evolution of paleooceanic structures and blocks with ancient continental crust, which were "involved" during the process of new consolidated crust formation. These blocks underwent younger granitic magmatism and metamorphism practically synchronous with the accretion-collision processes in the fold belt. Therefore, we do not show them in the scheme of tectonic zoning (see Fig. 18) but describe in the chapters of Explanatory Note devoted to the separated tectonic structures (fold belts and areas) or regional territories. For neogenic consolidated crust the most important complexesindicators reflected in the map are: (a) paleooceanic crust with different parts of ophiolite succession or mélange; (b) sedimentary-volcanogenic and magmatic assemblages of transitional (island arc) stage and complexes-indicators of final formation of the consolidated crust (granitic intrusions and batholiths) and molasses. Exactly the latter distributes all fold structures of Northern, Central and Eastern Asia by the age of consolidated crust. Summing all geological data which are available to the present time, we determine the following structures with consolidated crust among fold belts in the Russian and adjacent parts of the *Tectonic map of Northern-Central-Eastern Asia and Adjacent Areas*, which formed:

- to the end of the Neoproterozoic (Timan and Yenisei ridge);
- to the middle of the Devonian (Altay-Sayan and Sayan-Baikal mountain areas, Kazakhstan, Northern Tien Shan, Altun—Qilian—Northern Qinling and others);
- to the end of the Carboniferous—Early Permian (East Urals, Ob-Zaisan fold area, Southern Tien Shan, Central Kazakhstan and Junggar-Balkhash fold area, western part of the Mongol-Okhotsk fold belt and others);
- to the Middle—Late Triassic (Northern Pamirs, Kunlun ophiolite zones, Beishan—Solonker belt and others);
- to the beginning of the Cretacous (Oloy fold zone and South Anyui ophiolite zone, Sikhote-Alin);
- to the end of the Late Cretaceous (Koryak fold-thrust belt);
- in the Cenozoic (Kamchatka and Sakhalin).

We excellently understand that consolidated crust of the most above mentioned fold structures formed during enough long time and went through several phases. The best indicator showing the final formation of consolidated crust is the time of sedimentation starting and formation of sedimentary covers of young platforms and sedimentary basins.

Sedimentary basins principally distinguish form the sedimentary covers of platforms by their structure and evolution style. If in the history of platform evolution their cover in lateral direction was changed by shelf and slope facies of the surrounding paleooceans, then the sedimentary basin, as a rule, originated in the interior of fold or orogenic structures framing. This determined the properties of internal structure and sedimentation in basins. For the sedimentary basin the following features are predominant: heterogeneous and different-aged separated blocks of the basement, its relative tectonic instability; considerable and huge thicknesses in the central part of the basin; mainly terrigenous character of sediments; sometimes internal tectonic reconstruction in the sedimentary basin.

Thus, in the West Siberian Basin, since the Late Triassic, the sedimentary successions formed, the thickness of which exceeded the long-time accumulated cover of the adjacent East European and Siberian platforms. Heterogeneous basement of the West Siberian Basin (see Chap. 5) includes the pre-Riphean (i.e. pre-Mesoproterozoic) continental crust of the Siberian Craton, Caledonian consolidated crust of Northern

Kazakhstan, Variscan accretionary consolidated crust of the East Uralian zone and Ob-Zaisan fold area. This does not rule out that such heterogeneity and tectonic activity along boundaries of rheologically different blocks determined the conditions of sedimentation in the basins.

Considerable internal reorganization of the basin in the process of its evolution is peculiar only of the Mesozoic basins in the northeast and east of China: Songliao (at the Early and Late Cretaceous boundary) and Huabei (at the Paleogene and Neogene boundary). It was connected with the Yanshanian movements in the limits of North China (Sino-Korean) Platform and continuing collision between North China (Sino-Korean) and Yangtze platforms.

Turan Basin (or plate) represents an exception from the above mentioned examples, and it is the epi-Variscan young platform. Deformations in its basement and cover are caused by continuing collision of the Arabian Plate with the south of Eurasia and connected with simultaneous origination of tectonic structure and growth of the Kopet Dagh Mountain Ridge (see Chap. 11).

In the Russian part of the map, the main basins are West Siberian and Amur-Zeya ones, and in Chinese part—Songliao, Huabei, Sichuan, Junggar, Hami-Turpan and Qaidam basins.

References

Khain V, Leonov Y (eds-in-chief) (1982) International Tectonic map of the World, scale 1:15,000,000. Academy of Sciences of the USSR—Commission for the Geological Map of the World, 12 sheets

Khain V, Leonov Y (eds-in-chief) (1996) International Tectonic map of Europe, scale 1:5,000,000. CGMW-VSEGEI, 3rd edn, 5 sheets

Leonov MG (2008) Tectonics of consolidated crust. Nauka, Moscow (Transactions of Geological institute, Russian academy of sciences, 575), p 453 (in Russian)

Peive AV (1969) Oceanic crust of the geological past. Geotektonika 4:5–23 (in Russian)

Peive AV, Yanshin AL, Zonenshain LP, Knipper AL, Markov MS, Mossakovsky AA, Perfiliev AS, Pushcharovsky YM (1976) Formation of the continental earth crust in connection with creation of the new tectonic map. Geotektonika 5:6–24 (in Russian)

Peive A, Yanshin A (eds-in-chief) (1979) Tectonic map of Northern Eurasia, scale 1:5,000,000, composition. Depending on the distribution of plate complexes in the area, the map reflects, Moscow, 8 sheets

Petrov O, Leonov Y, Li T, Tomurtogoo O (eds-in-Chief) (2008) Tectonic Map of Central Asia and adjacent areas. In: Shokalsky S, Pospelov I, Koshkin V, Chen B, Hwang JH (Executive eds). VSEGEI, St. Petersburg, 9 sheets

Til'man S, Bogdanov N (1992)Tectonic map of Northeast Asia, scale 1:2,500,000. Committee on Geodesy and Cartography, Ministry of Ecology and Natural Resources, Russian Federation, Moscow, 2 sheets

Deep Structure Model

S. N. Kashubin, O. V. Petrov, E. D. Milshtein, E. A. Androsov, I. Y. Vinokurov, Y. M. Erinchek, T. P. Litvinova, V. N. Mukhin, Y. S. Golysheva, T. V. Kashubina, N. A. Krupnova, T. S. Sakulina, S. V. Suslova, O. A. Tarasova, T. M. Yavarova, and D. V. Vyatkina

Abstract The chapter provides results of the generalization and interpretation of deep geophysical studies. The model of the deep crustal structure is represented by a set of maps, showing the crustal thickness, and a 5400 km-long Geotransect, crossing major tectonic areas of northeastern Eurasia. A set of digital maps covering an area of 50,000,000 km^2 is compiled in a single projection and includes the following maps: the Moho depth; the thickness of main crustal units (sedimentary cover and consolidated crust); the anomalous gravity field and anomalous magnetic field used for zoning of the area; crustal types. The Geotransect crosses North-Eastern Eurasia and characterizes the vertical section of the earth's crust and upper mantle in the passive margin of the Eurasian continent (including submarine elevations of the Arctic Ocean and its shelf part), the active eastern continental margin and extends into the Pacific Plate.

Deep structure 3D model is represented by a set of maps showing crustal thickness parameters and by a geotransect. A set of digital maps covering an area of 50 million km^2 was compiled in a single projection and comprises the following maps: Moho depth, thickness of the main units of the Earth's crust (sedimentary cover and consolidated crust), gravity and magnetic anomaly maps used as the basis for zoning the area and a map of crustal types. All the available maps are reduced to a single coordinate system in Lambert Conformal Conic projection with a central meridian E 92° and standard parallels N 35° and N 65°. The geotransect extending for about 5000 km, crosses the eastern part of the project and characterizes the vertical section of the Earth's crust and upper mantle of the passive margin of the Eurasian

S. N. Kashubin (✉) · O. V. Petrov · E. D. Milshtein · E. A. Androsov · I. Y. Vinokurov · Y. M. Erinchek · T. P. Litvinova · V. N. Mukhin · Y. S. Golysheva · T. V. Kashubina · N. A. Krupnova · S. V. Suslova · O. A. Tarasova · T. M. Yavarova · D. V. Vyatkina
Russian Geological Research Institute (VSEGEI), 74 Sredny Prospect, St. Petersburg 199106, Russia
e-mail: Sergey_Kashubin@vsegei.ru

T. S. Sakulina
Institute of Earth Sciences of St. Petersburg University, 7–9 Universitetskaya nab., St. Petersburg 199304, Russia

O. V. Petrov and S. Dong (eds.), *Tectonics of Asia (Northern, Central and Eastern Asia)*, Springer Geology, https://doi.org/10.1007/978-3-030-62001-1_2

continent (including deepsea highs in the Arctic Ocean and its offshore part), the eastern continental margin proper and extends into the Pacific Plate. The Geological Survey of Russia (VSEGEI) was the coordinator and main contractor for the compilation of the 3D model. For compiling the set of maps, the data presented by participants of the International Project "Deep Processes and Metallogeny of North, Central and East Asia" were used.

1 Gravity and Magnetic Anomaly Maps

Gravity and magnetic anomaly maps were constructed by compilation and matching of digital data sets, both provided by the project participants and publicly available data. The base sets were GIS data of the Atlas of Geological Maps of Russia, the CIS Countries and Adjacent States, scale 1: 2,500,000 (Ch. Ed. Petrov O. V. 2008, https://vsegei.ru/ru/info/gis_cis). For the northern latitudes (from 70°), digital data of the Circum-Arctic Mapping Project were used (CAMP-M, Gaina et al. 2010) (CAMP-GM, https://earth-info.nima.mil/GandG/wgs84/agp/index.html). For expanding the maps in the area of the eastern marginal seas and the Pacific Ocean as well as general matching of digital data sets, the data of two global maps were used—the Earth Magnetic Anomaly Grid (EMAG2 https://www.geomag.us/models/emag2.html, Maus et al. 2009) and the development and evaluation of the Earth Gravitational Model 2008 (EGM2008, Pavlis et al. 2012, https://bgi.obs-mip.fr/data-products/Grids-and-models).

Magnetic anomaly map was constructed by compilation of the above maps and original data sets on the territory of China, Korea and Mongolia submitted by the project participants (Fig. 1).

Maps were compiled in several stages. At the first stage, maps having a significant commonality of territories (the territory of Russia and the CIS countries, northern latitudes from 70° and EMAG2) were matched according to the overlap system. The Magnetic Anomaly Map (ΔT) of Russia, the CIS countries and adjacent water areas was adopted as the base level (Petrov et al. 2008). Map levels were matched

Territory	Maps	Projection	Grid	Year
Russia, CIS states	Magnetic Anomaly Map	Krasovsky 1940 Equidistant Conic	2.5x2.5 km	2008
North of N 70°	CAMP-M	Stereographic_North_Pole	2x2 km	2009
EMAG2	Earth Magnetic Anomaly Grid	WGS 1984	2'x2'	2009
Korea	Magnetic anomaly map of Korea	WGS 1984	2'x2'	2013
China	Aeromagnetic anomaly map of China	Belge Lambert 2005	5x5 km	2016
Mongolia	Mongolia Aeromagnetic Mapping Project	Lambert Conical	1x1 km	2018

Fig. 1 Main cartographic materials used in compilation of the Magnetic Anomaly Map of Central and North East Eurasia

for the areas without intense magnetic anomalies. Digital sets of magnetic anomaly maps presented by the Geological Surveys of the Republic of Korea, the People's Republic of China and Mongolia were reduced to the selected level using overlaps with EMAG2.

As a result, the Magnetic Anomaly Map of Central and North East Eurasia was compiled at the same level with a grid size 5 × 5 km (Fig. 2).

Gravity Map was constructed in Bouguer reduction with the intermediate layer density 2.67 g/cm^3 based on compilation of three digital gravity maps (Fig. 3: (https://vsegei.ru/ru/info/gis_cis; EGM2008-https://bgi.obs-mip.fr/data-products/Grids-and-models; ArcGP-https://earth-info.nima.mil/GandG/wgs84/agp/index.html). Data on the area of the Arctic Ocean seas and EGM2008 were reduced to the normal field level calculated from Helmert's formula (1901) $\gamma = 9.78030 \times (1 + 0.005302 \times \sin^2\varphi - 0.000007 \times \sin^2 2\varphi) - 14$, *mGal*, where γ is normal field;

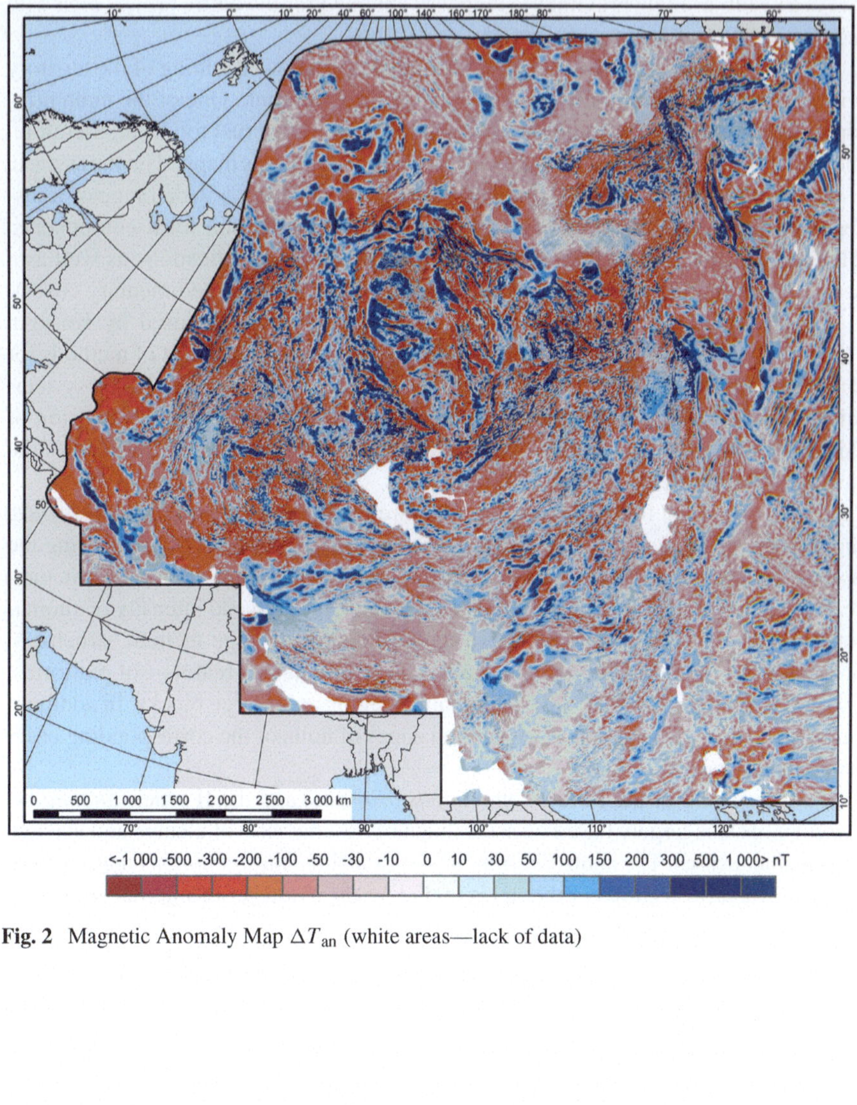

Fig. 2 Magnetic Anomaly Map ΔT_{an} (white areas—lack of data)

Maps	Projection	Grid	Year
Gravity Map of Russia, CIS Countries and Adjacent States	Krasovsky 1940 Equidistant Conic	2.5x2.5 km	2008
EGM2008	WGS 1984	2.5' x 2.5'	2008
ArcGP	Stereographic North Pole	10 x 10 km	2009

Fig. 3 Main cartographic materials used in compilation of the Gravity Map of Central and North East Eurasia

φ, latitude of the observation point. ArcGP and data on the territory of Russia are matched along the coastline.

Gravity maps were reduced to a single level and matched in overlap zones, while preference was given to more detailed maps constructed from the ground or on-board source data. Root-mean-square deviation (RMS) in the overlap zones was ± 5 mGal (without considering anomalous deviations).

As a result, a single gravimetric map was obtained for the entire project area with a 5×5 km digital array grid (Fig. 4).

Analysis of Gravity and Magnetic Anomaly Maps allowed zoning of the territory with distinguishing and outlining structures of different rank. One of the main principles of zoning was outlining structures at the crystalline crust top level; therefore, their boundaries may not coincide with geological boundaries mapped on the surface.

Zoning was based on a comprehensive analysis including a wide range of transforms: field upward continuation, calculations of regional and local components, TILT transformations (Miller and Singh 1994), various classifications as well as a set of deep structure maps (thickness of consolidated crust and sedimentary cover).

Zoning was based on the principles of tectonic zoning proposed by Kosygin (1975). According to these principles, zoning was considered as a set of methods for delimiting the space (including 3D version) in accordance with the chosen systematics of bodies (rank) in keeping with the rules for the complete exact division of this space without crossing the boundaries and individuality of characteristics of the distinguished elements (Voronin 2007).

Zoning provided for distinguishing of a three-rank system: *anomalous province, anomalous realm and anomalous region* (in decreasing order). When outlining the units of different orders, emphasis was placed on a combination of morphostructural features of the potential fields and, to a lesser extent, the sign and intensity of anomalies. The distinguished units reflect structural features of large tectonic structures: anomalous provinces—of continents and oceans; anomalous realms—of platforms, oceanic basins, large folded areas; anomalous regions—of large blocks. In addition, linear structures corresponding to junction zones of units of the corresponding order were distinguished.

Distinguishing of the I order (*anomalous province*) and the II order (*anomalous realm*) units was largely based on assessment of the character of change and average crustal thicknesses. The main criterion for distinguishing the *anomalous regions* were morphostructural features of the fields including the character of internal zoning.

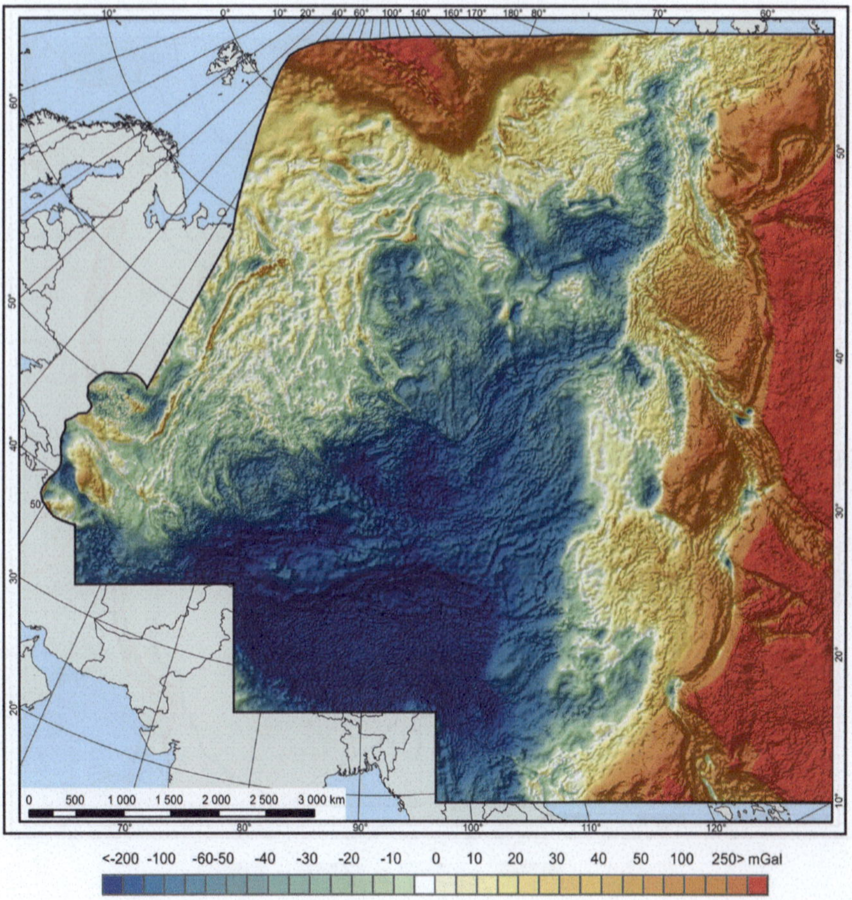

Fig. 4 Gravity Map in Bouguer reduction with intermediate layer density 2.67 g/cm^3

When outlining the units, a significant role was given to gravity and magnetic anomaly field transforms, in particular, the "TILT" transformation.

As a result, a zoning map of potential fields of the Central and North East Eurasia was compiled (Figs. 5 and 6).

Colour indicates provinces: 1—Eurasian, 2—North American, 3—Mid-oceanic ridges, 4—Pacific, 5—Subduction zones. Red lines—boundaries of provinces; blue lines, boundaries of realms; green lines, boundaries of regions.

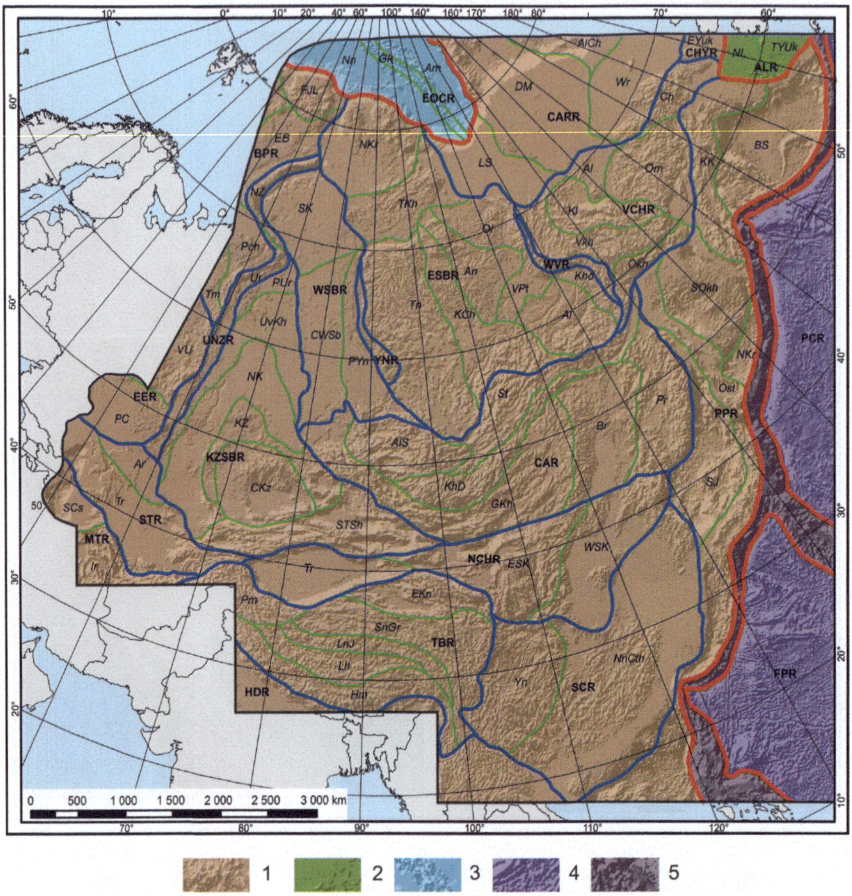

Fig. 5 Zoning scheme of Central and North East Eurasia after the character of potential fields

2 Set of Deep Structure Maps

The importance of information about the Earth's crust and upper mantle structure increases markedly with development of geological 3D mapping technologies (Petrov 2016; Kashubin 2016). This information is becoming more and more needed for resolving the fundamental problems related to the development of geotectonic models that determine understanding of the formation and mineragenic specialization processes of large geological provinces.

Currently, almost the entire territory of Eurasia is to some extent covered by deep seismic studies (Petrov et al. 2016). The common elements of deep structure distinguished on geological-geophysical lines are: (1) Moho depth (M discontinuity identified with the Earth's crust base); (2) base of sedimentary layer or basement surface (depending on the features of geological structure of the regions, the surface

Index in the scheme	Zoning of potential fields (units names)
	Eurasian Province
EER	**East European Realm**
VU	*Volgo-Ural Region*
PC	*Pre-Caspian Region*
BPR	**Barents-Pechora Realm**
EB	*East Barents Region*
FJL	*Franz Josef Land Region*
Tm	*Timan Region*
Pch	*Pechora Region*
WSBR	**West Siberia Realm**
SKr	*South Kara Region*
CWSb	*Central-West Siberian Region*
PYn	*Pre-Yenisei Region*
ESBR	**East-Siberian Realm**
NKr	*North Kara Region*
TKh	*Taimyr-Khatanga Region*
Tn	*Tunguska Region*
KCh	*Kotui-Chon Region*
An	*Anabar Region*
Ol	*Olenek Region*
Al	*Aldan Region*
Khd	*Khandyga Region*
VPt	*Vilyuy-Patom Region*
VCHR	**Verkhoyan-Chukchi Realm**
Vkh	*Verkhoyan Region*
Okh	*Okhotsk Region*
Kl	*Kolyma Region*
Om	*Omolon Region*
Ch	*Chukchi Region*
Al	*Alazeya Region*
CHYR	**Chukchi-Yukon Realm**
EYuk	*East Yukon Region*
CARR	**Central Arctic Realm**
LS	*Laptev Sea Region*
DM	*De Long-Makarov Region*
AlCh	*Alpha-Chukchi Region*
Wr	*Wrangel Region*
KZSBR	**Kazakhstan-Siberia Realm**
UvKh	*Uvat-Khanty-Mansi Region*
STSh	*South Tien Shan Region*
CKz	*Central Kazakhstan Region*
KZ	*Kazakhstab-Zaisan Region*
PUr	*Pre-Urals Region*
NK	*North Kazakhstan Region*
STR	**Scythian-Turan Realm**
Ar	*Aral Region*
Tr	*Turan Region*
YNR	**Enisei Realm**

Index in the scheme	Zoning of potential fields (units names)
	Eurasian Province
WVR	**West Verkhoyan Realm**
UNZR	**Urals-Novaya Zemlya Realm**
Ur	*Urals Region*
NZ	*Novaya Zemlya Region*
CAR	**Central Asia Realm**
KhD	*Khangai-Daurian Region*
Pr	*Primorye Region*
GKh	*Gobi-Khingan Region*
St	*Stanovoi Region*
AlS	*Altai-Sayan Region*
Br	*Bureya Region*
NCHR	**North China Realm**
WSK	*Western Sino-Korean Region*
ESK	*Eastern Sino-Korean Region*
Tr	*Tarim Region*
PPR	**Peri-Pacific Realm**
KK	*Koryak-Kamchatka Region*
SOkh	*Sea of Okhotsk Region*
BS	*Bering Sea Region*
Ost	*Ostrovnoi Region*
NKr	*North Kuril Region*
SJ	*Sea of Japan Region*
TBR	**Tibet Realm**
Pm	*Pamir Region*
EKn	*East Kunlun Region*
SnGr	*Songpan-Garse Region*
LnJ	*Lancang-Jang Region*
Lh	*Lhasa Region*
Hm	*Himalaya Region*
SCR	**South China Realm**
Yn	*Yangtze Region*
NnCth	*Nanhua-Cathaysia Region*
HDR	**Hindustan Realm**
MTR	**Mediterranean Realm**
Ir	*Iran Region*
SCs	*South Caspian Region*
	North American Province
ALR	**Alaska Realm**
NL	*Nest-Laurentian Region*
TYUk	*Tanana-Yukon Region*
	Mid-oceanic Ridge Province
EOCR	**Eurasian Oceanic Realm**
Gk	*Gakkel Region*
Nn	*Nansen Region*
Am	*Amundsen Region*
	Pacific Province
FPR	**Philippine Realm**
PCR	**Pacific Realm**

Fig. 6 Matching of letter symbols in the zoning scheme with the units identified

identified with either top of the consolidated (sedimentary metamorphosed) crust or with the top of the crystalline (upper) Earth's crust; (3) boundary of the upper and lower crystalline crust (of the "granite gneiss" and "basalt" layers). The position of M discontinuity is most reliably determined (from records of supercritical reflected PmP and refracted P-waves and velocities Vp > 7.8–8.0 km/s below this discontinuity) as well as the base of the sedimentary layer (by a characteristic change in wave pattern in the MCS time sections and Vp velocity jump from values lower than 3.5–4.5 km/s to 5.5 km/s and higher). Distinguishing of other deep structure elements is very ambiguous both due to a lower contrast of these elements in geophysical fields, and an uncertainty of their geological interpretation.

By now, a set has been compiled for the entire study area, which is a key element of deep structure maps of the international project "Deep processes and metallogeny of North, Central and East Asia". The set was compiled using ArcGIS tools and is represented by digital maps (sampling interval 5 × 5 km) in a single coordinate system using a topographic base of 1: 5 M scale in Lambert Conformal Conic projection with a central meridian E 92° and standard parallels N 35° and N 65°. The set comprises the following maps showing parameters of the main crustal units: (1) Moho depth (Fig. 7), (2) sedimentary cover thickness (Figs. 8 and 9), (3) consolidated crust thickness (Fig. 10). In addition, the set contains a map of comprehensive zoning of potential fields (see Fig. 5) and a map of crust types (Figs. 11, 12 and 13).

Moho (M) depth plays an important role in the study of deep structure of the Earth. In seismological and global geophysical constructions, the knowledge of this parameter is necessary for calculating the corresponding corrections, and in case of geological interpretation it is a basic element for both structural and geodynamic constructions. M depth is used to determine crustal thickness: on land, a correction is made for terrain height; in water areas, for sea depth. A change in crustal thickness in combination with velocity parameters serves as the main criterion for distinguishing the continental and oceanic types of the Earth's crust when studying the continent-ocean transition areas.

Crustal thickness is determined primarily by seismic methods. Deep seismic sounding (DSS) technique, when the Earth's crust base is identified with Moho (M) discontinuity calculated from the data of refracted and supercritical reflected waves, is generally accepted (Mooney 2007). Sometimes, the Earth's crust base is distinguished in seismic sections obtained by the reflected waves technique (MCS) (Models … 2007), and by distant earthquake converted-wave method (ECWM) (Zolotov et al. 1998). In the absence of seismic data, crustal thickness is estimated using correlation ratios between M depth, topography and Bouguer anomalies (Demenitskaya 1967; Kunin et al. 1987).

As the grid of deep geological-geophysical lines developed, starting in the 1990s, a number of researchers attempted to generalize the information on the Earth's crust base occurrence depths and the main interface within it by compiling the corresponding maps for individual regions (Druzhinin et al. 1990; Structure … 2006; Kostyuchenko and Morozov 2007; Models … 2007; Grad et al. 2009 et al.). Most of these constructions have not lost their relevance to this day. However, due to constant

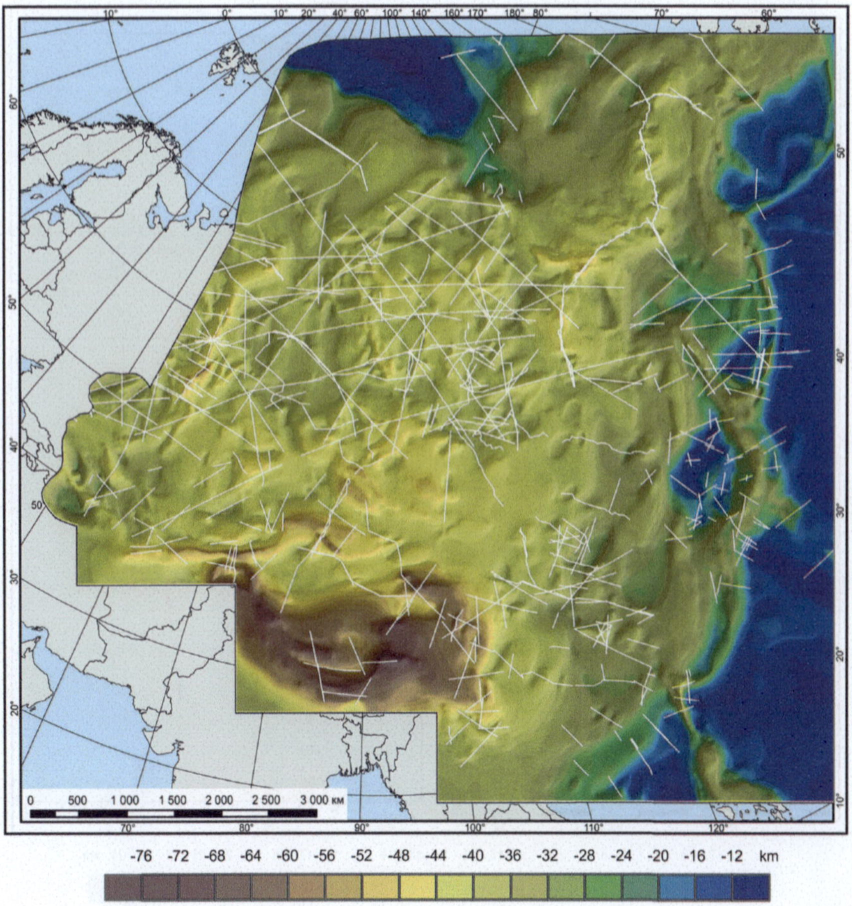

Fig. 7 M depth (white lines show the main deep seismic lines)

No.	Year	Author (Editor)	No.	Year	Author (Editor)
1	1981	Surkov V.S. et al.	13	2005	Kim B.I. et al.
2	1982	Semenovich V.V. et al.	14	2006	Koren T.N. et al.
3	1986	Krasny L.I et al.	15	2007	Surkov V.S. et al.
4	1988	Kostyuchenko S,L. et al.	16	2008	Larichev A.I. et al.
5	1988	Bogdanov N.A. et al.	17	2009	Surkov V.S. et al.
6	1999	Erinchek Yu.M. et al.	18	2009	Divins D.L. et al.
7	2000	Kostyuchenko S,L. et al.	19	2009	Grantz A. et al.
8	2001	Gramberg I.S. et al.	20	2010	Laske G. et al.
9	2001	Parfenov L.M. et al.	21	2010	Shokalsky S.P. et al.
10	2002	Erinchek Yu.M. et al.	22	2011	Sakulina T.S. et al.
11	2004	Gramberg I.S. et al.	23	2011	Poselov V.A. et al.
12	2004	Veselov O.V. et al.	24	2016	Petrov et al.

Fig. 8 Main cartographic materials used in compilation of the sedimentary cover thickness map of Central and North East Eurasia

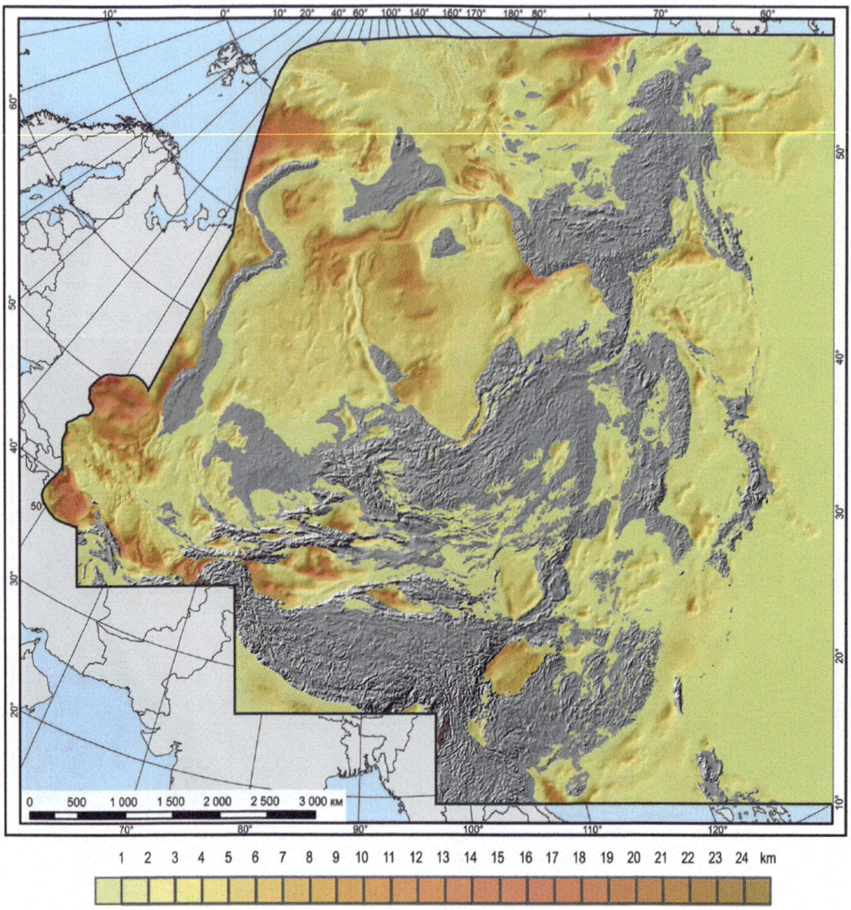

Fig. 9 Sedimentary cover thickness

accumulation of the information about deep structure, there is a need to update these maps and unify their compilation methods.

The map shown in Fig. 7 is compiled following the methodology described in detail in (Kashubin et al. 2011a, b, c). First, the values of M depth taken from seismic sections spaced at 25 km were put on the map of the actual material. Then, to fill in the depth values to Moho discontinuity of interline space and vast areas where seismic data are absent, digital maps of gravity anomaly, surface topography and ocean floor depth were used. Zm depths were calculated separately for the continental and offshore parts of the area with subsequent matching of contours in the area of their junction according to the values of Bouguer anomalies and elevations averaged over a radius of 100 km using the formulas given in (Kashubin et al. 2011a, b, c).

Interpolation error in recalculating Zm depth values into a uniform spacing was estimated by comparing the interpolated and initial values at points where depth

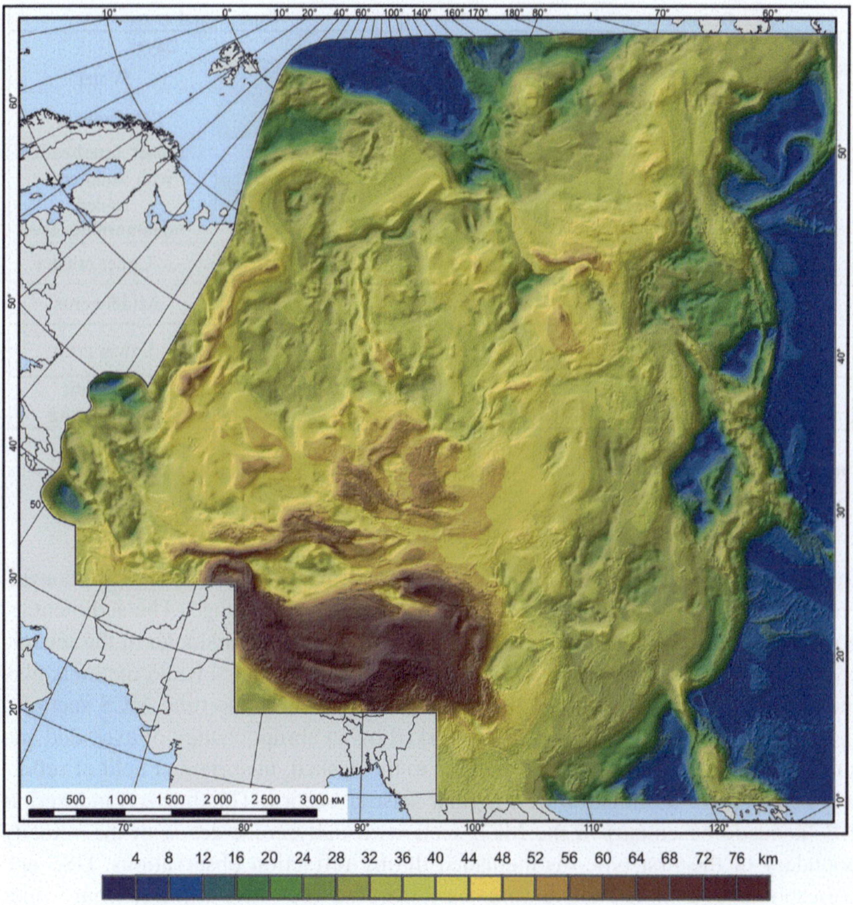

Fig. 10 Consolidated crust thickness

values were taken from seismic data. Standard deviation between the interpolated and initial values did not exceed ±2 km, and, therefore, the section between contours in the resulting map was taken equal to 4 km.

Map of sedimentary cover thickness plays an important role both for understanding the internal crustal structure, and for assessing the prospects of oil and gas presence in the study area.

Sedimentary cover is commonly regarded as slightly dislocated and usually non-metamorphosed rock strata characterized by gentle bedding and constituting the upper part of the crustal section. On continents, the sedimentary cover rests on consolidated crust; and in oceans, on the second oceanic layer. At the same time, intermediate complexes are distinguished in some sedimentary basins between the sedimentary cover and the crystalline basement represented by the formations that are slightly metamorphosed and/or dislocated to varying degrees. Sometimes, these

Oceanic crust			Vp,	Continental crust		
Layer		Vp/Vs	km/s	Vp/Vs		Layer
Water		-	1.45-1.50	-		Water
Sediments		2.1-2.5	2.0-4.5	2.1-2.5		Sediments
2nd oceanic layer		1.8-2.2	4.2-6.0	1.7-2.1		Basalt, interbedded with sediments / folded metamorphic layer
-	-	-	5.8-6.4	1.69-1.73		Upper crust
-	-	-	6.3-6.7	1.73-1.75		Middle crust
3^{d} layer of oceanic crust		1.81-1.87	6.6-7.2	1.75-1.77		Lower crust
Magmatic underplating		1.78-1.84	7.2-7.6	1.78-1.84		Magmatic underplating

Fig. 11 Generalized models of structure and velocity parameters of oceanic and continental crust (Kashubin et al. 2018a, b)

formations are assigned to the sedimentary layer, but more often they are considered as formations of the so-called intermediate structural stage. The sedimentary cover is confidently determined in seismic sections from the character of the seismic record and from values of elastic wave velocities; therefore, seismic methods play the leading role in the study of the sedimentary cover. In the time MCS sections, the sedimentary cover base is usually marked by an abrupt change of extended and subhorizontally oriented seismic features to the dashed, misoriented field of reflectors or complete termination of a regular seismic record. This horizon indexed as AB (acoustic basement) in the MCS sections usually coincides with the velocity boundary of the first type distinguished during refraction observations, DSS and corresponding, as noted above, to a sharp increase in P-wave velocity from values below 3.5–4.5 to 5.5 km/s and higher. As a rule, these features are used to construct the sedimentary cover base from seismic data.

Exploration maturity of the area based on seismic methods is extremely uneven. The West Siberian and East Siberian regions are most adequately covered by seismic studies. Total scope of the MCS surveys within oil and gas provinces (mainly on the shelf and in Western Siberia) exceeds hundreds of thousands of linear kilometers, while most of the data are summarized as medium-scale structural maps. Therefore, to compile a map of sedimentary cover thickness, there is no need to use primary seismic data, but rather use the results of these previous more detailed constructions.

It should be noted that the study area as a whole is covered by cartographic materials for compiling a sedimentary cover thickness map on 1:5 M scale. Information on the main materials used for compiling the sedimentary cover thickness map is given in Fig. 8.

Creation of the compilation map required unification of the data used and their matching in overlap areas of maps. The need for unification is due to the fact that the source cartographic materials are data files of various presentation forms: generalizing maps display both the sedimentary cover thickness and the relief of an uneven-aged basement. Matching of the available maps in their junction areas was carried out by revising the joints with regard for the character of potential fields. In overlap areas, preference was given to more detailed information. In the integrated map, isopaches with 1 km cross section represent the thickness of the sedimentary cover of different age (Fig. 9).

Map of consolidated crust thickness (Fig. 10) was compiled as difference between the crustal and sedimentary cover thickness maps.

Crustal types of Central, North East Asia, Far East and Arctic continent-ocean transition areas. The set of maps shown above in Figs. 5, 7, 9, 10 and the published deep geological-geophysical sections along the lines, the position of which is shown in Fig. 7, reflect the main features of deep crustal and upper mantle structure of Eurasia. Proceeding from these data and the notions about crustal types described in (Belousov and Pavlenkova 1989; Mooney 2007; Kashubin 2013, 2018a, b), a sketch map of crustal types of Central, North East Asia and continent-ocean transition areas presented in Fig. 12 was compiled. All these typifications of the Earth's crust are based on differences in oceanic and continental crust in terms of thickness, internal structure, and composition.

According to these parameters, the characteristic features of continental crust are: a great thickness (as a rule, over 30 km) and the occurrence in the consolidated part of the crust of a thick (to 10 km or more) upper layer with P-wave velocities of 5.8–6.4 km/s, often called a "granite gneiss" layer. Oceanic crust, in contrast to the continental crust, is thin (as a rule, less than 10 km); there is no "granite gneiss" layer in it, and it is almost completely represented by rocks with P-wave velocities over 6.5 km/s. Detailed seismic studies covering the active and passive margins of continents and oceanic rises showed that, in addition to continental and oceanic crust, crust with intermediate parameters is often found. It is characterized by a thickness from 10 to 30 km and the granite gneiss layer in it is substantially reduced or absent. The assignment of this crust to the oceanic or continental type is often ambiguous; therefore, it is often proposed to define such crust as transitional (Belousov and Pavlenkova 1989, etc.).

Oceanic and continental crust differs markedly in thickness and velocity characteristics. These differences are especially evident on velocity models constructed from the multiwave seismic data using Vp/Vs ratios as an additional characteristic of the medium (Ljones 2004; Breivik et al. 2005; Mjelde 2009; Kashubin 2016, 2018a, b). In the crystalline crust of the continents, the Vp/Vs ratio rarely exceeds 1.77, while in the 2nd and 3rd layers of oceanic crust Vp/Vs is usually 1.85–1.90. Moreover, in the sedimentary layer, both in oceanic and continental crust, Vp/Vs varies in a wide range generally exceeding 1.9–2.0. Considering the relationship between total silica content in crystalline rocks and the Vp/Vs ratio (Aleinikov et al. 1991), these differences seem to be quite natural and indicate different basicities of oceanic and continental types of the Earth's crust.

Figure 11 summarizes data on the models of structure and velocity parameters of the main layers of the oceanic and continental crust, which, along with data on the thickness of different layers and crustal thickness as a whole, serve as the basis for typification of the Earth's crust. As can be seen from the Table, the difference between the continental and the oceanic crust is that in the latter there is no upper and middle crust. This is most reliably recorded by differences in Vp/Vs ratios.

As type columns of the Earth's crust for the blocks shown in the sketch map of crustal types (Fig. 12 (beginning)), actual columns compiled on the basis of the published seismic models are given (Fig. 13 (end)) in accordance with the generalized velocity parameters in Fig. 11. It should be noted that this scheme based solely on geophysical data bearing information about deep structure, in some cases, can significantly differ from the current geological and tectonic notions. Moreover, according to the authors, the data presented can and should stimulate the appearance of various geological tectonic constructions in the region including the alternative ones.

As can be seen from Fig. 12, the Earth's crust in the region is highly diverse. Blocks with both thin (less than 5–6 km), virtually 2-layer oceanic crust, and with very thick (over 70 km) Himalayan crust and the 4-layer consolidated Urals crust are distinguished here. Three main types of the Earth's crust are distinguished in the scheme: oceanic, transitional and continental, each of which, in turn, is divided into a number of subtypes (see the end of Fig. 13).

Two subtypes of **oceanic crust** (subtypes 1 and 2 in Fig. 12) differ primarily in crustal thickness. Thin crust (less than 5–6 km) is represented by two layers (the 2nd and 3rd layers of oceanic crust in Fig. 11) overlain by thin sediments (Iwasaki et al. 2013). It occurs in the deep part of the Pacific and the Eurasian Basin of the Arctic Ocean. In the area of the Bonin Ridge in the Pacific, a rise is identified noted for a much thicker (over 20 km) oceanic crust similar in its properties to the thickened crust of the East Pacific Rises and oceanic plateaus. In this case, thickness increases due to appearance of a crustal-mantle complex with P-wave velocities of 7.4–7.6 km/s in the lower crust (Iwasaki et al. 2013). Despite a significant thickness and similarity of this parameter with the continental crust, the crust of this Rise by its geological position belongs to the oceanic type.

Crust of the Far East continent-ocean transition area is distinguished as a special, transitional type. It is represented by an extended linear subduction zone (subtype 3 in Fig. 12), in which the crust of the Pacific Plate submerges beneath the continental margin of Eurasia (Nakanishi 2009).

DSS data (Iwasaki et al. 2013; Nakanishi 2009) confidently record the velocity parameters of the suprasubduction zone with a rather typical continental crust. The subduction zone proper is usually distinguished after seismological data as a seismic focal zone with a dip about 35°–45° towards the continent. Projections of the earthquake sources on the day surface in this zone representing the transition from the Pacific Plate to the Eurasian continent form a 150–200 km wide band extending for several thousand kilometers (Kanao et al. 2015). Areal parameters of the zone correspond to projection onto the Earth's surface of the sections of continental crust and oceanic crust underthrust beneath it the combined along the Benioff-Zavaritsky zone. A specific structure and extent of this geological phenomenon motivated the

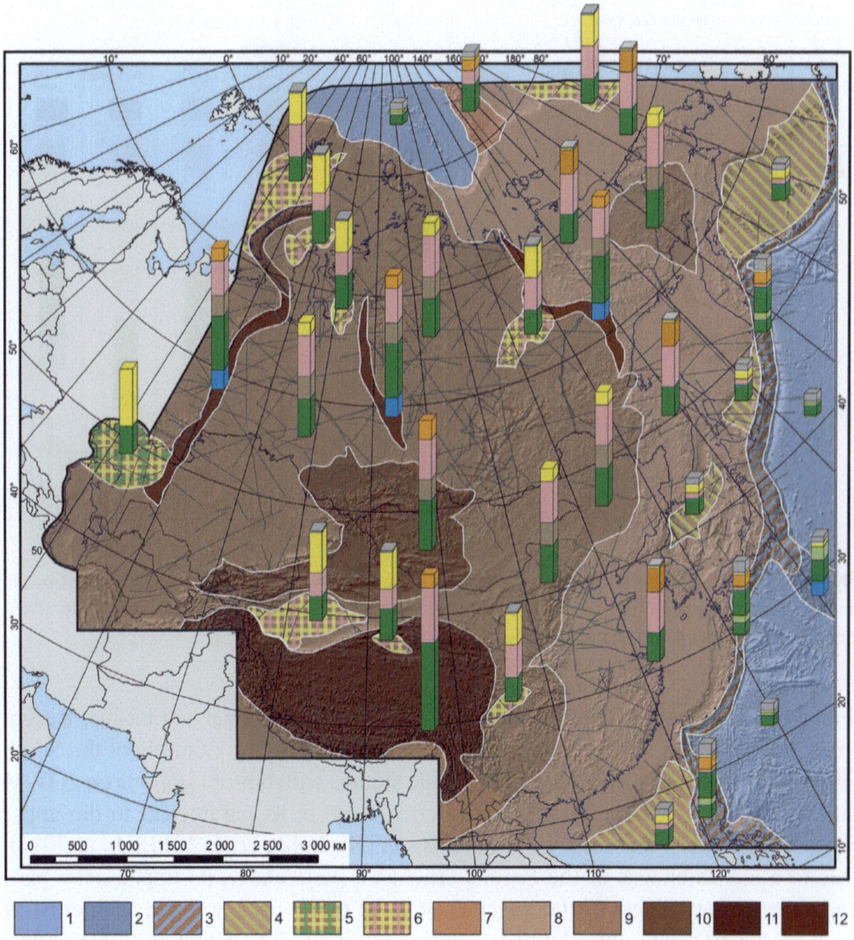

Fig. 12 (Beginning). Sketch map of crustal types ***1, 2—oceanic crust***: 1—oceanic basins, 2—oceanic plateaus; ***3, 4—transitional crust***: 3—subduction zones, 4—back-arc basins; ***5–12—continental crust***: 5—crust of deep sedimentary basins (cis-Caspian type, "granite-free"), 6—crust of deep sedimentary basins (paleorift type), 7—crust of submarine ridges and rises, 8—crust of marginal folded areas and shelf seas, mainly with a two-layer crystalline crust, 9—crust of platforms and median massifs, 10—crust of intracontinental folded areas, mainly with a three-layer crystalline crust, 11—crust of folded-thrust areas (Himalayan type), 12—crust of intracontinental fold belts and boundary zones. Grey lines show refraction, DSS lines. Type columns of the Earth's crust based on seismic data see at the end of the Figure

authors to distinguish a separate subtype of the subduction zone crust (subtype 3 in Fig. 12).

Another contrasting crustal subtype in this continental-ocean transitional area (subtype 4 in Fig. 12) was a markedly thinned crust of deep-sea troughs of back-arc basins characterized by a reduced thickness of the crystalline crust and an increased

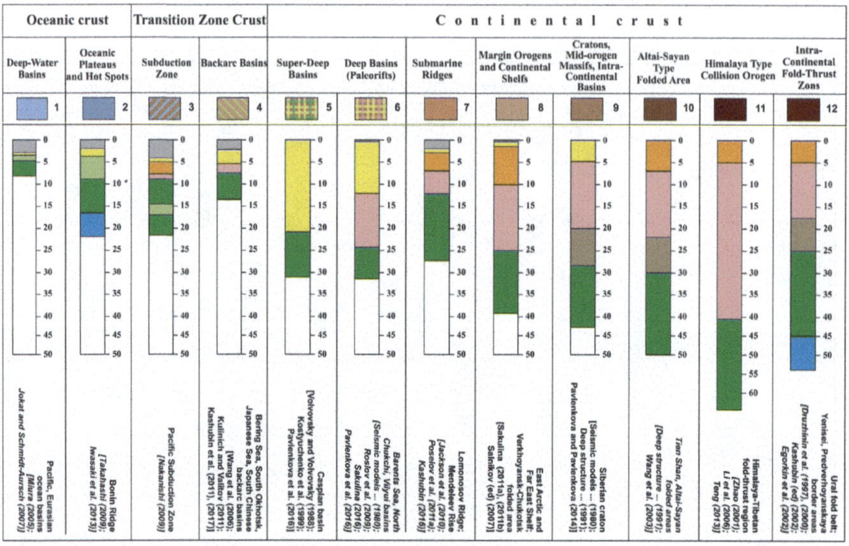

Fig. 13 (End). Type columns of crustal structures of Central, North East Asia and continental-ocean transition areas compiled on the basis of published seismic models in accordance with generalized velocity parameters given in Fig. 11

thickness of sediments from the rest—shelf—part of marginal sea basins. Recent studies of the Earth's crust of this subtype have shown on the example of the South Okhotsk deep-sea basin (multi-wave seismic studies with bottom stations on 2DV-M line in the Sea of Okhotsk), that a thin crust close in its parameters to the upper continental crust is recorded within this Basin (Kashubin et al. 2011a, b, c, 2017). Due to a low thickness of the Earth's crust and, particularly, its upper sialic part, and velocity parameters of the lower crust close to the oceanic type, such crust of back-arc basins can be considered as "suboceanic crust". The extent of its occurrence in the transition area requires clarification, since the detailed materials justifying this crustal subtype are so far available only for the Sea of Okhotsk marginal basin.

Continental crust covering most of the study area, comprises the remaining subtypes (subtypes from 5 to 12 in Fig. 12) combined into four main groups.

The first group comprises the 5th and 6th subtypes of the Earth's crust of deep sedimentary basins. Both subtypes are characterized by a reduced crust due to a decrease of its consolidated part and the thickest (to 18–20 km) sedimentary filling (see Fig. 9). Subtypes differ in the number of layers distinguished in the crystalline crust. The 5th subtype comprises sedimentary basins with a "granite-free" (or "suboceanic") crust, such as the cis-Caspian Basin (Volvovsky and Volvovsky 1988; Pavlenkova et al. 2016; Kostyuchenko et al. 1999; Seismic models … 1980).

Other deep sedimentary basins distinguished in the map (North Barents, South Kara, South Chukotka, Vilyui, etc.) are underlain by a 2-layer crystalline crust (Pavlenkova et al. 2016; Sakulina 2016; Seismic models … 1980; Roslov et al. 2009).

Riftogenic structures occur, as a rule, at the base of most basins of this group.

Folded-metamorphic layer ("transitional complex") is found in the upper part of the basement of some basins, which occur mainly within young plates (South Kara, Pur-Gydan, North Chukotka basins). In others, for example, the North Barents or Vilyui basins usually lying within ancient platforms with a Proterozoic basement, the folded complex can be absent.

The second group in the study area is represented by only one subtype (subtype 7 in Fig. 12), corresponding to the crust of subsea rises forming in the environment of rift passive continental margins. Such rises are, for example, the Lomonosov Ridge and the Mendeleev Rise.

Continental nature of the Lomonosov Ridge crust is recognized by almost all Arctic researchers (Jackson et al. 2010; Poselov et al. 2011a). Currently, the Lomonosov Ridge is considered as an underwater extension of the margins of the Eurasian continent and Greenland joining in the form of the Lomonosov Ridge in the deepsea part of the Arctic Basin.

Earth's crust of the Alpha-Mendeleev Rise system recently studied along several DSS lines (Lebedeva-Ivanova et al. 2006; Poselov et al. 2011a, b; Funck et al. 2011; Kashubin 2016; Kashubin et al. 2018a, b), also corresponds to the continental type, but differs from the typical continental crust by a greater thickness of its lower part (according to V. V. Belousov, due to "basification"). Now it is presumed that this feature of deep structure of the above Central Arctic Rises is due to basitic reworking of the sialic continental crust and, in particular, underplating, formation of peripheral basalt foci in the crust and intrusion of mafic complexes during the formation period of the vast Cretaceous magmatic province HALIP—High Arctic Large Igneous Province—in this part of the Arctic (Morozov et al. 2013; Buchan and Ernst 2006; Embry 1991; Estrada et al. 1999; Petrov 2016).

The third group combines the 8th and 9th subtypes characterized by the normal continental crust of platforms and marginal folded areas framing them on land and shelf. They cover most of North East Asia and differ in the number of layers distinguished in the crystalline crust. Thus, subtype 8 in Fig. 12 is represented by the Verkhoyansk-Chukotka and Amur folded areas with adjacent East Arctic and Far East shelves. Crustal thickness for this subtype does not exceed 35–40 km, and the upper and lower crystalline crust are usually distinguished in the crystalline crust (Sakulina et al. 2011a, b; Structure and composition … 2007). West Siberian Plate, Siberian Craton, Omolon Block, Kazakhstan folded area and other regions of Central and East Asia, the crust of which is characterized by a total thickness of 40–45 km and, as a rule, a 3-layer structure, are assigned to subtype 9 in Fig. 12.

The fourth group characterizes crustal subtypes of extended intracontinental areas (collision folded belts) and boundary zones with a markedly thickened crust (subtypes 10–12 in Fig. 12). All three subtypes differ in the number of layers distinguished in the crystalline crust. Subtype 10 combines the Tien Shan and Altai-Sayan folded areas with a 3-layer (upper-middle-lower) crystalline crust with a thickness of 50 km or more (Deep structure 1991; Wang et al. 2003). Subtype II comprises the Himalayan–Tibetan folded-thrust area with the thickest crust over 70 km (Li et al. 2006).

Urals folded belt with total crustal thickness of about 55 km, within which under the 3-layer crystalline crust a crustal-mantle layer is consistently distinguished, is assigned to subtype 12 ("Granite" Geotraverse… 2002; Druzhinin et al. 1997, 2000 etc.) as well as the Yenisei and Pre-Verkhoyansk "boundary zones" distinguished along the western and north-eastern boundaries of the Siberian Craton (Egorkin et al. 2002).

Thus, the seismic studies in the Central and North East Asia as well as the Far East and Arctic continent-ocean transition areas revealed a significant heterogeneity of crustal structure, which allows distinguishing its definite types and subtypes. The accomplished constructions showed a regular decrease in total thickness of the Earth's crust from the central part of Eurasia to its Far East margin and further to the Pacific. A decrease in consolidated crust thickness is associated with transition from predominantly 3-layer crystalline crust in the center of the continent to a 2-layer consolidated crust at the margin of the continent and within the shelf seas. The established pattern and the distinguished crustal types and subtypes can be used to improve the geological and prognostic mineragenic constructions in Central and North East Asia and in the adjacent Far East and Arctic continent-ocean transition areas.

3 Geotransect

One of the elements of studying deep structure of large lithosphere blocks is running super-long geotransects that reflect the relationships of regional tectonic structures with each other and allow illustrating the geological evolution history of the study area. Modern geotransect is an element of the 3D model of deep structure of lithosphere and is a series of geophysical monomethod sections of the Earth's crust and upper mantle of the Earth along the selected line and a set of interpretation models including the geological-geophysical model.

Composite geotransect "Arctic Ocean—North East Russia—Pacific" extending for 5400 km crosses essentially different tectonic areas: Arctic passive continental margin, Mesozoic folded area, active continental margin (Fig. 14). It is supported to the fullest extent possible by actual deep geophysical observations and represents a continuous junction of four reference geological-geophysical lines: Arctic-2005, 5-AR, 2-DV, 2-DV-M—and their extension into the Pacific Ocean Plate.

Geotransect model is represented by a series of monomethod geophysical sections (composite seismic sections (MCS, DSS), density section) that provide the factual basis for all further constructions, and the interpretation geological-geophysical model based on their combination, which ensures the most complete and visual demonstration of the original factual data on deep structure, confirming the geological-geophysical interpretation.

Over its length, the geotransect intersects a number of large tectonic structures (from north to south): Mendeleev Rise, Vilkitsky Basin (North-Chukchi Basin), Novosibirsk-Chukotka fold system as part of the Wrangel cratonic block (median

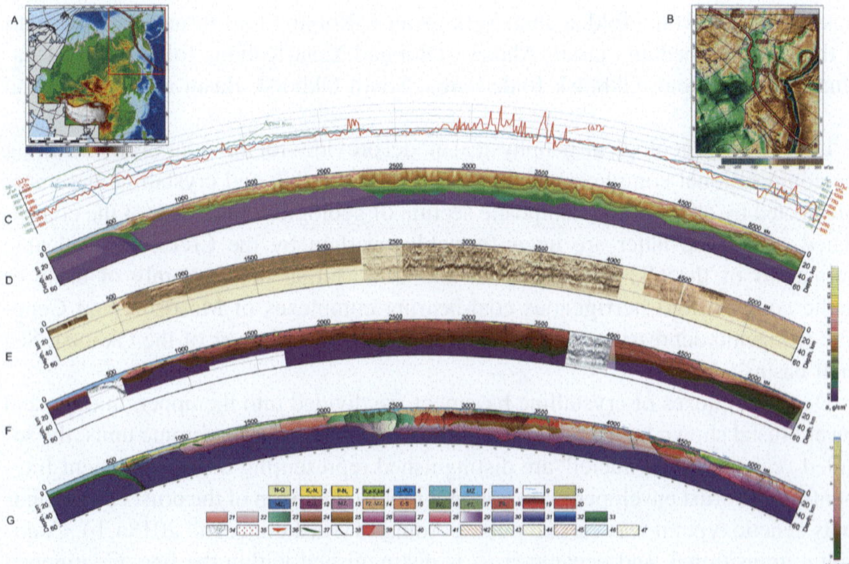

Fig. 14 Geotransect "Pacific Ocean—North East Russi—Arctic Ocean" A—line location scheme, B—map of gravity anomalies (combined free-air-Bouguer reduction), C—diagrams of gravity anomalies in free-air-Bouguer and Bouguer reductions (2.67 g/cm3) and magnetic anomalies, D—density model, E—MCS seismic section, F—DSS seismic section, G—geological-geophysical model 1–8—non-dislocated SCC (crustal structural and compositional complexes) of sedimentary cover: 1—Neogene-Quaternary terrigenous, 2—Upper Cretaceous—Lower Miocene terrigenous, 3—Paleogene-Miocene volcanogenic, 4—Lower Brookian terrigenous, 5—Beaufort (synrift) terrigenous, 6—Late Mesozoic sedimentary volcanogenic, 7—Mesozoic sedimentary volcanogenic (undivided), 8—Mesozoic terrigenous (undivided); 9—clinoform complexes; 10—complexes of an accretionary prism; 11–12—dislocated SCC of the Arctida platform cover: 11—Middle Mesozoic terrigenous, 12—Ellesmerian carbonate terrigenous (undivided), 13–18—folded metamorphic complex of consolidated crust: 13—Lower Mesozoic terrigenous, 14—Upper Paleozoic-Lower Mesozoic terrigenous carbonate 15—Franklin terrigenous carbonate, 16—Lower-Middle Paleozoic terrigenous carbonate, 17—Paleozoic terrigenous carbonate, undivided, 18—Wrangelian (metavolcanites, metasandstones, schists); 19–24—upper crust (after consolidation time): 19—Early Archean, 20—Late Archean, 21—Early Proterozoic, 22—Late Proterozoic, 23—Late Proterozoic-Early Paleozoic, 24—Early Paleozoic; 25—middle crust of cratonic blocks (crystalline massifs); 26–33—lower crust (by crust type): 26–29—continental crust: 26—crust of platforms (crystalline massifs), 27—crust of folded areas, 28—crust of deep sedimentary basins, 29—crust of volcano-plutonic belts; 30–32—transitional crust: 30—crust of back-arc basins, 31—crust of island arcs, 32—crust of subduction zone; 33—oceanic crust; 34–37—areas of thermal reworking of crust: 34—basification, 35—granitization, 36—intrusive bodies of acidic composition, 37—intrusive bodies of basic composition; 38–39—block boundaries: 38—I order, 39—intrablock; 40–41—faults: 40—established, 41—presumed; 42–44—areas of folded deformations and metamorphism shows: 42—Middle Paleozoic, 43—Late Mesozoic, 44—Cenozoic; 45–47—areas of destructive crustal processes shows: 45—Middle Paleozoic, 46—Middle-Late Mesozoic, 47—Cenozoic

massif) and Chukotka folded area, Verkhoyansk-Kolyma fold-thrust system as part of the Omolon median massif, Alazeya-Oloi and Yana-Kolyma folded areas, Uda-Murgal fold system, Okhotsk folded area, South Okhotsk Basin and Kuril island arc.

Legend to the geological-geophysical model provides for the symbols for denoting the compositional complexes of the folded metamorphic and crystalline basement intersected by a line of the composite section of geological structures of the consolidated crust. The latter are unconformably overlain by the Cretaceous volcanic formations of the Okhotsk-Chukotka and Uda-Murgal belts (mainly of basic or acidic composition), terrigenous coal-bearing complexes of Mesozoic and Cenozoic riftogenic depressions and the forming sedimentary cover of the Okhotsk Sea shelf basin.

Deep complexes of crystalline basement are divided into the upper, middle, and lower crustal shown by a gradient vertical filling. Within large tectonic units, the so-called "consolidation nuclei" are distinguished representing the most ancient fragments of the crust enveloped with younger blocks. Division of the crust corresponds to its genetic type; it was carried out according to (Kashubin et al. 2018a, b). Continental, transitional, and oceanic crust is distinguished within the line. Continental crust, in turn, is divided into 4 types: crust of platforms (median massifs), folded areas, deep sedimentary basins and volcanogenic plutonic belts. Three types are distinguished within the transitional crust: back-arc basins, island arcs and subduction zones. These characteristics, to a greater extent, correspond to the lower crust, which is, in our opinion, bearing the basic genetic information on the structure of large tectonic units, to which they are fully assigned in the section.

Upper crust within the selected blocks is differentiated by the consolidation age of the Earth's crust in accordance with geological data on large tectonic elements and their environment. Geological-geophysical boundaries such as the base of the sedimentary cover, the base of the folded basement, the top of the lower crust and the Moho discontinuity are shown with special signs in the section. Additional signs in the section show the velocity and density parameters for different crustal levels. In the process of interpretation, an analysis of tectonic environment was performed for revealing magmatic formations; zones of high tectonomagmatic activity in the upper part of the section are distinguished and classified by their specialization: granitization and basification.

According to the consolidated crust structure in the geotransect band, several large blocks (from north to south) are clearly distinguished: Mendeleev Rise crust characterized by an increased thickness of the lower layer; crust of the North Chukchi Basin subject to destruction with a thick sedimentary cover; thick, presumably, ancient crust of the Wrangel-Herald Rise (Wrangel cratonic block or median massif); low-velocity crust of the Chukotka folded area, which is practically devoid of sedimentary cover; thick ancient three-layer continental crust of the Omolon cratonic block (median massif) and its northern margin; two-layer crust of its southern margin subject to destruction; crust of the Uda-Murgal block characterized by an increased thickness of the lower layer; thin continental crust of the Sea of Okhotsk Basin subject to the

Late Cenozoic stretching; transitional crust of the "back-arc basin—volcanic arc—subduction zone" zone with highly variable characteristics and oceanic crust of the Pacific Plate.

Section of the Earth's crust within the geological-geophysical model based on geotransect "Arctic Ocean—North East Russia—Pacific Ocean" has been studied through its entire thickness. Moho discontinuity is traced at depths of 10–52 km. The simplest structure of the upper mantle was recorded under the structures of the Verkhoyansk-Kolyma and Okhotsk folded areas; the most complex one, under the Wrangel median massif, the North Chukchi Basin and the Uda-Murgal block, where the rise of a relatively high-velocity and high-density mantle is simultaneously distinguished, and at the northern extremity of the Omolon Massif and within the South Anyui zone, where a mantle fragment with an abnormally low density and velocity (to 7.8 km/s) is segregated directly below the M discontinuity. Under the South Okhotsk Basin, features of the upper mantle stratification are recorded. At the southern extremity of the line, a slab of oceanic crust plunging beneath the continental crust goes into the mantle.

Thickness of the Earth's crust varies widely from 10 to 12 km within the South Okhotsk Basin and 28–30 km within the North and South Chukchi and Magadan riftogenic basins to 40–42 km in the Wrangel cratonic block and 50–52 km in the Uda-Murgal FS and in the northern Omolon Massif displaying a general tendency to decrease southwards to the ocean.

Structural image of the crust and the obtained DSS velocities, with a certain conditionality, allow within the ancient Wrangel cratonic block and the Omolon median massif and its margins to divide it into three parts—lower, middle and upper. In other blocks, the crust is two-layer.

Within the Arctic passive margin, two ancient cratonic blocks are distinguished within the geotransect: Wrangel and Mendeleev separated by the North Chukchi Basin. They differ in deep structure, age of cratonization and geological evolution history.

Wrangel cratonic block (Wrangel-Herald Rise) is the oldest one as regards the crust age, and can be considered as a rather large fragment of the Siberian craton torn-off from the latter in the course of the Early Caledonian events to the area of the future accretion-collision system of the Early Mesozoides. It has a crust 30–35 km thick decreasing northwards to the North Chukchi Basin. Moreover, the thickness of its crystalline part varies from 18 to 25 km with the same tendency to decrease northwards. In the south of the block, the lower and upper layers have approximately equal thicknesses of 15 km, whereas in the north, the lower layer is 10 km, while the upper layer decreases to 8 km. In our opinion, this is evidence of the riftogenic processes superimposed on the northern flank of the Wrangel Rise, first at the end of the Devonian—the beginning of the Carboniferous, and then in the middle of the Jurassic. Density model demonstrates a pronounced vertical and lateral heterogeneity of the consolidated crust within the massif. It can be assumed that lateral heterogeneity is due to both fault tectonics and formation changes. The southern boundary of the massif can be drawn from abrupt changes in the occurrence depth of the upper crust top and an increasing crustal thickness due to its crystalline part.

The northern boundary of the massif passes under the western flank of the North Chukchi Basin. Most likely, boundaries of the massif are of tectonic nature.

Another interesting fact is a northward increase in P-wave propagation velocities in the lower crust from 6.7 to 7.0 km/s. This can be explained both by riftogenic processes and mantle uplift, as well as by a rather ancient age of the crust not only in the central Wrangel Rise (Archean-Lower Proterozoic), but also in its peripheral parts (Neoproterozoic).

Mendeleev cratonic block, apparently, has a basement of Neoproterozoic age and can be considered as a fragment of the ancient Arctida mainland. Velocity and density parameters of the crust allowed presenting the Mendeleev Rise in the section as a block of the two-layer Late Precambrian continental crystalline crust to 30 km thick and suggesting the presence in its acoustic basement of the Late Proterozoic (?), Paleozoic and Early Mesozoic sediments of the Arctida craton cover. The lower crust here is characterized by higher, in comparison with southern blocks, P-wave propagation velocities from 6.8 to 7.2 km/s and a sufficiently large and consistent along the strike thickness to 20 km with total crustal thickness of 32–34 km.

North Chukchi Basin, which is part of the Vilkitsky Basin system, is considered to be riftogenic with two cycles of riftogenic processes: post-Ellesmerian and Mesozoic. Under it, an extended crust with total thickness of 28–30 km is recorded. Of these, up to 14–16 km is a sedimentary cover; 2–4 km, supra-crustal (transitional) complex; and less than 10–12 km, crystalline crust. Thickness of the reworked upper crust in the central part of this block, which is clearly recorded from the P-wave propagation velocity of 6.2 km/s and Vp/Vs ratio of 1.71–1.73, is reduced to 3–5 km. P-wave propagation velocities in the lower crust increase northwards to 6.9–7.0 km/s.

South of the Wrangel cratonic block, Mesozoides of the Chukotka folded area, which is the southern extremity of the Novosibirsk-Chukchi fold system, occur. Proceeding from the specific features of geological structure, two extended sublatitudinal folded zones of the Mesozoides are distinguished (from north to south): Chaun and Anyui and the superimposed Rauchuan Trough separating them. In the south, its boundary is overlapped by volcanics of the Okhotsk-Chukchi belt.

Geotransect allowed investigating the structure of the Verkhoyansk-Kolyma and Novosibirsk-Chukotka folded-thrust systems junction zone. Transition from the first to the second system is accompanied by a rapid decrease in total thickness of the crust—from 48 to 40 km (at a 40-km distance from the eastern boundary of the South Anyui zone along the line) and further to 35 and even 30 km in the Chaun zone. In this case, a relatively dense lower crust is thinned primarily (decreasing to 7–8 km under the South Chukchi Basin), while in most cases the average thickness of the upper folded-metamorphic "layer" and the granite metamorphic crystalline crust, which experience sharp, coordinated thickness fluctuations, is preserved, due to which the thickened crystalline crust in areas the folded layer thinning can come close to the Earth's surface and even create bedrock exposures and be destroyed. The South Anyui zone is the most important element of recent paleotectonic reconstructions accepted by many researchers as a collision (suture) zone—a trace of the closed Late Paleozoic-Early Mesozoic Protoarctic Ocean. It should be stated that, despite its obvious sutural border character (an abrupt change in crustal thickness and composition), no sites and high-energy reflections extending into the mantle were recorded within the geotransect on the MCS section.

Alazeya-Oloi FS lying south is a transformed northern extremity of the Omolon Massif. Structural image of the crust, the obtained DSS velocities and density model with a certain conditionality within it and within its continental margins make it possible to divide the crust into three parts—lower, middle and upper. This once again emphasizes the ancient cratonic character of both this structure and its reworked margins (Alazeya-Oloi FS and Sugoi zone of gentle dislocations) making it one of the centers of cratonization and accretion activity in the region. It has a three-layer crust with a thickness to 48–52 km decreasing southward to its southern margin to 36–40 km. Moreover, the thickness of its crystalline part within the massif varies from 35–40 km in the northern part to 18–20 km on its southern margin with a clear tendency to decrease southwards. Crust structure of the massif clearly indicates a pronounced vertical and lateral heterogeneity. Under its northern part, according to the DSS data and density modeling, a low-velocity mantle with velocities of 7.8 km/s and densities of 3.2 g/cm^3 is distinguished.

Southwestern boundary of the massif along the upper horizons is drawn in the subsidence zone of the Paleozoic formations cropping out on the surface on the South Omolon Rise after deep crustal complexes, in the zone of an abrupt upthrust of the lower crust reflection systems with increasing the thickness of the lower crust in the Sugoi zone of gentle dislocations (PK 2590-2780) lying at the margin of Omolon, to 14–16 km, with total crust thickness of 40–44 km. A locally developed lens of crustal-mantle mixture with enclosing substance densities 3.15 g/cm^3 is confined to this boundary in the lower crust.

Sugoi zone of gentle dislocations from the south is directly adjoined by a narrow (along the profile line) Buyunda-Balygychan folded-block zone of Yano-Kolyma FS proper separated by the Omsukchan deep fault zone (PK 2780-2900). It is this fault, as it seems to us, that separates two large crustal blocks: Yana-Kolyma block (in the south) and Omolon block (in the north).

Obviously, the Buyunda-Balygychan folded-block zone is a boundary structure with the Uda-Murgal FS in the south representing the southeastern narrow fragment of the folded zones of the Mesozoides of the Yano-Kolyma system, which is rather clearly seen on the line. Probably, this block formed as a result of repeated complex shear-thrust-upthrust processes resulting in local compression of the Earth's crust under the influence of convergence of large multidirectional megastructures. This can account for the upper crust thickness increase to 24 km with total thickness of the crystalline crust within the given structure of the Yana-Kolyma and Uda-Murgal FS recorded on the southern margin of this block.

Thus, the Buyunda-Balygychan folded-block zone of the Yana-Kolyma FS and the Uda-Murgal FS appeared to be squeezed between the ancient Omolon median massif and the structures of the Okhotsk block, in our opinion, the reworked water area margin of the Okhotsk Massif. A complex internal structure of this zone shown in the section along the geotraverse, is due to repeated reworking of the crust both under compression and under stretching with repeated shows of intense processes of its fluid-magmatic reworking.

Uda-Murgal block is regarded by us as a junction zone of crustal blocks that are fundamentally different in their deep geological structure. The crustal section starts with a lens of crustal-mantle mixture with a density of 3.15 g/cm^3, which represents an almost isosceles triangle in projection onto the vertical plane of the section. Structure of the block is asymmetric; total thickness of the crust in the northern (continental) part of the block reaches 50 km, while at the southern boundary under the Magadan Basin it stretches to 25–27 km. Moreover, the thickness of its crystalline part is 36–38 km in the north decreasing to 22–24 km southwards. In the north, much more than a half (to 22 km) is the lower crust with velocities of 6.6–6.8 km/s decreasing southwards to 7–8 km; the upper crust with velocities of 6.2–6.4 km/s is more evenly distributed reaching thicknesses of 16–18 km. We distinguished such a crust into a separate type of continental crust of volcanogenic plutonic belts.

With transition from the continent to the Sea of Okhotsk water area, a decrease in crustal thickness was noted. The destructive transformations that led to this, most likely, were caused by the rise of a mantle diapir and caused an extension of the continental crust, a decrease in its thickness (to 20–22 km and less in the southern parts of the water area in the vicinity of the South Okhotsk Borderland), and the formation of an arched uplift (Central Okhotsk Rise) and the belt of the Cenozoic troughs framing it (Magadan, for example, superimposed in the Cenozoic on a deep suture between the Uda-Murgal and Okhotsk FS).

The Central Okhotsk Rise, together with the structures of the South Okhotsk Borderland (PK 3720-4250), has an ancient continental crust, which underwent various destructive transformations, which led to its thinning to 25 km within the Rise and 20 km in the Borderland area. The interpretation of the Sea of Okhotsk Plate by some representatives of the geological community as part of the Upper Cretaceous Kula oceanic plate does not correspond to the modern geological-geophysical information and contradicts the materials on this composite geotransect.

A trend of a consistent decrease in crustal thickness and, especially, its consolidated part, is even more intensified when approaching the Kuril Islands. The main changes in the geological structure of the Sea of Okhotsk occur on the northern slope of the South Okhotsk Basin, which is a kind of a morphological benchmark that separates this deepsea basin from the Central Okhotsk Arch. Total thickness of the crust south of this benchmark is reduced to 10 km, with changes affecting both the sedimentary and the consolidated parts of the crust. The delamination process associated with intrusion of a mantle diapir and causing extension of the continental crust in the area of modern basin, vertical accretion of the consolidated crust of adjacent structures and intense magmatic activity can be considered as the main factor determining the evolution geodynamics of this region starting from Miocene. Despite a reduced thickness of the consolidated part, the Earth's crust of this structure retains a thin upper crust.

Southern flank of the South Okhotsk Basin is separated by a system of composite faults with a clearly shear component from the Kuril island arc. Specific features of its modern deep structure are largely determined by the volcanic processes. Thickness of the crust within it is 20–22 km, in which the lower crust layer is markedly predominant (to 13–15 km). A very remarkable behaviour of the upper crustal layer has been established: it is developed unevenly, its thickness sharply increases southwards 3 times from 2 to 6 km, along with a density increase from 2.78 to 2.88 g/cm^3 at velocities of 6.0–6.2 km/s. This may indicate the ongoing stretching processes of the crust and its transformation due to modern basic volcanism.

Geotransect clearly characterizes the structure of the subduction zone of the Pacific Plate under the Eurasian Plate. We distinguish the subduction zone proper by seismological data as a seismic focal zone with a dip of about 35°–45° in the direction of the continent. Projections onto the section line of the earthquake sources in this zone, which represents a transition from the Pacific Plate to the Eurasian continent, form a fairly wide band. Its width is approximately correlated with thickness of combined in this zone sections of continental crust and oceanic crust underthrust under it. Slab thickness swell to 15 km is actually explained by the wish of the authors to take into account the data on earthquake sources. Elements of an accretionary prism, a geological body forming in the course of submersion of oceanic crust of the Pacific Plate into the mantle in the frontal part of the overlying Eurasian Plate, are identified. Relatively high velocities (about 5.0 km/s) within it are accounted for by multiple layering of sedimentary rocks of both plates and a strong deformation of the piled material destroyed by endless thrusts. An accretionary prism on the geotraverse line lies between a deep-sea trench and the fore-arc South Urup Trough (Basin).

It can be concluded that all the SCC of the continental crust (despite their different genetic types) are traced continuously from the Arctic continental margin through the Mesozoides of North East Russia to the Sea of Okhotsk water area and are recorded gradually decreasing in thickness along the profile to the Kuril (South Okhotsk) Basin.

References

Aleinikov AL, Nemzorov NI, Kashubin SN (1991) Rock type determination from seismic data. Author's certificate No 1642416 A1 cl. G 01 V1/30

Belousov VV, Pavlenkova NI (1989) Types of the Earth's crust in Europe and North Atlantic. Geotectonics 3, pp 3–14

Belousov VV, Pavlenkova NI, Kvyatkovskaya GI (eds) (1991) Deep structure of the USSR/Publ. M.:Nauka, p 224

Breivik AJ (2005) Caledonide development offshore–onshore Svalbard based on ocean bottom seismometer, conventional seismic and potential field data. In: Breivik AJ, Mjelde R, Grogan P, Shimamura H, Murai Y, Nishimura Y (eds) Tectonophysics 401, pp 79–117

Buchan KL, Ernst RE (2006) The high arctic large igneous province (HALIP) evidence for an associated giant radiating dyke swarm. LIP of the Month. https://www.largeigneousprovinces.org/06apr

Demenitskaya R.M. Crust and mantle of the Earth. M.: Nedra, 1967

Druzhinin VS (1997) The main features of the interface between the crust and the upper mantle in the Middle Urals (in vicinity of the deep drillhole SG-4). In: Kashubin SN, Kashubina TV, Kolmogorova VV, Parygin GV, Rybalka AV, Tiunova AM (eds) Tectonophysics, vol 269, pp 259–268

Druzhinin VS, Egorkin AV, Kashubin SN (1990) New data on deep structure of the Urals and adjacent areas after DSS. In: Proceedings of the USSR academy of sciences, vol 315, N 5, pp 1086–1090

Druzhinin VS, Karetin YS, Kashubin SN (2000) Deep geomapping of the Urals region based on DSS data. Region. In: Geology and metallogeny, vol 10, pp 152–161

Egorkin AV, Razinkova MI (1980) Cis-Caspian Basin. Seismic models of main geostructures of the USSR. In: Zverev CM, Kosminskaya IP (eds) - M.: Nauka, pp 90–96

Egorkin AV, Akinshina LV, Artemenko LS et al (2002) Crystalline crust structure in Siberia along the Khanty-Mansiysk—Lena line. Explor Prot Mineral Resour 2:33–35

Embry AF (1991) Mesozoic history of the Arctic Islands. In: Trettin HP (ed) Geology of the Innuitian orogen and Arctic Platform of Canada and Greenland, Geology of Canada 3 (also Geological Society of America) pp 371–433

Estrada S, Henjes-Kunst F, Höhndorf A (1999) Cretaceous volcanites from the Canadian Arctic islands: magmatism related to the opening of the Arctic Ocean. Eur J Mineral 11(1):66

Funck T, Jackson HR, Shimeld J (2011) The crustal structure of the Alpha Ridge at the transition to the Canadian Polar Margin: results from a seismic refraction experiment. J Geophys Res Solid Earth 116:1–26

Gaina C, Werner S, Saltus R, Maus S and the CAMP-GM group (2010) Circum-Arctic Mapping Project: New Magnetic and Gravity Anomaly Maps of the Arctic. In: AM, Gautier D, Stoupakova A, Embry A, Sørensen K (eds) 'Arctic Petroleum Geology' Spencer, Geological Society, London, Special Publications. https://earth-info.nima.mil/GandG/wgs84/agp/index.html

Grad M, Tiira T, Behm M, Belinsky AA, Booth DC, Brückl E, Cassinis R, Chadwick RA, Czuba W, Egorkin AV, England RW, Erinchek YM, Fougler GR, Gaczyński E, Gosar A, Guterch A, Hegedüs E, Hrubcová P, Janik T, Jokat W et al (2009) The Moho depth map of the European plate. Geophys J Int 176(1):279–292

Iwasaki T, Levin V, Nikulin A, Iidaka T (2013) Constraints on the Moho in Japan and Kamchatka. Tectonophysics 609:184–201

Jackson HR, Dahl-Jensen T, the LORITA working group (2010) Sedimentary and crustal structure from the Ellesmere Island and Greenland continental shelves onto the Lomonosov Ridge, Arctic Ocean. Geophys J Int 182:11–35

Jokat W, Schmidt-Aursch MC (2007) Geophysical characteristics of the ultraslow spreading Gakkel Ridge, Arctic Ocean. Geophys J Int 168:983–998

Kanao M, Suvorov VD, Toda S, Tsuboi S (2015) Seismicity, structure and tectonics in the Arctic region. Geosci Front 6:665–677

Kashubin SN (ed) (2002) "GRANITE" geotraverse: East European Platform—Urals—Western Siberia (crustal structure based on results of integrated geological-geophysical studies). Ekaterinburg, p 312

Kashubin SN (2013) Crustal types of Circumpolar Arctic. In: Kashubin SN, Pavlenkova NI, Petrov OV, Milshtein ED, Shokalsky S, Erinchek YM (eds) Region. Geol Metallogeny 55:5–20

Kashubin SN (2016) Deep structure of the Earth's crust and upper mantle of Mendeleev Rise based on Arctic-2012 DSS line. In: Kashubin SN, Petrov OV, Artemieva IM, Morozov AF, Vyatkina DV, Golysheva YS, Kashubina TV, Milshtein ED, Rybalka AV, Erinchek YM, Sakulina TS, Krupnova NA (eds) Region. Geol Metallogeny 6:16–35

Kashubin SN, Petrov OV, Androsov EA, Morozov AF, Kaminsky VD, Poselov VA (2011a) Map of crustal thickness of circumpolar arctic. Regional Geol Metallogeny 46:5–13

Kashubin SN, Petrov OV, Androsov EA, Morozov AF, Kaminsky VD, Poselov VA (2011b) Crustal thickness in the Circum Arctic. In: ICAM VI: proceedings of the international conference on arctic margins VI, Fairbanks, Alaska, p 1–17

Kashubin SN, Sakulina TS, Pavlenkova NI, Lukashin YP (2011c) Specific features of the P- and S-wave fields during deep seismic studies in water areas. Seismic Survey Technol 4:88–102

Kashubin SN, Petrov OV, Rybalka AV, Milshtein ED, Shokalsky SP, Verba ML, Petrov EO (2017) Earth's crust model of the South-Okhotsk Basin by wide-angle OBS data. Tectonophysics 710–711:37–55

Kashubin SN, Petrov OV, Milshtein ED, Androsov EA, Vinokurov IYu, Shokalsky SP (2018a) Crustal types of Central and North East Asia, Far East and Arctic continent—ocean transition areas. Reg Geol Metallogeny 73:6–18

Kashubin SN, Petrov OV, Artemieva IM, Morozov AF, Vyatkina DV, Golysheva YS, Kashubina TV, Milshtein ED, Rybalka AV, Erinchek YM, Sakulina TS, Krupnova NA, Shulgin AA (2018b) Crustal structure of the Mendeleev Rise and the Chukchi Plateau (Arctic Ocean) along the Russian wide-angle and multichannel seismic reflection experiment "Arctic-2012". J Geodyn 119:107–122

Kostyuchenko SL, Morozov AF (2007) Geological and geophysical images of the Earth's crust and upper mantle of the territory of Russia in maps and models. In: Models of the Earth's crust and upper mantle based on results of deep seismic profiling. Materials of the International Scientific and Practical Seminar. Rosnedra. VSEGEI. SPb.: VSEGEI Publishing House, pp 82–85

Kostyuchenko SL, Egorkin AV, Solodilov LN (1999) Structure and genetic mechanisms of the Precambrian rifts of the East-European Platform in Russia by integrated study of seismic, gravity, and magnetic data. Tectonophysics 313:9–28

Kosygin YA (1975) Basics of tectonic zoning. In: Principles of tectonic zoning. Vladivostok, pp 8–24

Kulinich RG, Valitov MG (2011) Crustal thickness and types in the Sea of Japan according to marine satellite gravimetry. Pac Geol 30(6)3–13

Kunin NY, Goncharova NV, Semenova GI et al (1987) Map of mantle surface relief in Eurasia. M.: IFZ USSR Academy of Sciences, Mingeo RSFSR

Lebedeva-Ivanova NN, Zamansky YY, Langinen AE, Sorokin MY (2006) Seismic profiling across the Mendeleev Ridge at 82°N: evidence of continental crust. Geophys J Int 165:527–544

Li S, Mooney WD, Fan J (2006) Crustal structure of mainland China from deep seismic sounding data. Tectonophysics 420, 239–252

Ljones F (2004) Crustal transect from the North Atlantic Knipovich Ridge to the Svalbard Margin west of Hornsund. In: Ljones F, Kuwano A, Mjelde R, Breivik A, Shimamura H, Murai Y, Nishimura Y (eds) Tectonophysics 378:17–41

Maus S et al (2009) EMAG2: A 2-arc min resolution Earth Magnetic Anomaly Grid compiled from satellite, airborne, and marine magnetic measurements, Geochemistry. Geophys Geosyst 10 (Q08005). https://www.geomag.org/models/emag2.html

Meissner R, Brown L, Durbaum HJ, Frauke W, Fucks K, Seifert E (eds) (1991) Continental Lithosphere: deep seismic reflection. Geodynamic Series, vol 22, Am. Geophys. Union, Washington, DC

Miller HG, Singh V (1994) Potential field tilt—a new concept for location of potential field sources. J Appl Geophys 32(2–3):213–217

Miura S (2005) Structural characteristics of Miyagi forearc region, the Japan Trench seismogenic zone, deduced from a wide-angle reflection and refraction study. In: Miura S, Takahashi S, Nakanishi A, Tsuru T, Kodaira S, Kaneda Y (eds) Tectonophysics 407:165–188

Mjelde R (2009) Crustal structure and evolution of the outer More Margin, NE Atlantic. In: Mjelde R, Raum T, Kandilarov A, Murai Y, Takanami T (eds) Tectonophysics 468:224–243

Models of the Earth's crust and upper mantle based on results of deep seismic profiling (2007) Materials of the International Scientific and Practical Seminar. Rosnedra. VSEGEI. SPb.: VSEGEI Publishing House, p 245

Mooney WD (2007) Crust and Lithospheric Structure—Global Crustal Structure. Romanowicz B, Dziewonski A (eds) Treatise on geophysics. Vol. 1: Seismology and structure of the earth. Elsevier, pp 361–417

Morozov AF (2013) New geological data substantiating continental nature of the area of Central Arctic Rises. In: Morozov AF, Petrov OV, Shokalsky SP, Kashubin SN, Kremenetsky AA, Shkatov MY, Kaminsky VD, Gusev EA, Grikurov GE, Rekant PV, Shevchenko SS, Sergeev SA, Shatov VV (eds) Region. Geol Metallogeny 53:34–55

Morozov AF, Mezhelovsky NV, Pavlenkova NI (eds) (2006) Structure and dynamics of lithosphere of Eastern Europe. Research Results under EUROPROBE Programme. Issue 2. M: GEOKART, GEOS, p 735

Nakanishi A (2009) Crustal evolution of the southwestern Kuril Arc, Hokkaido Japan, deduced from seismic velocity and geochemical structure. In: Nakanishi A, Kurashimo E, Tatsumi Y, Yamaguchi H, Miura S, Kodaira S, Obana K, Takahashi N, Tsuru T, Kaneda Y, Iwasaki T, Hirata N (eds) Tectonophysics 472:105–123

Pavlenkova NI, Pavlenkova GA (2014) Structure of the Earth's crust and upper mantle of Northern Eurasia based on seismic profiling with nuclear explosions. M.: GEOKART GEOS, p 191

Pavlenkova NI, Kashubin SN, Pavlenkova GA (2016) Earth's crust of deep platform basins of Northern Eurasia and nature of their formation. Phys Earth 5:150–164

Pavlis NK, Holmes SA, Kenyon SC, Factor JK (2012) The development and evaluation of the Earth Gravitational Model 2008 (EGM2008). J Geophys Res 117 (B04406):1–38. https://bgi.obs-mip.fr/data-products/Grids-and-models. (Pavlis et al. 2012)

Petrov O (2016) Crustal structure and tectonic model of the Arctic region. In: Petrov O, Morozov A, Shokalsky S, Kashubin S, Artemieva IM, Sobolev N, Petrov E, Ernst RE, Sergeev S, Smelror M (2016) Earth-Sci Rev 154:29–71

Petrov OV et al (Editor-in-Chief) (2008) GIS Atlas of Geological Maps of Russia, the CIS countries and adjacent states, scale 1: 2 500 000. SPb, VSEGEI. https://vsegei.ru/ru/info/gis_cis

Petrov OV, Shuwen D, Kiselev EA, Morozov AF (eds) (2016) Atlas of geological maps of Asia and adjacent areas. SPb.: VSEGEI Publishing House, p 48

Poselov V, Butsenko V, Chernykh A et al (2011a) The structural integrity of the Lomonosov Ridge with the North American and Siberian continental margins. In: ICAM VI: proceedings of the international conference on arctic margins VI (Fairbanks, Alaska, May 2011) 2014, pp 233–258

Poselov VA, Avetisov GP, Kaminsky VD et al (2011b) Russian Arctic geotraverses. SPb: FGUP "VNIIOkeangeologia" named after I. S. Gramberg, p 172

Roslov YV, Sakoulina TS, Pavlenkova NI (2009) Deep seismic investigations in the Barents and Kara Seas. Tectonophysics 472:301–308

Sakulina TS (2011a) Geological model of the Sea of Okhotsk Region from 1-OM and 2-DV-M reference lines. In: Sakulina TS, Kalenich AP, Atakov AI, Tikhonova IM, Krupnova NA, Pyzhyanova TM (eds) Explor Prot Mineral Resour 10:11–17

Sakulina TS (2011b) Integrated geological-geophysical studies on 5-AR reference line in East Siberian Sea. Sakulina TS, Verba ML, Kashubina TV, Krupnova NA, Tabyrtsa SN, Ivanov GI (eds) Explor Prot Mineral Resour 10:17–23

Sakulina TS (2016) Deep structure of the Earth's crust and upper mantle of North Chukchi Basin along the DSS Dream-line. In: Sakulina TS, Kashubin SN, Petrov OV, Morozov AF, Krupnova NA, Dergunov NT, Razmatova AV, Tabyrtsa SN, Kashubina TV, Yavarova TM (eds) Region. Geol Metallogeny 68:52–65

Salnikov AS (ed) (2007) Crustal structure and composition in Magadan sector of Russia after geological-geophysical data. Nauka, Novosibirsk, p 173

Takahashi N (2009) Structural variations of arc crusts and rifted margins in the southern Izu-Ogasawara arc–back arc system. In: Takahashi N, Kodaira S, Tatsumi Y, Yamashita M, Sato T, Kaiho Y, Miura S, No T, Takizawa K, Kaneda Y (eds) Geochem Geophys Geosyst 10. Q09X08. https://doi.org/10.1029/2008GC002146

Teng J (2013) Investigation of the Moho discontinuity beneath the Chinese mainland using deep seismic sounding profiles. In: Teng J, Zhang Z, Zhang X, Wang C, Gao R, Yang B, Qiao Y, Deng Y (eds) Tectonophysics 609:202–216

Volvovsky BS, Volvovsky IS (1988) Structures of continents with "granite-free" crust type. Geodynamic Studies. Problems of Deep Geology in the USSR. M., 12:169–187

Voronin AY (2007) Zoning of territories based on artificial intelligence and pattern recognition in environmental management tasks. Synopsis of Doctoral Thesis in Engineering Science. Moscow, p 44

Wang Y, Mooney WD, Yuan X, Coleman RG (2003) The crustal structure from the Altai Mountains to the Altyn Tagh fault, northwest China. J Geophys Res 108(B6):2322. https://doi.org/10.1029/2001jb000552

Wang TK, Chen M-K, Lee C-S, Xia K (2006) Seismic imaging of the transitional crust across the northeastern margin of the South China Sea. Tectonophysics 412:237–254

Zhao W (2001) Crustal structure of central Tibet as derived from project INDEPTH wide-angle seismic data. In: Zhao W, Mechie J, Brown LD, Guo J, Haines T, Hearn T, Klemperer SL, Ma YS, Meissner R, Nelson KD, Ni JF, Pananont P, Rapine R, Ross A, Saul J (eds) Geophys J Int 145:486–498

Zolotov EE, Kostyuchenko SL, Rakitov VA (1998) Tomographic sections of lithosphere of East European Platform. In: Mitrofanov FP, Sharov NV (eds) Seismological model of lithosphere of Northern Europe: Barents Region. KSC RAS, Apatity, Pt 1, pp 71–79

Zverev SM, Kosminskaya IP (eds) (1980) Seismic models of lithosphere of main geostructures of the USSR/Publ. M.: Nauka, p 84

Tectonic Domains of Northern Asia

O. V. Petrov, I. I. Pospelov, V. N. Puchkov, K. S. Ivanov, V. A. Koroteev, A. N. G. Bulgatov, I. V. Gordienko, and S. D. Sokolov

Abstract The chapter gives a brief description of the geology and tectonic evolution of large geological structures in Northern Asia, such as orogenic systems of the western part of Central Asian Orogenic belt—the Urals Mountains, Sayan-Baikal mountain area, Verkhoyansk-Kolyma and orogenic structures of the Paleozoic basement of the West Siberian Plate. The history of the regional studies, data on the magmatic and sedimentary complexes as well as isotope-geochronological datings are provided.

1 Urals: Main Tectonic Features

The Urals is one of the most famous examples of fold belts with a complete cycle of evolution. A vast body of publications including numerous monographs is devoted to problems of geological structure and evolution of the Urals (Peive et al. 1976; Koroteev et al. 1979; Puchkov 1979, 2000; Ivanov et al. 1986; Sobolev et al. 1986; Savelieva 1987; Mizens 1997; Ivanov 1998; Yazeva and Bochkarev 1998; Deep composition… 2001; Ruzhentsev and Degtyarev 2005; Brown et al. 2002; Morozov 2006; Kashubin et al. 2006; Fershtater et al. 2009; and many others). In the last 10–15 years, several major geological-geophysical projects were carried out

O. V. Petrov (✉) · I. I. Pospelov
Russian Geological Research Institute (VSEGEI), 74 Sredny Prospect, St. Petersburg 199106, Russia
e-mail: vsgdir@vsegei.ru

V. N. Puchkov · K. S. Ivanov · V. A. Koroteev
The Zavaritsky Institute of Geology and Geochemistry of the Ural Branch of the Russian Academy of Sciences (IGG UB RAS), 15 Akademika Vonsovskogo str., Ekaterinburg 620016, Russia

A. N. G. Bulgatov · I. V. Gordienko
Geological Institute of Siberian Branch of the Russian Academy of Sciences (GIN UB RAS), 6a Sakhyanova str., Ulan-Ude 670047, Russia

S. D. Sokolov
Geological Institute, Russian Academy of Sciences (GIN RAS), 7 Pyzhevksy per., Moscow 119017, Russia

O. V. Petrov and S. Dong (eds.), *Tectonics of Asia (Northern, Central and Eastern Asia)*, Springer Geology, https://doi.org/10.1007/978-3-030-62001-1_3

(Project “Uralides—Variscides” under the EUROPROBE Program, Project “MinUrals” and others). These projects are based on seismic and structural profiles across the Southern and Middle Urals (URSEIS-95 and ESRU-SB: position of the profiles is shown in Fig. 1). These profiles were instrumental in further understanding the deep structure of the Uralian orogen.

The Uralian fold belt extending for 2500 km is an extreme northwestern branch of the huge Paleozoic Ural-Mongolian mobile belt. The main folding, as well as tectonic overthrusting, metamorphism and granitization in the Urals took place during the Late Paleozoic (with maximum during the Late Carboniferous—Early Permian). Frequently Uralides are identified with Variscides (Hercynides) of Western Europe. But results of studies during last ten years have clearly shown the existence of principal differences between the Variscides and Uralian fold belt (Puchkov 2003). Uralides had more long-time history of orogeny with the last orogenic phase in the Early Jurassic (to say nothing about the neo-orogenic stage).

The Urals is characterized by a number of peculiarities such as an exceptionally wide distribution and good preservation of ophiolites and island arc complexes, presence of platinum-bearing intrusive belt and Uralian belt of high-pressure metamorphism. Geophysical data show the presence of a “cold”, isostatically equilibrated “mountain root” under the central part of the Uralian bi-vergent structure.

Identification of the Urals as Variscides in the “Tectonic map of Central Asia and Adjacent Areas” is a tribute to a long-term tradition which winds down. It is possible to expect that the legends for tectonic maps will contain in future the “Uralides” as an independent tectono-chronological category equal in rank to the older Variscides in which the folding has finished mostly in the Carboniferous.

Geographically, the Urals is divided into five segments: Southern (including Mugojary), Middle, Northern, Cis-Polar and Polar. Northwards, the western Uralian structural zones continue as the Pai-Khoy—Novaya Zemlya fold belt.

Tectonically, the Urals is subdivided into several meridional structural zones (megazones) subparallel to the margin of the East European Platform. The western (mega) zones are traced at the earth surface along the whole extent of the belt; the eastern ones are exposed only in the Southern and Middle Urals and disappear gradually to the north under the Mesozoic-Cenozoic sedimentary cover of the young West Siberian Basin. The South Tien Shan fold belt is a most probable prolongation of the Urals to the southeast. Experience of comparative study of the Urals and the Tien Shan indicates that analogues of many Paleozoic complexes of the Urals are situated in the South Tien Shan (although between the structures of these two regions there are quite essential distinctions); at the same time, Uralian Late Paleozoic rock assemblages (starting from Famennian—Tournaisian Zilair flysch series) are traced to the Caucasus through the Pre-Caspian depression.

The Urals fold belt usually is subdivided into megazones (Fig. 1). First three of them (the westernmost ones) form the Uralian paleocontinental sector—a former margin of Baltica/Laurussia paleocontinent, and the last three—a paleooceanic sector, a collage of ophiolites, island arc and microcontinental terranes of Paleo-Uralian Ocean. The boundary between these sectors is represented by a huge ophiolite-hosting suture zone of the Main Uralian Fault (Fig. 1).

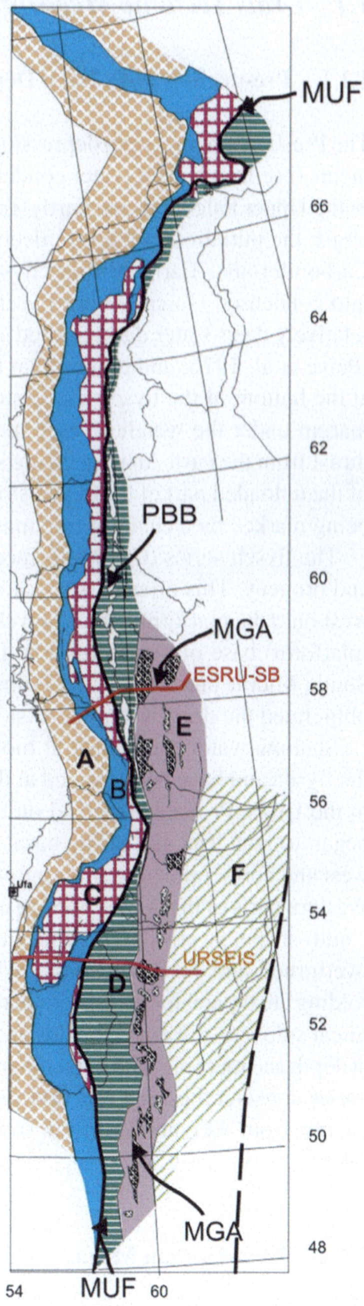

Fig. 1 Tectonic megazones of the Urals (explanations in the text)

1.1 Urals Tectonic Megazones

1.1.1 Pre-uralian Fore-Deep Depression

The Pre-Uralian fore-deep depression (**A**, see Fig. 1) is filled up with terrigenous sediments (pre-flysch deep-water condensed sediments, evaporates, flysch and molasse) of the Upper Paleozoic and partly Triassic age up to 5–6 km thick. In the south of the Urals, the initiation of the fore-deep was accompanied by the accumulation of Late Carboniferous—Early Permian flysch series. Westwards, the flysch is transformed into condensed (so-called pre-flysch) series composed mostly of the alternation of relatively deep-water dark-colored shale, marl and limestone. Usually it is supposed (Peive et al. 1976; and others) that the formation of the condensed preflysch series at the bottom of the flysch succession was connected with a depression continental margin under the weight of tectonic sheets composed of island arc complexes and thrust from the east. An east-facing structural step has formed along the western side of the unloaded part of the depression. The step was migrating with time to the west, being marked by a chain of reef massifs.

The flysch series (C_2–P_1) formed during the epoch of active collision, thrusting and orogeny. This process was accompanied by a migration of thedepression to the west onto the platform, which is well determined by the age change of the carbonate (platform) base of the depression. During the Kungurian, the flysch was partly (in South, Middle and North Urals) substituted by evaporates which finally filled up and obliterated the deep-water depression.

Shallow-water and subaerial molasse (P_2–T) migrated to the west in relation to the flysch and have been formed at the stage of attenuation of the orogenic processes in the Urals. Probably the load on the platform margin decreased and the platform began to rise. The age of the basal layers of the foredeep becomes younger to the west and north, reflecting a migration of the basin westward and probably northward. Western parts of the depression are mainly characterized by mild platform structures. Comb-shape, swell-like, boxlike and more compound compressed folds (including overturned and isoclinal ones) of Uralian strike and reworked by Uralian linear folding are typical for the eastern parts. The above-described folds are complicated by linear sulphate-salt diapiric folds and thrusts, dipping to the east, flattening gradually at depth and merging with detachment surfaces. The depth of the Pre-Uralian foredeep varies considerably, and it is divided by transversal uplifts into a system of isolated basins: from Aktyubinsk in the south to Kara one in the north (Mizens 1997; and others).

1.1.2 West-Uralian Megazone

West-Uralian megazone (**B**, see Fig. 1) was during the Paleozoic a passive (Atlantic-type) margin of the East European Platform, i.e. transitional area from the platform to the Uralian Paleoocean situated to the east. There, two regional structuraland-formational zones from the Ordovician to Carboniferous are identified: the western,

Belsk-Elets zone is composed of paleoshelf terrigenous-carbonate series and the eastern, Zilair-Lemva zone consists of terrigenous-siliceous-shale accumulation interpreted (Puchkov 1979, and others) as an area of passive continental margin. It is only during the Early—Late Carboniferous that a new structure (above described Pre-Uralian fore-deep depression) has been originated. Shelf complexes overlying in the external (Belsk-Elets) zone are almost undestroyed but somewhat thinned crystalline basement shows platform features. The presence of reef, bioherm, organogenic-detrital and other limestone as well as dolomite, quartz sandstone with features of coastal-marine origin is common there. The sections of this zone are characterized by greater thickness, but in principle they do not differ from sections of the sedimentary cover in the adjacent East European Platform. The sections begin with the Lower Ordovician terrigenousoligomictic series overlying graben facies or directly the crystalline basement. Upwards it gives way to the Middle–Upper Ordovician terrigenous-limestonedolomite series. In northern regions, the Late Ordovician is characterized by the first appearance of carbonate reefs at the shelf periphery and evaporates in its internal parts. Silurian–Carboniferous deposits are mainly represented by shallow-water stratified and reef limestone with layers of well-sorted quartz sandstones (Emsian, C_1h series and others). Lower Devonian reef limestone (up to 1500 m thick) forms a barrier reef extending along almost the whole margin of the East European paleocontinent.

Zilair-Lemva structural zone consists of six areas of bathyal complexes. Three stages of evolution of this zone are distinguished: (1) initiation (riftogenic); (2) mature passive continental margin: (3) pre-orogenic (graywacke flysch stage). For the rift stage, shallowwater terrigenous molasse-like series with alkaline and alkaline-basalt volcanic rocks are most typical. The stage of passive continental margin in the major part of the Urals started in the Middle Ordovician when a continental slope was formed and started to descend. This is shown in the accumulation of siliceous–shaly sediments which are frequently condensed. Early Silurian deposits are represented everywhere by phtanite—black shale series. Colored chert, cherty breccias and sandstone with rare horizons of deep-water clayey limestone are deposited in the southern areas until the Frasnian but in the north till the Bashkirian inclusively. The upper part of the sections in the Southern Urals is represented by a thick series of greywacke flysch (the Famennian—Tournaisian Zilair series) which resulted from erosion of tectonic sheets of volcanogenic-sedimentary formations. In the northern areas, the flysch appeared somewhat later—during the Early Carboniferous. The compression which began at that time caused a complete closure of the Zilair-Lemva zone. This process was accompanied by the generation of west-vergent fold-and-thrust structures above detachment surfaces and the westward overthrusting of the deposits formed in Zilair-Lemva zone. Overthrusting of bathyal complexes onto the shelf is proved by structural drilling in the western slope of the Middle and Polar Urals. In the Sakmara, Kraka and some other areas, ophiolites, volcanogenic-sedimentary and intrusive island arc formations are overthrusted onto sedimentary complexes of the passive continental margin.

1.1.3 Central Uralian Megazone

The Central Uralian megazone (**C**, see Fig. 1) is mainly composed of metamorphosed Precambrian and Early Paleozoic rocks which form axial, the most uplifted part of the Ural Mountains. At the same time, slightly metamorphosed Precambrian rock complexes are exposed in large uplifts of the Southern and Middle Urals (Bashkirian and Kvarkush meganticlines). So, the Bashkirian Meganticline of the Southern Urals is composed of Riphean shallow-water terrigenous-carbonate series (up to 15 km thick) with moderate amount of subalkaline volcanic rocks and intrusions of a rift origin, approximately at the levels of 1800–1650, 1380–1350, 720–700 Ma.

In the legend for the Tectonic map, the Riphean sedimentary complex with the age older than 1350 Ma that covers the Archaean-Paleoproterozoic crystalline basement is indicated as Grenvillian. But it does not mean that the Grenvillian folding took place there. One may speak only about a rifting event which can be approximately correlated with the Grenvillian orogeny.

The crystalline basement of the East European Platform covered with Riphean deposits is exposed in the Taratash block of the Bashkirian Meganticline. Rock assemblages metamorphosed there mostly in granulite and amphibolite facies are correlated with Karelides although include more ancient (Archean) complexes.

During the Vendian (Ediacaran, 600–540 Ma), eastern parts of the Bashkirian and Kvarkush meganticlines underwent collision and folding. These events were expressed in angular unconformities between the Neoproterozoic and the overlying Paleozoic, metamorphism of the Vendian age and formation of molasse, a result of erosion of internal parts of the orogen. These structures are traced to the Timan Ridge and represent externides of a large folded belt of Timanides (Puchkov 1997, 2003, 2006; Gee and Pease 2004). In the legend of the map, the Timanides are correlated with Baikalides and Cadomides.

The internides of the Timanides are exposed in the northern uplifts of the Central Uralian zone and form the crystalline basement of the Timan-Pechora Basin of the East European Platform. There, in contrast to the externides, in addition to riftogenic formations, rare and dismembered Late Riphean ophiolites are also developed, as well as the Late Riphean—Vendian suprasubduction intrusive and volcanogenic complexes. Molasse there is of the same Vendian age similar to the externides.

A considerable part of internides formed in the course of the Late Riphean—Vendian accretion and continental collision of blocks separated previously during the Riphean as a result of a break-up of Archaean-Paleoproterozoic supercontinents—Nuna and Rodinia. Notwithstanding strong subsequent reworking, the relicts of these blocks (covered by thick Riphean and Vendian series) appear as gneiss complexes in the cores of Lyapin and Kharbey dome-like uplifts. "Granilite-type" zircons found in these complexes are Paleoproterozoic (Pystin and Pystina 2008).

Complexes shown in Lyapin Anticline as Caledonian are not products of Caledonian folding in this area. Recently A-granite intrusions were established there; they have a narrow spectrum of compositions (leuicocratic varieties predominate), form contrasting associations with gabbro (rhyolite comagmatic to thc granites make a contrasting association with basalts). Petrologically, they are related to magmatic formations of divergent geodynamic environments. Age dating of zircons covers an interval of 564–498 Ma—the end of the Vendian and most of the Cambrian (Kuznetsov et al. 2005). It is possible to suggest that convergent environments of the Late Riphean—Vendian in the northern part of Timanides have been altered in the Cambrian by extension; this caused a collapse of orogen to form contrasting formations.

Ural-Tau antiform has a peculiar position. Till mid-80s of the last century, a viewpoint predominated that it is a Precambrian structure, the lower part of which is represented by Middle Riphean eclogite-glaucophane-schist Maksiutovo complex and the upper part, by thick Upper Riphean–Vendian terrigenous Suvanyak complex. As a result, the Ural-Tau was confidently regarded as part of a Precambrian structure of the Central Uralian zone (and in fact was its synonym). At present, numerous occurrences of Paleozoic fossils and absolute age datings (mostly Paleozoic) make us to abandon this viewpoint particularly while interpreting the southern part of Ural-Tau Ridge.

The above-described three zones formed during the deformation of the passive continental margin of the Laurussia continent (East European Platform). Only some allochtones (Sakmara and Kraka ones of the Southern Urals, Nyazepetrovsk nappe of the Middle Urals) were transported tectonically from the east, i.e. from paleo-island-arc part of the Uralian fold belt.

1.1.4 Main Uralian Fault

Zone of the Main Uralian Fault (**MUF**, see Fig. 1) represents a typical ophiolite suture of variable width (sometimes up to 20 km and more) as a relic of the disappeared fore-arc trench of the Paleo-Uralian Ocean. Wide distribution of serpentinite mélange and tectonic megabreccias traced at hundreds km (Sakmara-Voznesensk zone of the Southern Urals, Ray-Iz–Kharamatalou zone of the Polar Urals and others) is typical for MUF. Zones of dislocational metamorphism, blastomylonite are also typical of this fault. The all-Uralian belt of eclogite–glaucophaneschist metamorphism is traced along the fault as a discontinuous belt. The reflected waves method research has shown that the fault surface dips eastward usually at an angle of 35–55° (up to 90° at the latitude of the Ufimian promontory) and divides the complexes of the ancient continent, and Paleozoic oceanic and island-arc rock assemblages thrust over them. Old complexes including Archaean—Paleoproterozoic crystalline basement are traced in the depth eastward (gradually thinning out) to the axial part of the Magnitogorsk zone where the relics of the island arc are developed at the surface.

Formation of MUF was a long-term and multistage process. Initially it apparently was a riftogenic extension fault (a normal fault in the upper, fragile part of the Earth crust, substituted by zones of plastic flow in the middle and lower parts). The time of full break-up of the continental crust in the Urals is just before the Late Arenig, i.e. 480 Ma. During the Middle Paleozoic, MUF experienced a stage of quiescence separating the passive continental margin and oceanic basin. A collision of the Late Paleozoic Magnitogorsk island arc and paleocontinent that started in the Early

Famenian in the Southern Urals (in the northern areas approximately 30 Ma later) transformed MUF into a thrust-fault. Structural, paleomagnetic, paleogeographic and other data testify that the collision between the East European Platform and Uralian island arc terrains was not frontal but oblique (Puchkov 2000; Ivanov et al. 2000, and others). Thus, in the Middle Paleozoic (i.e. before the collision of the Magnitogorsk island arc with the paleocontinent), the Main Uralian Fault was a typical subduction zone which structure has been preserved till the present.

1.1.5 Tagil-Magnitogorsk Megazone

The Tagil-Magnitogorsk megazone (**D**, see Fig. 1), also known as an area of "greenstone synclinoria", "main volcanogenic zone of the Urals", is located immediately eastward of MUF and is not an all-of-a-piece structure. The megazone is distinctly subdivided into two zones (island arc terrains) of different age but similar composition: older Tagil and younger Magnitogorsk zones. The Tagil zone has been originated during the Middle Ordovician and the Magnitogorsk zone during the Early Devonian.

The Tagil zone is widely distributed in the northern part of the Middle Urals, in the Northern, Cis-Polar and Polar Urals. But the problem of appearance of its fragments in the Southern Urals (although extremely reduced) is a matter of discussion.

The Magnitogorsk zone constitutes the largest part of the eastern slope of the Southern Urals. Its dislocated fragments are exposed in the Middle Urals (Koroteev et al. 1979 and others) to the east of the Tagil zone. In addition, the eastern part of the Tagil Synclinorium (shown on the tectonic map as a Variscan structure) can be recognized as the extension of the suprasubduction zone of the Magnitogorsk island arc superimposed onto the Tagil island arc structure during the Middle Devonian (Puchkov 2009).

In the Tagil and Magnitogorsk zones, tholeiitic low-potassic pillow basalts of 1.5–2.5 km thick constitute lower parts of volcanogenic successions. The underlying formation is represented by poorly exposed sheeted-dyke diabase complex. As a rule, basalts are only associated with altered hyaloclastics and thin layers of jasper. The latter contains sufficiently abundant complexes of Middle—Upper Ordovician conodonts in the Tagil zone and Emsian ones in the Magnitogorsk zone. In general, the successions of the Tagil and Magnitogorsk zones consist of the following greenstone-altered formations (upwards): 1. Sodium basaltic, 2. Sodium rhyolite-basaltic, 3. Andesitedacitic, 4. Andesite-basaltic, 5. Andesitic, 6. Basalt-trachyte-trachyrhyolitic. However, the age difference between the formations in both sequences is preserved from the bottom to the top. Subalkaline volcanogenic formations of the Tagil zone are covered by the Lower—Middle Devonian bauxite-bearing limestone. In the Magnitogorsk zone, subalkaline volcanogenic formations appeared only in the Upper Devonian; they are facially replaced by flysch and covered with limestone with tholeiitic and subalkaline volcanic rocks (already without features of suprasubduction origin). The area of widespread Tagil-type formations is shown on the Tectonic Map as Caledonian (with termination in the Early Devonian) whereas the area of the Magnitogorsk-type formations as Variscan (with termination in the Early Carboniferous).

Along the western border of the paleo-island arc sector of the Urals, in the western part of the Tagil zone, near the MUF, a unique Platinum-bearing belt (**PBB**, see Fig. 1) extends for more than 900 km. It is one of the largest and most important Uralian geological complexes. This huge geological object is represented in the Middle, Northern and Cis-Polar Urals by a chain of thirteen concentric-zonal isometric or tectonically stretched massifs. The latter are composed of dunite, clinopyroxenite, olivine and two-pyroxene gabbro, granitoids and are typical representatives of zonal mafic–ultramafic massifs (socalled Ural-Alaskan type). The considerable difference of this Ural-Alaskan type association from ophiolites (and other oceanic associations) is the absence of harzburgite, sheeted dyke complex and basalts and relatively high contents of Fe and Sr (over 300 ppm). Owing to gravity and seismic methods, Platinumbearing massifs are traced to sufficiently large depths (6–8 km) as steeply eastward-dipping bodies. It is shown (Ivanov 1998; Volchenko et al. 2007 and others) that the rock associations of the Platinumbearing belt are the products of island arc processes, i.e. melting in different depths above the subduction zone (suprasubduction magmatism has finished there 415–420 Ma ago). Their island arc origination is proved by resemblance with gabbroid and ultramafic xenoliths from volcanic rocks of recent island arcs, by geochemical characteristics and other data.

Widespread island arc magmatism in the Tagil and Magnitogorsk zones called forth a diversity of intrusive complexes corresponding to effusive rocks. Island arc intrusive complexes are widespread. Intrusions of M-type tonalites and plagiogranites associated with hornblende gabbro (gabbro-plagiogranite and gabbrotonalite complexes) are related to initial episodes of island arc magmatism. Later they were altered by gabbro-granitoid associations with I-type granites which geochemistry has been changed from potassicsodium calc-alkaline to potassic subalkaline type in the process of island arcs evolution.

Volcano-sedimentary series of the Tagil-Magnitogorsk megazone in the Northern and Southern Urals are usually folded in relatively simple structures. Volcanic series there preserve frequently their primary textures; relics of volcanic buildups are often present. In the Middle Urals, volcanic series are more intensively tectonizied, crushed against rigid block of the Ufimian promontory; this was accompanied by formation of the back-thrusts in the eastern part of the Middle Urals. The absence of uplifts of the Precambrian sialic metamorphic rocks is a peculiar feature of the whole megazone.

One of main differences between the Tagil and Magnitogorsk zones (besides their ages) consists in the fact that the former was an ensialic island arc and the later, ensimatic (except its northern extension). Devonian volcanic series of the Magnitigorsk zone have geochemical features typical of suprasubduction complexes (Ivanov et al. 1986; Yazeva and Bochkarev 1998; Kosarev et al. 2005, 2006), such as negative anomalies of Nb, Ta, Zr, Hf, Y and enhanced concentrations of LIL elements (K, Rb, Ba, Cs) and LREE. They do not show any signs of contamination with the continental crust and can be considered as ensimatic island arc complexes formed above an east-dipping subduction zone.

Petrologic and geochemical analyses of the Late Ordovician—Silurian volcanic formations of the Tagil zone (data of V. N. Smirnov and others) show that the evolution of volcanism consisted in a gradual transition from calc-alkaline rocks to rocks of subalkaline trend. Eruption of low volume subalkaline volcanics took place already at the initial stage of zone formation. Later on, their share increased gradually; the final Tura trachybasalt-trachyte volcanic complex is already completely composed of subalkaline varieties. There are no volcanic associations with tholeiitic trend of differentiation typical of ensimatic island arcs in the Tagil zone. Spider diagrams of basalts show all peculiarities of island arc associations: well expressed maxima of Sr, Ba content; distinct Nb, Th and weak Ce minima. Thus it is possible to consider the Tagil paleo-island arc as an analogue of recent ensialic island arcs. According to gravimetry data, in the central part of the Tagil zone, volcanogenic series are underlain by relatively "light" rocks which can be identified as rocks of old continental crust. Occurrences of xenogenic garnets (similar in composition to garnets of metamorphic rocks) and zircons with Precambrian datings in volcanic rocks of the Tagil zone can be also regarded as an evidence of contamination of island arc magmas with a substance of the old crust from the island arc basement.

To be unbiased, it is necessary to mention contrary opinions of some other researchers who believe that the Tagil arc was ensimatic (Narkissova 2005; Puchkov 2006; and others).

The idea of early mobilistic publications that formations of an oceanic stage (of mid-oceanic ridges and oceanic plateaus) are widespread in the Tagil-Magnitogorsk megazone has not been confirmed. Series of initial tholeiitic basalts, sheeteddyke complexes etc., described as oceanic, were later attributed to initial island arc (mainly back-arc) formations (Ivanov et al. 1986, and others). Rocks of the oceanic stage of evolution in the Urals are almost lacking because the paleooceanic crust was very easily absorbed in a subduction process.

It should be noted that recently (Puchkov 2006; Savelieva et al. 2006; Tessalina et al. 2007; Popov et al. 2008; Fershtater et al. 2009) there appeared a tendency to use ancient (mostly Late Precambrian (and Cambrian) datings (510–885 Ma by Sm–Nd and U–Pb data after zircons from the Uralian gabbroultramafic massifs, both of alpine-type and Platinumbearing associations). In relation to ophiolite, it is not clear how this dating correlates with sufficiently numerous (tens) and very reliable Ordovician (mainly Late Arenig—Middle Ordovician) age determinations (based on representative conodont complexes from syngenetic jasper layers) of ophiolitic basalts (Ivanov et al. 1986; Ivanov 1998; Puchkov 2000; Ruzhentsev and Degtyarev 2005; and others) evidently conjugated with gabbro-ultramafic complexes. Incidentally, there is no reliable dating (performed using different methods with compatible results) of all members of ophiolite association in the same complex or massif. Recent data suggest different ages of ophiolite members which constitute a single complex (lower ultramafic-gabbro parts of ophiolite sections being probably considerably older than tholeiitic basalts and sheeted-dyke complex). Thus it is possible to suppose different ages of different layers of ophiolites. Gabbro-ultramafic ophiolitic complexes (Puchkov 2006) are (in overwhelming majority) unevenly retrometamorphosed and repeatedly partially remelted (depleted) rocks of ancient mantle.

There were both restites and melts which did not reach the earth surface and so they can keep the "isotopic memory" about previous Wilson's cycles.

The eastern boundary of the Tagil-Magnitogorsk megazone passes along the East-Magnitogorsk mélange zone in the South Urals and along the Serov-Mauk fault controlled by mélange serpentinites in the central and southern parts of the Northern Urals.

1.1.6 East-Uralian Megazone

This part of the Urals (**E**, see Fig. 1) differs from the Tagil-Magnitogorsk megazone adjoining on the west in wide distribution of granitoids and gneisses, occurrence of microcontinent blocks with Pre-Paleozoic(?) crystalline sialic-type crust and generally continental type of the crust with well-developed granitic layer. It is just this megazone that includes the so-called "Main Granitic Axis" (**MGA**, see Fig. 1) of the Urals where the greatest amount of granites is concentrated in this region. Magmatism of marginal (back-arc) basins is limited by gabbro-tonalite M-type association (the most typical example is the Reft gabbro-tonalite complex in the eastern part of the Middle Urals). Next stage of marginal-continental evolution is characterized by intrusion of enormous (up to 100 km long) I-type granodiorite and tonaitegranodiorite batholiths. The Verkhne-Isetsk tonalitegranodiorite massif of the Middle Urals is a typical example of these batholiths. At the later stages, the calc-alkaline tonaite-granodiorite magmatism gave way to subalkaline magmatism, but areal extent of subalkaline rocks of this stage is negligible. The collision stage of the Urals evolution is characterized by the formation of batholith-like bodies of crustal anatectic granites. Isotopic data for the granite of the "Main Granitic Axis" in the South Urals suggest its formation above the Late Paleozoic subduction zone (Ivanov et al. 1986, and others); later stages of this process are related to palingenesis in the continental crust, which was thickened as a result of intensive approaching crustal blocks by the eastward movement of the thrust system, inherited from the Early Carboniferous—Bashkirian subduction zone. This system is reliably identified at the URSEIS-95 and SB-ESRU seismic profiles (Deep composition… 2001; Kashubin et al. 2006; and others).

The East-Uralian megazone (uplift) as a whole is mainly composed of intrusive and metamorphic rocks of low to medium pressure, i.e. by formations of the lower and middle parts of the Earth's crust. This crust completely formed during the Late Paleozoic (Early Permian). Judging by the degree of metamorphic alterations which accompany the intrusion of granites, their upward movement toward the surface caused the erosion of the 10–15-km upper crust. This may be due to the mountainous relief and associated "mountain root".

The Kartaly fault is the eastern marginal boundary of the East-Uralian megazone.

1.1.7 Trans-Urals Megazone

The Trans-Urals megazone (**F**, see Fig. 1) is the easternmost structure of the «opened» (exposed) Urals and has probably the accretionary nature. Paleozoic volcanogenic and sedimentary series of variable composition are common in the Trans-Urals megazone. Following complexes of different rock assemblages have been identified in Pre-Carboniferous formations: (1) blocks of Pre-Paleozoic(?) crystalline schists; (2) Ordovician terrigenous-volcanogenic rift complexes; (3) Middle-Upper Ordovician ophiolite that constitutes submeridional zones (Denisov, Varna); (4) Silurian volcano-sedimentary island arc complexes; (5) Middle-Upper Devonian deep-water siliceous shale series. All these series are covered with the Early Carboniferous suprasubduction andesitic basalt of the post-accretion complex. Carboniferous deposits are subdivided (from east to west) into Valerianovka, Borovoy and Ubagan zones. All the zones are covered with Meso-Cenozoic platform sediments and differ in composition. According to drilling data, mainly Carboniferous, particularly Lower Carboniferous complexes are present here. Sedimentary deposits are represented by shallow-water limestone and sandy-shale series. Volcano-sedimentary and volcanogenic series, represented by andesitic and andesite-basaltic porphyry lavas and their tuffs, are widespread. Volcanic rocks together with basic and intermediate intrusives form an integral volcano-plutonic association.

The Trans-Urals megazone is characterized by very complicated tectonics; thrusts, strike-slip faults, zones of mélanged serpentinite and blastomylonite are abundant there. According to geological mapping, structural studies and interpretation of the URSEIS-95 seismic profile, major structural elements of the megazone are of eastward vergence.

The relationships between the most important geological rock assemblages of the Southern Urals are shown in Fig. 2.

1.2 Main Features of the Deep Structure in the Urals

URSEIS-95 is the complex seismic profile crossing the Southern Urals. It was made by international team of geophysicists from Russia, Germany, Spain and the USA. This 465-km profile is one of the most representative lines in the world. According to results of URSEIS-95 (Fig. 3) (Echtler et al. 1996; Deep composition… 2001; and others), the Uralian fold belt consists of three large domains (segments): western, central and eastern.

The western domain includes the eastern margin of the East European Platform together with Pre-Uralian fore-deep basin (Fig. 3), West-Uralian and Central Uralian megazones and is limited from the east by the Main Uralian Fault (MUF). The central domain corresponds to the Magnitogorsk megazone and East Uralian Uplift. The eastern domain is situated to the east of the East-Uralian Uplift with the boundary between them represented by the Kartaly fault zone (Fig. 3). The Moho discontinuity, well fixed in the western and eastern domains, dips gradually to central part of

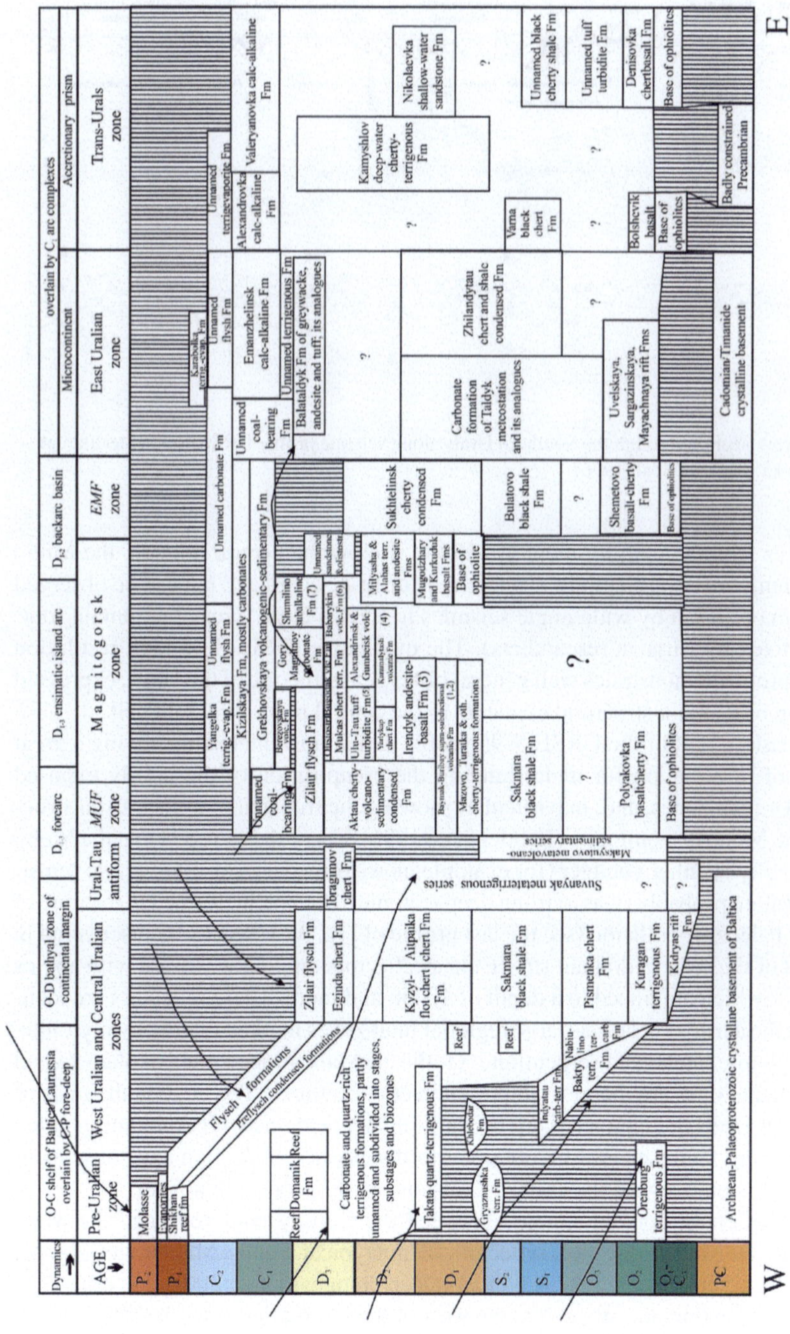

Fig. 2 Main rock assemblages of the Southern Urals. The terrigenous provenance direction is represented by the arrows

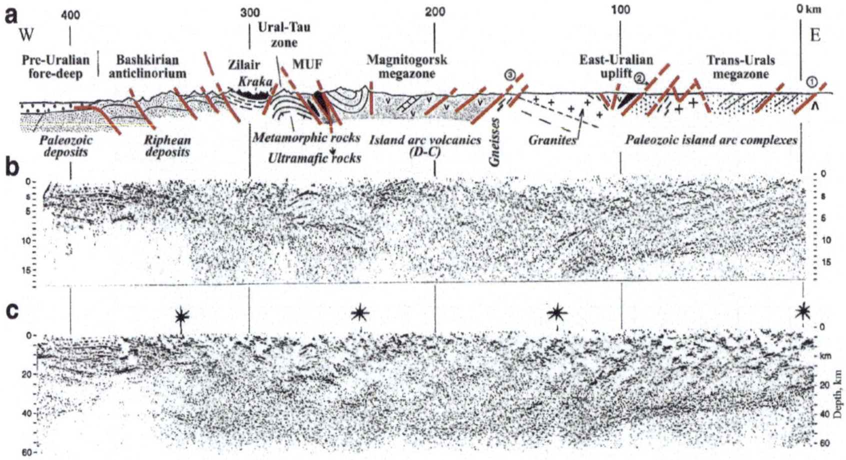

Fig. 3 Cross-section through the Southern Urals along seismic profile URSEIS-95 (Sterlitamak—Nikolayevka) (Echtler et al. 1996)

the orogen from a depth of about 40–55 km. Below the central domain, the Moho discontinuity loses its "legibility" and has a "diffuse" character, but can be observed at a depth of 58 km by wide-angle seismics to form a clear "crustal mountain root' (earlier fixed by Russian researchers). The qualitative URSEIS-95 high-resolution seismic profile demonstrates well general bi-vergent structure of the Urals, expressed in the dip of its main structural elements to the central part of the fold belt.

The eastern part of the URSEIS-95 profile is the most interesting, giving a great volume of information for understanding the composition of the poorly exposed eastern Uralian zones, and, in particular, showing the truncation of the crustal structure at the Moho discontinuity (Echtler et al. 1996). These data suggest that the Moho discontinuity is either younger (for example, new-formed as a result of phase transition) or (more probably) was a strong deep tectonic detachment, formed in the time of the Late Paleozoic collision (see the last profile of Fig. 3). URSEIS-95 discovered in the crust of the eastern domain strong clear reflectors dipping westward with angles of 30–40° which are traced to a depth of 40 km and more. These reflectors are identified on the earth's surface as large regional fault zones marked mainly by mylonite, blastomylonite, mélanged serpentinite. Of the abovementioned, the Nikolaevka and Kartaly fault zones are the most important tectonic sutures. The large fault zones of the eastern Urals are characterized by multiple dislocations with dominating sinistral strike-slip faults, caused by oblique (northwestern) subduction and subsequent terrane collision between the eastern Urals and East European Platform.

For the Central domain situated between the Kartaly fault zone and the Main Uralian Fault (MUF), long clear reflectors are not typical. For the Magnitogorsk zone, diffuse character of reflections is typical (Deep composition… 2001; and others).

The western domain, situated to the west of the Main Uralian Fault (MUF, which dips eastward at an angle of ≅ 45° and is confidently enough traced to a depth of about 30 km) is characterized by highly reflective layered crust dominated by the west-vergent structural elements. An exception is a clearly expressed antiform of metamorphic rocks of the Ural-Tau zone which is separated from both sides by thrusts and normal faults. Below and to the west of this antiform, a series of long, eastward dipping reflectors is traced to a depth of 30 km. Beneath the Bashkirian Megaanticlinorium, four series of such type are exposed (Echtler et al. 1996; Deep composition… 2001). They steeply dip eastward near the surface and more gently at depth. At the western end of the profile (380–465 km) there is visible a highreflecting weakly-deformed layered convexo-convex lens of Paleozoic, Vendian and Riphean sedimentary rocks (total thickness of 20 km) which corresponds to the southeastern end of the Kama-Belsk Aulacogen. Sedimentary rocks cover the faintly-reflectant Pre-Riphean crystalline basement of the East European Platform having the thickness of about 24 km.

The studies along the SB-ESRU Middle-Uralian transect (Kashubin et al. 2006; and others) resulted in obtaining numerous new interesting data, principally specifying ideas concerning the lithospheric structure of the Urals. It is shown that the main Uralian structures, West-Uralian, Central Uralian, Tagil and considerable part of East-Uralian megazones are situated in allochthonous position, i.e., are represented by non-root structures, tectonically torn off the basement. In general, the Uralian crust has a bi-vergent composition. The bi-vergency axis in the upper and middle crust is situated in the Tagil zone. The lower crust is also bi-vergent, but its axis in the Middle Urals is shifted westward for 50–60 km relatively to the axis in the upper crust. The upper mantle can be called mono-vergent, though it has practically no seismic reflectors, except for one gently westward-dipping reflecting zone traced to a depth of about 80 km. This tectonic structure dips into mantle beneath the Central Uralian megazone and is traced at depth to the central part of the area, above which the Pre-Uralian fore-deep basin is situated and where the western topographic slope of the Ural Mountains is almost terminated. Probably, it is above this presently(?) activated structure that the present-day Ural Mountains formed. The fragments of the gneiss-amphibolite and granulite complexes of the East-Uralian megazone accessible for observation are of Paleozoic (Devonian–Permian) age and formed from the heterogenous substrate (Kashubin et al. 2006; and others).

1.3 Tectonic Evolution

The above-mentioned and other numerous facts show that in the Riphean (Meso-Neoproterozoic)—Mesozoic the Urals underwent two complete cycles of geodynamic evolution. The first one took place during the Riphean and Vendian and finished in the formation of Timanides; and the second (Paleozoic—Early Mesozoic) caused the formation of the Uralides. **The first cycle** was preceded by poorly studied, continuous and probably polycyclic period of formation of the Archean-Paleoproterozoic crust of the Baltica paleocontinent, observed in the Taratash uplift in the Southern Urals. According to the recent opinion (Bogdanova et al. 2008), the paleocontinent formed during the late Paleoproterozoic as a result of amalgamation of three smaller continents: Fennoscandia, Sarmatia and Volgo-Uralia in a process of supercontinent Nuna assemblage. The Riphean (Meso-Neoproterozoic) evolution of the Ural-Timan continental margin was connected with episodes of riftogenesis, of which the Mashak (Middle Riphean) rifting was the strongest. It is possible that this event (and/or the following the Late Riphean rifting) caused the formation of the passive continental margin in the east and northeast of Baltica paleocontinent, and also the oceanic crust and some microcontinents along its periphery. During the Late Riphean the subduction processes began in the future Timanides, and were changed finally by continental collision and growth of the Timanian orogen.

The second cycle (Fig. 4) is studied better and permits identification of several stages, partly overlapping, as a result of diachroneity of processes of the same type.

Continental rifting (Cambrian—Early Ordovician). This stage includes the uplifting of the whole region followed by formation of the main Uralian rift. The volume of volcanic rocks gradually increases up the section (and from west to east) while their alkalinity decreases.

Oceanic spreading (Middle-Late Ordovician). Spreading and ophiolite formation in the Urals started in the Late Arenig (Dapingian of the International Geochronological scale). This is proved by conodont findings in the Akay, Sugrala, Lower Polyakovкa, Denisovka and other series of tholeiitic basalts (thickness up to 2500 m) where jasper interlayers with conodont complexes contain *Periodon flabellum—P. aculeatus zgierzensis*; *Periodon aculeatus aculeatus—P. aculeatus zgierzensis—Pygodus serrus*; *Periodon aculeatus aculeatus—Pygodus anserinus*. These complexes allow considering that the ocean floor spreading continued 25–30 Ma and the width of the Ordovician Uralian paleoocean was not less than 600–800 km.

Island arc stage (Late Ordovician—Early Carboniferous). The eastern sector of the Urals consists of two main island arc terranes of different ages: the Tagil (Ordovician—Lower Devonian) and Magnitogorsk (Lower Devonian—Carboniferous) ones characterized by rather similar composition. In the Tagil terrane, in its western part, deep-crustal magmatic suprasubduction complexes are exposed and represented by massifs of the Platinum-bearing belt; the presence of non-exposed gabbro belt may be suggested, according to gravity data, also in the Magnitogorsk terrane.

Early collision stage (Late Devonian—Early Carboniferous). This stage corresponds to a collision of the Magnitogorsk island arc and the Laurussia passive continental margin. Direction of the collision was oblique (northwestern).

Late subduction stage: subduction of a relic oceanic basin crust of the Paleo-Uralian Ocean (Early Carboniferous—Bashkirian). Accretion of the Trans-Urals zone and subduction of oceanic crust took place near and under the Eastern Uralian zone; it resulted in the formation of the tonalite-granodiorite association of the "Main Granitic Axis" of the Urals and intensive suprasubduction volcanism in eastern zones of the Urals. Apparently, the cessation of suprasubduction magmatism and paleogeographic changes in the Bashkirian time marked the transition from subduction to collision.

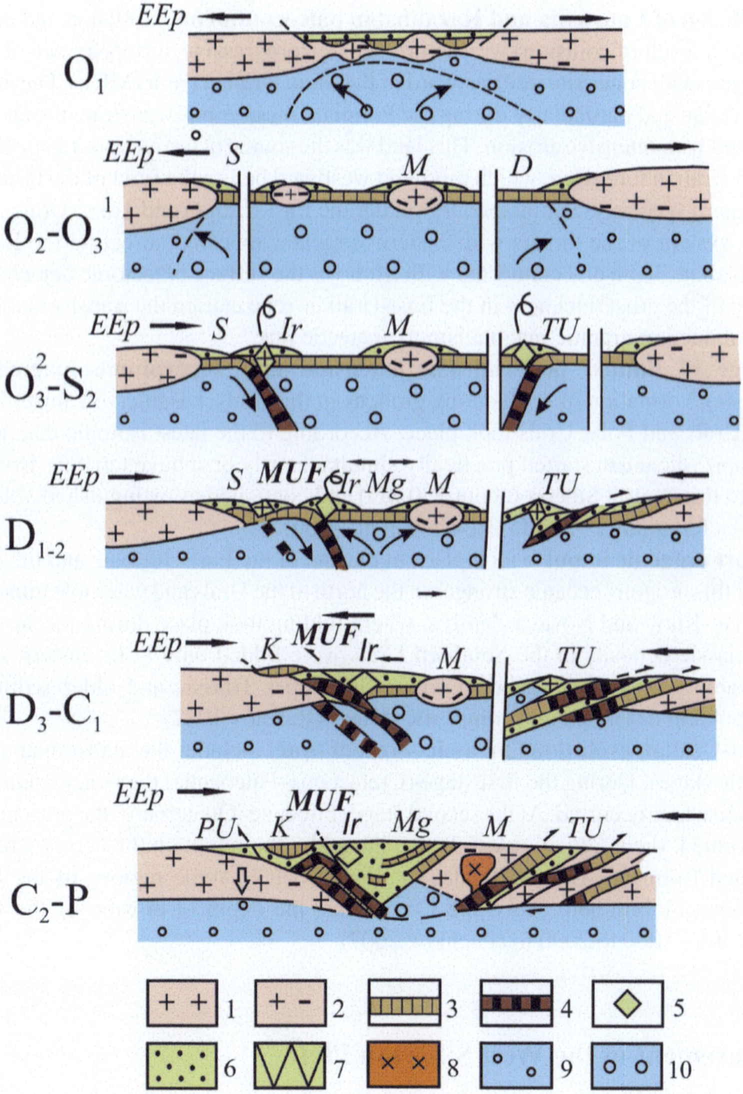

Fig. 4 Scheme of geodynamic evolution of the Southern Urals during the Paleozoic. 1—continental crust; 2—the same, intensively deformed; 3—basalts and other rocks of the oceanic crust, and also andesitoids of the island arcs; 4—the same, metamorphosed mantle in subduction zones; 5—island arcs; 6—sedimentary complexes; 7—accetionary prisms; 8—granites of the 'Main Granitic axe" of the Urals; 9–10—mantle: 9—normal, 10—anormal. Letters on the picture: EEp—East European Platform; Kc—Kazakhstan continent; S—Sakmara zone; M—ugodzhary (East-Uralian) microcontinent; D—Denisov volcanogenic zone; TU—Trans-Urals megazone; MUF—Main Uralian Fault; Ir—Irendyk island arc; Mg—Magnitogorsk volcanogenic zone; K—Kraka allochthon; PU—Pre-Uralian fore-deep depression. Arrows show the direction of movements

Collision of Laurussia and Kazakhstan paleocontinents. Collision and orogenesis (as a result of collision) were expressed as a progressive disappearance of sedimentation in all zones situated eastward of the Main Uralian Fault (MUF). During the Pre-Permian and particularly during the Permian, these zones were transformed into a dry land with intensive erosion. This land was the source of terrigenous material into the Pre-Uralian fore-deep, which migrated westward before the front of the thrusts of the western vergency. Simultaneously under the East Uralian and Trans-Urals zones a thick system of the thrusts with eastern vergence appeared, affecting the Earth's crust to all its depth and connected with Moho as the surface of tectonic detachment. Growth of the crust thickness in the East-Uralian zone caused the transformation of suprasubduction granitic magmatism to anatectic one.

Stage of limited post-collision extension and superplume magmatism (Triassic). Formation of coalbearing grabens in the Urals, basaltic volcanism in the Trans-Urals and Polar Urals took place. According to the latest isotopic data intensive trapp volcanism started practically simultaneously on a huge territory from the Urals to the Central Siberia (about 250 Ma) and continued as extinguished volcanic impulses for another ~20 Ma (Reichov et al. 2009).

Short orogenic impulse took place at the end of the Early Jurassic and the influence of this orogeny became stronger in the north of the Urals and was most important in the Pai-Khoy and Novaya Zemlya, where folding took place during the Jurassic. The Triassic deposits in the Southern Urals were folded only in its eastern sector (Chelyabinsk and other grabens) where the Upper Triassic and older sediments correspond to deformed and composite thrusts (Rasulov 1982).

Post-Uralian evolution (Jurassic—recent time) includes the platform and neoorogenic stages. During the first stage (Cretaceous–Paleogene) the peneplanation of the Uralian area occurred. At the second stage (Pliocene–Quaternary) the present Ural Mts. formed; their formation can be explained by the remote influence of stresses, expanded from the Alpine-Himalaya collision belt to some regions of the Ural-Mongolian mobile belt. Since the Cretaceous, the depth of erosion of the Urals ranged from 1000 to 2000 m (Puchkov 2007).

2 Basement of the West Siberian Plate

Structure of the West Siberian Plate is divided into three structural stages:

(1) Folded basement composed by rock complexes of almost exclusively Paleozoic age;
(2) Riftogeneous structural stage composed by the Early Triassic basalts (sometimes basalts with rhyolite) covered by terrigenous series of the Middle and Upper Triassic;
(3) Platform cover formed by the Jurassic and younger practically non-deformed sedimentary complexes. The latter contain almost all deposits of oil and gas

in Western Siberia. Thickness of sedimentary cover increases to the north and reaches 6 and more km.

The first two lower stages usually represent the basement of the West Siberian Plate.

In present time, Paleozoic and Triassic rock complexes of the Pre-Jurassic basement of the West Siberian Plate are penetrated by more than 5000 boreholes situated mostly in the southern and central areas of West Siberia.

There are several schemes of zonation (Surkov and Trofimuk 1986; Yolkin et al. 2001, 2008; Bochkarev et al. 2003; Klets et al. 2007; Surkov and Smirnov 2008; Ivanov et al. 2013 and others) and prolongation into the basement of West Siberia of the Paleozoic fold belts and their structural-formational zones surrounding sedimentary plate. The basement of western part of the West Siberian Plate is composed by structural zones of the eastern sector of the Urals, and basement of eastern part of the plate—by complexes of the Siberian Craton and its folded framing. General feature of zonation schemes for the West Siberian Plate is the presence of a huge block of Kazakhstanides situated to the east from Uralides and pinching out to the north (Fig. 5). These main megazones (or domains) are divided by major ophiolite sutures—Valerianovsk and Chara ones.

Siberian Platform with surrounding fold systems is the core of the Siberian domain. Three primary megazones are reconstructed for them (Yolkin et al. 2001, 2008). The latter ones form a united facial line and characterize sedimentation environments on Siberia paleocontinent and its margin with gradual deeping to the west. So megazone II (Fig. 5) is composed mostly by shallowwater terrigenous-carbonate series of the uppermost Cambrian—lowermost Upper Carboniferous, which continued the shelf of Siberia paleocontinent. Megazone III is composed mostly by more deep shelf and continental slope facies and also volcanogenic complexes. It is supposed (Kontorovich et al. 2003) that all three described megazones (together with the Siberian Platform) are united by single Pre-Cambrian (Proterozoic—Lower Riphean) crystalline basement. Sufficiently numerous relict zircons of 2051 + /–23 Ma are determined in a subvolcanic body of the Early Permian potassic rhyolite-granite (Ivanov and Erokhin 2010). This is the evidence that granite magma interacted with ancient granite-metamorphic basement. Starting from the Late Riphean throughout the Cambrian, extension and fragmentation of the Siberian domain took place (Yolkin et al. 2008). The first rupture of the whole continental crust with its separation was at the Early and Middle Riphean boundary with formation of turbidite complexes and ophiolites. The evidence of the last impulse of riftogenesis probably is the basaltic volcanic rocks petrologically similar to back-arc basin and middle-ocean ridge basalts (Sarayev et al. 2004).

To present time, there are many disputable issues on the problems of the Kazakhstanides evolution (megazone IV, Fig. 5). Either their belonging to Kazakhstanides separated structures composed by different complexes (among which andesite and its tuffs, also the Devonian—Carboniferous carbonate-slate complexes and et al. are predominate), or the whole evolution history of this structural unit of West Siberia is discussed. Generally, it is considered (Bochkarev and Krinochkin 1988; Yergaliev

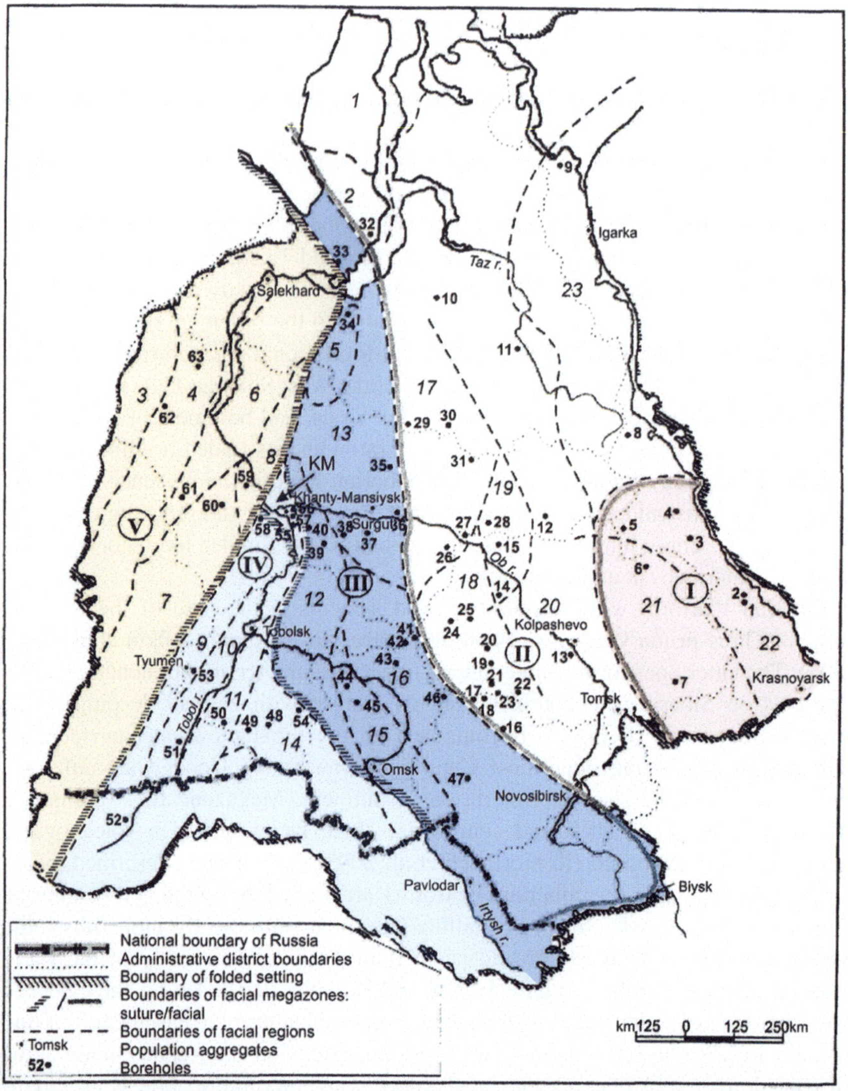

Fig. 5 Facial regions and megazones in structure of the West Siberian Plate basement (Yolkin et al. 2008)

et al. 1995) that the southern part of this megazone represents the northern deepening of the Kokshetau Massif. To the north from it, Krasnoleninsky dome is situated. Probably, both of these Pre-Cambrian blocks were united in the Frasnian. Until now, the supposition about the presence of the Pre-Cambrian in Krasnoleninsky dome has not been proved. Granitoid intrusions of this (Kazakhstan) domain are considerably older (440–450 Ma by SHRIMP-II U/Pb dating) than in the Pre-Uralian area (280–290 Ma). Subplatform environments in limits of the Kazakhstanides were settled to the end of the Late Devonian (Yergaliev et al. 1995).

To the west of domain in the Early Carboniferous evidently sufficiently narrow shelf was situated, where accumulation of terrigenous-carbonate sediments of huge thickness took place (Yolkin et al. 2001).

Kazakhstanides are separated from the Pre-Uralian part of the West Siberian Plate by Valerianovsk suture which is traced by geophysical data to depth no less than 20–30 km (Dyakonova et al. 2008).

Chara suture is situated between the Kazakhstanides and Siberian domain. Age of ophiolites is determined as the Visean and Serpukhovian boundary of the Early Carboniferous. Most of these and other regional faults of the West Siberian Plate basement have a strike-slip origin (amplitude of displacement reaches hundreds km). It is connected (apparently by paleomagnetic data) with a clockwise rotation of the Siberian domain with respect to the East European Craton (Buslov et al. 2003; Kazansky and Metyolkin 2008).

Pre-Uralian part of the West Siberian Plate (megazone V, Fig. 5) is composed by complexes of eastern island arc sector of the Urals (Peive et al. 1976; Ivanov 1998; Ivanov et al. 2006). As a result of mapping of large segments in this area, a new scheme of structural-formational zones of the western part of West Siberian plate basement was created and a simplified (scale 1:1,000,000) geological map for the Pre-Uralian part of the Pre-Jurassic basement of the West Siberian Plate was compiled (Fig. 6) (Ivanov et al. 2009).

Normal fault along the western flank of North-Sos'va graben is the ingenuous boundary between the Urals and West Siberian young platform. This fault-boundary extends along the "opened" (i.e. outcropping) Urals in submeridional direction to the Pre-Polar part 350 km in length. Study of magmatic and metamorphic complexes, also volcanogenic (including ophiolitic ones) terrigenous-slate and carbonate series of the Urals and western part of West Siberia show their obvious resemblance. Structure of the plate basement has much general features with the "opened" Urals. As in the Urals, in West Siberia basement two periods of ophiolitic magmatism are determined—Ordovician and Devonian (by Sm/Nd dating, conodonts and radiolarians from jasper interlayers). By geochemical characteristics, mafic complexes formed in island arc (probably back arc) conditions.

At the same time, there are considerable differences between the Urals and West Siberia basement. Within the outcropping Urals (with diversity of disjunctive fault system), the Late Paleozoic submeridional sinistral strike-slips predominate over others. In the basement of the western part of the West Siberian megabasin, a system of dextral sublatitude strike-slip faults was revealed. These faults have W–N-W strike with amplitudes of 6–16 km and caused an echeloned displacement of the main structures. Strike-slips divide the basement into several blocks about 40–50 km long; and each more northern block is displaced eastward (and is frequently lowered) with respect to more southern one. This strike-slip system was formed in the Middle—Late Triassic (somewhere later) probably as a result of a sublatitudinal extension of crust and deepening of its northern part. This caused, at first, origination of the

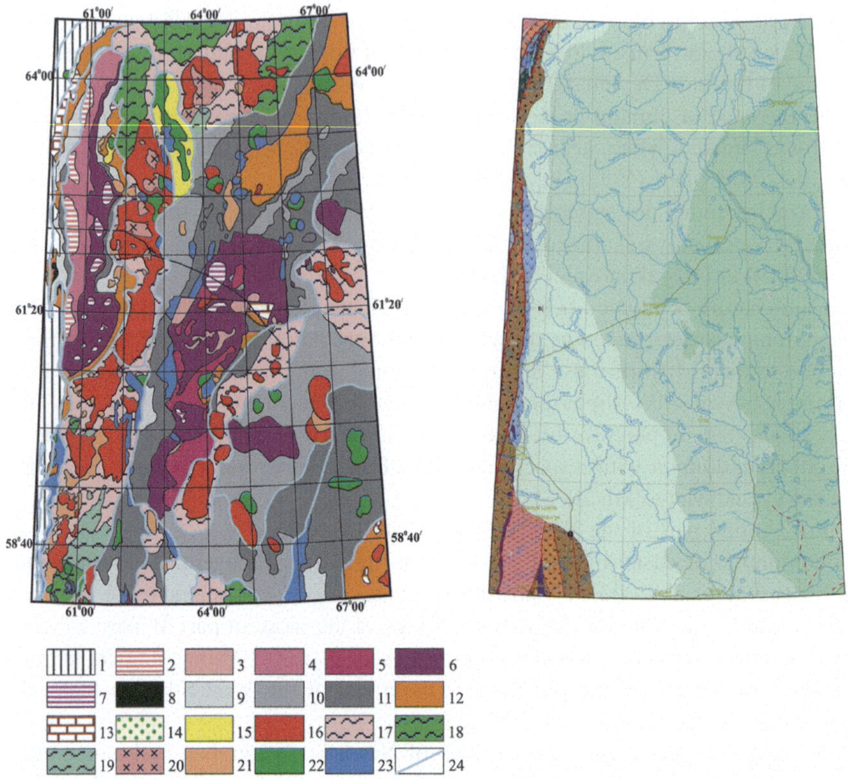

Fig. 6 Schematic geological map of the pre-Jurassic basement of the western part of the West Siberian Plate (Ivanov et al. 2009) (left) and fragment of the *Tectonic map of Northern-Central-Eastern Asia and Adjacent Areas* for the western part of the West Siberian Plate (right). 1—Paleozoic basement outcrops (Ural Mts.), 2—Triassic in troughs and brachysynclines, 3—Triassic rhyolite, 4—Triassic basalt-terrigenous formation, 5—Triassic effusives with predominance of tuffs, 6—Lower Triassic basalts, 7—Triassic gabbro-dolerite, 8—Carboniferous gabbrodiabase, 9—terrigenous-carbonate deposits (C_1t-v), 10—Carboniferous terrigenous-slate formation, 11—Effusives (D_3–C_1), 12—Devonian volcano-sedimentary series, 13—Limestones (D_{2-3}), 14—Silurian sedimentary-volcanic series, 15—Lower Silurian andesitic basalts, 16—Granitoides, 17—Sialic gneiss and crystalline schists, 18—Amphibole schists, 19—Epidote-amphibole schists, 20—Quartz diorite and magnetic sialic gneiss, 21—Plagiogranite, quartz diorite, granodiorite, 22—Gabbro, 23—Serpentinite, ultramafic rocks, 24—Boundaries of structuralformation zones, 25—Shear zones and sutures

Triassic graben system filled by volcanogenic and terrigenous-volcanogenic series, and thereupon the whole West Siberian oil-and-gas megabasin. Wide distribution of the Triassic volcanogenic complexes in the West Siberian Plate basement is the principal difference in comparison with the Urals.

Ophiolites and other mafic–ultramafic complexes are widely distributed in the West Siberian Plate basement (especially in its central and western parts). These complexes are situated along large faults dividing structural-formational zones of different type (Surkov and Trofimuk 1986; Dobretsov 2003). Ophiolites are often represented by elements of dismembered section tectonically aggregated with other rocks. Most representative Paleozoic ophiolite complex is composed by mélanged serpentinite, gabbroid, plagiogranite, basalts with jasper interlayers containing the Late Ordovician radiolarians and conodonts (Ivanov et al. 2007). This ophiolite complex was described within the Shaim gas-and-oil area of the Pre-Uralian megazone. It is the most ancient complex in basement structure in the western part of West Siberia. In all probability, spinel lherzolite is a relict of melanocratic basement of the Early Paleozoic (Uralian?) paleoocean.

Completion of the Paleozoic geodynamic evolution in these regions appeared as a result of collision accompanied by folding, tectonic aggregation, intrusion of granite plutons, metamorphism and formation of newly-originated crust of continental type. The time of these events consolidating Paleozoic complexes of basement of the future West Siberian megabasin is determined as Early Permian for the Pre-Uralian part of the plate. Relatively low $^{87}Sr/^{86}Sr$ ratio in granites ($I_{sr} = 0.7046–0.7047$) of the western part of West Siberia indicates that substratum for melting of these granites probably consisted of the Paleozoic complexes with a considerable portion of mantle, i.e. oceanic and island arc, material. Three types of granites are distinguished. Among them, rocks of monzodiorite-granite series predominated (age 280 Ma); they are similar to analogous from the eastern sector of the Urals.

Triassic period was the most important in the postPaleozoic evolution of the West Siberia basement. In Triassic (mostly in the beginning of Triassic (Medvedev et al. 2003) riftogenesis, formation of graben system, also uplift of intrusive and metamorphic complexes forming "anticlinorium" cores took place. Emplacement of the megablocks composed by deep-crust complexes to the surface took place as a result of their uplift to the level of the upper crust during rupture and/or extension of the latter. Probably, extension began in the Early Triassic. Triassic volcanism is a result of dispersed riftogenesis during the Triassic post-collision sublatitudinal extension of the Urals and origination of the West Siberian megabasin which are closely associated to each other.

By K/Ar method data, phases of tectonic activization in the Mesozoic are determined (Fedorov et al. 2004). It is known that the Triassic basalts and rhyolite with unconformity are covered by the Jurassic sediments. Therefore, a considerable part of datings for volcanic rocks (younger than 230 Ma) shows not the time of their origination, but the time of their secondary alteration which is connected with phases of tectonicthermal activity of the region. The following phases of endogenic activities are distinguished:

1. Late Permian—Early and Middle Triassic (with peak at 250–230 Ma)—riftogenesis and intense volcanism;
2. Early Jurassic (201–200 Ma)—short but intense growth of tectonic activity accompanied by uplift of territory;
3. Middle Jurassic (180–160 Ma)—alternating uplifting and downwarping movements of the territory, accumulation of continental sediments;

4. Early Cretaceous (with peak at 130–120 Ma)—a new phase of tectonic activity, formation of sandclayey marine clinoform series;
5. Late Cretaceous—Early Paleogene (with peak at 80–70 Ma)—tectonic activity with slow attenuation.

Accumulation of separated horizons of clayey deposits (end of the Early—beginning of Middle Jurassic, Late Jurassic—beginning of the Early Cretaceous, Early Aptian) was connected with phases of the calmest tectonic activities.

3 Fold Systems of the Sayan-Baikal Mountain Area

In the Sayan-Baikal mountain area (SBMA), there are following geographical regions (from north to south): Patom and North Baikal uplands, Middle Vitim, Angara-Barguzin, East-Sayan and Dzhida mountain areas, West, East and South Pre-Baikalia, Vitim tableland and West Transbaikalia. Fold systems of this area from west, north and east are limited by the Siberian Craton and from southeast by MongolOkhotsk fold belt (Fig. 7).

Fold systems with continental crust consolidated during the second half of the Neoproterozoic and during the Late Cambrian—Early Ordovician existed in the SBMA.

3.1 Fold System with Consolidated Crust Formed During the Second Half of Neoproterozoic

It involves the northern, central and south-western parts of SBMA. The boundary between the Siberian Craton and the Fold system with consolidated crust that formed during the second half of the Neoproterozoic runs along the Main Sayan Fault (dextral) in the southeastern part of the East Sayan and corresponds to the 150-km Primorsky sinistral strike-slip—reversefault in West Pre-Baikalia. Northward, at distance of about 1000 km, the structural suture that separated the Siberian Platform from the SBMA folded system is tectonically covered with Precambrian complexes overthrusted onto the platform margin. The boundary of their extension is discontinuous due to different amplitude of horizontal offset of the fold system complexes onto platform deposits. Maximal amplitude of replacement (about 50–80 km) has been registered in the area of the Akitkan Ridge in the Northern Pre-Baikalia where the Akitkan orocline is known, and in the Patom uplands where clearly expressed orocline of the same name formed (Ivanov and Ryazanov 1992). Between the above mentioned oroclines, the amplitude of horizontal replacement of overthrusts and reserved faults ranges from first km to first tens km (Alexandrov 1990).

The Patom orocline to the east is limited by the Zhuya border shear zone of 10–12 km wide composed of some oblong tectonic sheets, divided by reversed faults

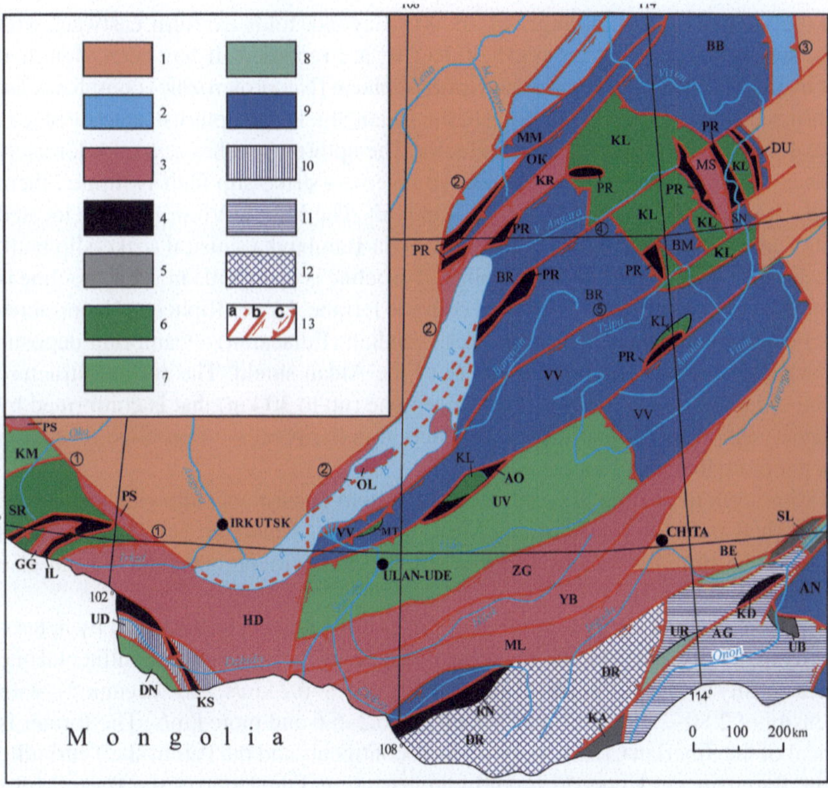

Fig. 7 Map of tectonostratigraphic terranes of the Sayan-Baikal mountain area and adjacent areas. 1—Siberian Craton; 2—Patom fold-thrust belt (passive continental craton margin). Other terranes of the southeastern folded framing of the Siberian Craton: 3—*Meso-Neoproterozoic and Paleozoic metamorphic*: (GG—Gargan, HD—Khamar-Daban, KR—Kichera, ML—Malkhan, MS—Muya, OL—Olkhon, PS—Proto-Sayan, YB—Yablonoviy, ZG—Zagan); *oceanic*: 4—*Neoproterozoic* (IL—Ilchir, PR—Parama, AO—Abaga-Olan), 5—*Vendian (Ediacaran)—Early Paleozoic* (KS—Khasurta, KN—Kunaley, KD—Onon-Kulinda); *Silurian—Devonian* (KA—Kiran, UB—Ust'-Borzya, SL—Pre-Shilka); *island arc*: 6—Neoproterozoic (KL—Keliana, MT—Meteshikha, SR—Sarchoi), 7—Vendian (Ediacaran)—Early Paleozoic (UV—Uda-Vitim, DN—Dzhida, KM—Khamsara); 8—Devonian—Early Carboniferous (BE—Bereya, UR—Urtuy); *turbidite basins:* 9—Meso-Neoproterozoic (AN—Argun, BB—Bodaibo, BM—Bambui, BR—Barguzin, DU—Delyun-Uran, MM—Mama, OK—Olokit, SN—Shaman, VV—Verkhne-Vitim); 10—Vendian (Ediacaran)—Early Paleozoic (UD—Upper Dzhida); 11—Devonian—Early Carboniferous (AG—Aga), 12—Devonian—Late Carboniferous (DR—Daur); 13—faults: **a** conventional faults, **b** strike-slip faults, **c** upthrusts and nappes. Faults names (arabian figures in circles): 1—Main Sayan, 2—Primorsky, 3—Zhuya, 4—Tompuda-Nerpa, 5—Argoda-Bambuy. *Note* Overlapping Meso-Cenozoic structural-formational complexes are not shown

and dipped eastward at an angle of 70–90°. Beyond, folds overturn eastward with axial surfaces dipping at an angle of 70–50° are recorded. It has been identified that by the Zhuya border shear zone Upper Riphean (Neoproterozoic) complexes are upthrusted onto the Lower Cambrian of the Aldan Shield and upthrust tectonic blocks were deformed by dextral strike-slip faults. The upthrow reaches several kilometers (Ivanov and Ryazanov 1992). The Zhuya reverse—strike-slip fault is limited there by the Tompuda-Nerpa sinistral strike-slip fault. The Middle Vitim orocline located southward is limited in the south by the Argoda-Bambuyka sinistral strike-slip fault. The distance between the extreme points of orocline is up to 150 km. This orocline is composed of complexes of the Muya cratonic terrane, Upper Riphean (Neoproterozoic) oceanic and island arc complexes, Vendian (Ediacaran)—Cambrian deposits obducted onto the crystalline basement of the Aldan shield. The surface structure of the orocline is well-correlated with deep one (up to 30 km) that is conformed by seismic methods. The amplitude of obduction in the process of orocline formation was not more than 10–20 km.

From north to south, the described fold system consists of following structures: Patom shelf-turbidite basin, Baikal-Muya ophiolite—island arc belt, Barguzin and Upper Vitim turbidite basins. In these structures, microcontinents have been also identified (See Fig. 7).

Upper part of the Earth's crust under this fold system is characterized by heterogeneity of density and velocity. There are two types of the upper crust: continental one with a density of 2.60–2.80 g/cm^3 and velocity of 5.8–6.2 km/s, and "oceanic"—with a density of 2.80–2.90 g/cm^3 and velocity of 6.2–6.6 and more km/s. The former is typical of the Siberian Craton margin, microcontinents and the Patom shelf-turbidite basin, Barguzin and Upper Vitim turbidite basins, and the latter, of the BaikalMuya ophiolite-island arc belt (Bulgatov and Gordienko 1999; Bulgatov et al. 2004).

Microcontinents are composed by the Early Precambrian stratified complexes metamorphosed in granulite and amphibolite facies and granitoides. They underwent substantial structural and thermal reworking during the Late Riphean (Neoproterozoic) and Phanerozoic. The Chernogriva series amphibolite ages (eastern coast of Lake Baikal) are 3022 and 3086 Ma (U–Pb, SHRIMP-II) (Rudenko et al. 2010), and the age of granulite from the western coast (Bibikova et al. 1990) of Lake Baikal is 1880 and 1877 Ma (U–Pb). But at present, there are many determinations confirming the Early Paleozoic age of this complex. Detrital zircons from interbedded marbles of quartzite layers (Sliudianka series of the Khamar-Daban microcontinent) are dated to 3232–3140 Ma (grain's nucleus) and older than 2500 Ma (grain's peripheries) and only in two cases it is younger—2409 and 1209 Ma (U–Pb, SHRIMP-II). The granitoids of microcontinents can be subdivided into three age groups: 2366, 2070–2020 and 1906–1846 Ma (U–Pb).

Patom shelf-turbidite basin was filled with Late Riphean (Neoproterozoic) terrigenous and carbonates sediments up to 12–13 km thick which accumulated in shelf, continental slope and its foot environments. Relatively early formations are represented by coarse-terrigenous sediments and metabasalts of E-MORB-type including the interlayers of ferrous quartzite related to the Late Riphean (Neoproterozoic) local spreading zone of the oceanic crust.

Ophilites of the **Baikal-Muya belt** constitute tectonic sheets, lenses and wedges and often are situated along boundaries of different tectonic unites, marking structural sutures, and rarely—among island arc complexes. The ophiolites are represented by siliceous-basaltic (E-MORB-type) formations which are sometimes associated with protrusive restite ultramafic rocks (Parama, Shaman and others). Trustworthy cumulative and dyke complexes are not determined in ophiolite composition. Geochronological data of ophiolitic basalts are the following: 1135 and 927 Ma (Rb–Sr), 1035 Ma (Sm–Nd).

Island arc complexes of the Baikal-Muya belt are represented by tuffs, tuffites, rhyolite-andesitebasalt-boninite lavas and by ultramafic–mafic, gabbro, tonalite and trondiemite massifs of the Keliana island arc. The rhyolite, tonalite and trondiemite age is 830–812 Ma (from eight U–Pb analyses) and gabbro age is 835 Ma (Sm–Nd) (Tsygankov 2005).

Turbidite complexes of the **Barguzin and Upper Vitim basins** are represented by sandy-slate (bottom of succession) and limestone deposits with the thickness up to 12–13 km, metamorphosed from green-schist to amphibolite facies. The bottom of the succession includes the basalts formed in the local spreading zones. Such zones are identified in the northwestern margin of the Barguzin turbidite basin and in the southwestern and northwestern margins of the Upper Vitim Basin. The basalt belongs to E-MORB-type and has absolute age dating. The earliest basalts are dated to 1021, 972, 934 Ma (U–Pb, SHRIMP-II) (Gordienko et al. 2010).

Complexes of the Patom shelf-turbidite basin, Baikal-Muya ophiolite-island arc belt, Barguzin and Upper Vitim turbidite basins have been accreted to the Siberian Craton during the second half of the Late Riphean (Neoproterozoic). Formation of depressions and grabens filled with molasses, basic and acid volcanic rocks, intrusion of peridotite-gabbro, gabbro and granitoid plutons was connected with accretioncollision processes. Volcanic rocks in depressions and grabens have the age of 735–700 Ma (6 U–Pb analyses), 765–670 Ma (6 Rb–Sr analyses), peridotite-gabbro and gabbro massifs—723 Ma (U-Pb), 774–612 Ma (10 Sm–Nd analyses), 774–733 Ma (3 Rb–Sr analyses), 787–650 Ma (3 Ar-Ar analyses), granitoid plutons—790–663 Ma (17 U-Pb analyses, U–Pb analyses, SHRIMP-II, 4 Rb–Sr analyses).

Late Riphean (Neoproterozoic) complexes of different geodynamic origin accreted to the Siberian Craton, and syn-collisional and syn-accretion complexes of the same age are covered with unconformity with Vendian (Ediacaran) terrigenous and Cambrian carbonate non-metamorphosed deposits. They are similar to synchronous sediments of the Siberian Platform and formed in the passive continental margin of the Siberian paleocontinent (Bulgatov 1983).

During the Neoproterozoic, fold systems with consolidated crust are typical of the southwestern part of SBMA within eastern part of the East Sayan. They are separated from the Siberian Craton by a system of the Main Sayan dextral fault. Early Precambrian complexes of Sharadzhalga ledge (Siberian Craton) and of the Gargan microcontinent are distinguished there. In the southeastern part of the Eastern Sayan, at the first stage of Baikalides evolution (1020 Ma), the Sayan oceanic basin and Dunzhugour ensimatic island arc formed (Kuzmichev 2004). Later, (about 800 Ma) this arc was obducted onto the Gargan massif and was broken by collisional tonalite of Sumsunur complex (785 Ma) and became its extinct. As a result of these processes, a dynamic continental margin occurred along which the Sarkhoi Andean-type island arc system with the age of volcanics of 782 ± 7 Ma and the Shishkhid ensimatic island arc (rhyolite age of 800 Ma) (Kuzmichev and Larionov 2011) formed. At the beginning of the Vendian (Ediacaran), the dynamic continental margin underwent rifting and was separated from the Siberian Craton. Due to these processes, an oceanic basin extending from East Tuva to West Transbaikalia was originated.

3.2 *Fold Systems with Consolidated Crust Formed During Late Cambrian—Early Ordovician*

Such systems in SBMA are distinguished in the southern Pre-Baikalia (Dzhida zone) and western Transbaikalia (Uda-Vitim zone). Tectonic history of these structures is strictly connected with the formation, evolution and closing of Paleoasian Ocean which dynamic evolution took place 630–540 Ma, i.e. during the Vendian (Ediacaran)—Early Cambrian. At this time MOR spreading zones, ensimatic and ensialitic island arcs with extensive subduction zones, fore-arc, back-arc basins and intra-arc spreading marginal seas formed.

3.2.1 Dzhida Zone

Dzhida zone is represented as a construction of overthrusted or upward faulted tectonic sheets with replacement in southern direction (in present-day coordinates). As a result, the Khamar-Daban orocline of southward extension formed in the southern frame of wedge-shaped ledge of the Siberian Craton. The strike of the structure in the Khamar-Daban orocline gradually changes its orientation from northeastern (Sayan direction) to northwestern (Baikal direction); a structural conformity between complexes of different ages is recorded there. But this conformity is secondary; it is caused by formation of arc-shaped Khamar-Daban orocline. It is suggested that the formation of the Khamar-Daban orocline was caused by southward movement of Siberian Craton and its underthrusting beneath the folded frame.

Tectonic structures of the Dzhida zone form ensimatic type island arc system of the same name and following complexes have been identified there: ophiolite (of oceanic crust), guyot, island arc, back-arc marginal basin and collisional. But the upper Earth's crust has relatively low density (2.70–2.80 g/ cm^3). This discrepancy can be explained by the fact that ophiolite and guyot complexes compose the nonrooted tectonic sheets in the Khamar-Daban thrustfolded structural arc.

Ophiolite complex is represented by restite ultramafic rocks, ultramafic-gabbro layered series, sheeted dyke complex, and N-MORB-type basalts. OIB-type subalkaline basalts and hawaiites, tuffs, tuffites and limestone, which belonged to guyots, are considered to be intra-plate oceanic complex.

Island arc complex is formed by two subcomplexes: ensimatic island arc and near-arc basins. The former is mainly composed of basalts and andesitic basalts, but some part—by andesite, boninite, dacite and rhyolite, their tuffs and gabbro-dioriteplagiogranite massifs with the age of 504 and 506 Ma (U–Pb). Subcomplex of near-arc basins is represented by tufa-sandstone, tufa-gravelite, andesibasaltic, andesite, dacite and rhyolite tuffs.

Carbonate-terrigenous flyschoid series (sandstone, siltstone, clayey shale and limestone with gravelite and conglomerate interbeds) is a complex of marginal basin. Island arc and gyuots were the sources of terrestrial fragments for sedimentary rocks. These sediments are characterized by Early Cambrian fossils and intruded by collisional tonalite-plagiogranite ptutons of 489 and 490 Ma (U–Pb), 476 Ma (Ar-Ar) ages. Granitoids associated with collision processes are determined in the northern framework of the Dzhida zone of the Khamar-Daban microcontinent. Different granitoids have the following absolute age datings (by U–Pb method): the first phase of Zun-Muren massif—494 Ma, the second phase—468 Ma; monzonite-syenite of Malo-Bystrinsky massif—474 and 470 Ma; pegmatite veins—477, 460 and 447 Ma. Tectonic event and crust growth appear to be strongest in the Early–Middle Ordovician when the collision between Dzhida complexes and Siberia paleocontinent came to an end (Gordienko et al. 2007).

3.2.2 Uda-Vitim Zone

The Uda-Vitim similar to the Dzhida zone forms massive island arc system of West Transbaikalia. It stretches from the lower reaches of the Selenga River in the northeast to midstream of the Vitim River having the length of 500 km and the width of 150–160 km. On northwest, southeast and southwest, the zone is limited by strike-slip faults. Ophiolite complex is not determined there, although more than twenty small massifs of ultramafic and serpentinite bodies are dispersed in this zone. The density of the upper crust is mainly 2.70–2.75 g/cm^3, some local isometric parts have the density of 2.65–2.70 g/ cm^3 and 2.75–2.80 g/cm^3; this fact indicates the continental type of the crust.

In this zone, island arcs, near-arc basins and shelf complexes are determined. The first complex is mainly composed by plagiorhyolite, plagiodacite, andesite and their tuffs. Basalts, tuffite and limestones are less common. Volcanism has antidromic character: it is resulted from eruptions of acid rocks with formation of central-type volcanoes, and then there were eruptions of andesite and basaltic lavas with their tuffs. Volcanism appeared in subaeral and shallow-water marine environments. Island arc volcanic rocks form a differentiated order from basalts to rhyolite and belong to calc-alkaline series. In REE distribution and deficit of Nb, Zr, Ti, the island arc volcanic rocks of the Uda-Vitim zone are correlated with those of the Kuril-Kamchatka island arc.

A complex of near-arc basins is identified in the central part of the zone. It is composed of tufa-sandstones, tufa-aleurolite, phyolite, dacite and andesite tuffs, limestone and marl with interbeds of rhyolite, dacite and andesite lavas.

A shelf complex is known at northwestern (Turka River basin) and northeastern (riverhead of the Bolshoy Amalat River) limits of the zone; in all likelihood, the shelf complex fixes zone boundaries. The complex is composed of sandstone with fewer amounts of siltstone, gravelite, conglomerate, limestone. The age of the limestone of all the complexes is determined based on Early Cambrian fossils, and that of volcanic rocks, by absolute age dating (U–Pb, SHRIMP-II): rhyolite—535 and 530 Ma, andesite—530 Ma; rhyiolite tuffs—526 Ma; plagiogranite-porphyry—515 Ma (Gordienko et al. 2010).

Collision and accretion of the Uda-Vitim island arc to paleocontinent margin (which in the Vendian (Ediacaran)—Early Paleozoic was a passive continental margin) took place at the end of the Early Paleozoic. It is accompanied by large-scale displacements along the strike-slip fault of submeridianal and NE strike, folded and thrust-faulted deformations, gabbro and granitoid magmatism of 500–460 Ma age (10 determinations by U–Pb method) in northwestern and northern frame of the Uda-Vitim zone.

3.3 Structures and Complexes of the Sayan-Baikal Mountain Area Formed After Crust Consolidation

After consolidation of the SBMA crust in the second half of Paleozoic the volcano-plutonic belts and riftogenous belts formed in continental environment.

The Sayan-Transbaikalia belt ($D–C_1$) is coupled with the Khangay-Khentey-Daur Basin filled with turbidites located south of the belt (in present-day coordinates). Within the eastern part of East Sayan and Dzhida, this belt consists of thick subalkaline and alkaline, including bimodal, volcanic and sedimentary complexes and alkali-gabbro and alkali-granitoid massifs comagmatic to them.

Still, in West Transbaikalia, the belt is mainly occurred as the Late Devonian—Early Carboniferous marine strike slip depressions of pull-apart type (Uakit, Bagdarin, Urma etc.) which are relic basins.

Absence of the Devonian granite in Transbaikalia shows that the accretion of tectonic structures to Siberian continent occurred later (in the Late Paleozoic) (Gordienko 2006).

Selenga-Vitim volcano-plutonic belt ($C_2–T_1$) is represented by the following complexes: trachyandesiterhyolite with molasse (C_2), trachyrhyolite with molasse, subalkaline granitoid (P_1), trachyrhyolite-trachybasaltic with comendite and molasse, alkaline granitoid ($P_2–T_1$). This belt extends to Central Mongolia, in Russia, to the basin of the Vitim—Olyokma rivers. Its length is 2000 km and width is 250–300 km.

Formation of rift structures of the Selenga-Vitim volcano-plutonic belt resulted in further extension of the lithosphere due to mantle and reworked crustal matter rising isostatically along tensile zones. In case, riftogenesis and accompanying volcanism was concentrated not only in narrow tectonic zones but also in enormous surrounding terrains where the Angara-Vitim (320–290 Ma) and Khangay (260–250 Ma) huge granitoid batoliths formed (Tsygankov 2005; Gordienko 2006).

Mongol-Transbaikalia rift system (T_2–K_1) is extensive as well. About 200 rift depressions filled with continental molasse and volcanogenic complexes are recorded within its limits. In addition, massifs of alkaline granite and syenite, nepheline syenite are present there. The initial volcanic cycle is fixed in time interval of 240–210 Ma; volcanism was of trachyandesite-trachybasalt and trachybasalttrachyrhyolite composition. The molasse accumulation in the depressions occurred at the next stage. The second volcanic cycle took place in time interval of 160–113 Ma and was impulsive. Molasse accumulated in depressions during intervals between volcanic impulses. Volcanic rocks of the second cycle are of bimodal systems from trachybasalt-trachyrhyodacite and trachybasalt-trachyandesite series. Evolution of riftogenous depressions finished in molasse accumulation at the end of the Early Cretaceous.

4 Tectonics of the Verkhoyansk-Kolyma Fold Area

Conception of the "**Verkhoyansk-Kolyma fold area (VKFA)**" arose in the late 30-ies—early 40-ies of the twentieth century in connection with implementation of regular geological prospecting works. The first state geological mapping works started in the 50-ies of the last century finally assert this conception. By idea of that time, the Verkhoyansk-Kolyma folded area consisted of two main geotectonic elements—Verkhoyansk fold system and Kolyma central massif (Bogdanov 1963).

Middle of the 80s was a crucial moment in structural conception of the Verkhoyansk-Kolyma area. Kolyma-Omolon region represents a system of accretional complexes of different ages: microcontinents with the pre-Phanerozoic continental crust, Paleozoic oceanic and island arc formations (Natapov and Stavsky 1985; Stavsky 1984), and Paleozoic blocks of Kolyma loop were the boundary of Siberian Craton (Natapov 1988). Later on, some researchers described that the Verkhoyansk-Kolyma terrane collage formed as a result of relatively fast accretion of terranes in the Late Jurassic—Early Cretaceous (Parfenov 1984; Oxman 2000).

At first, the present conception of tectonic zoning for the Verkhoyansk-Kolyma fold area was reflected in the *Tectonic map of Northeast Asia, scale 1:5,000,000* (Bogdanov and Til'man 1992). This map shows the compound heterogeneous structure of Kolyma part of VKFA, composed of different-type microcontinents, uneven-aged accretion zones and collision sutures, collected together rock assemblages with age from Neoarchean to Paleogene. Unfortunately, the explanatory note to this map does not contain the spatial geodynamic model of mosaic structure formation for NE Asia and origin of so-called "Kolyma loop".

4.1 Tectonic Position of the Verkhoyansk-Kolyma Fold Area

Verkhoyansk-Kolyma fold area (Fig. 8) in present structure of NE Russia has clearly marked tectonic boundaries. From NE, it is limited by South Anyui ophiolite suture and from SE—by West Koryak fold-thrust system with superimposed Okhotsk-Chukotka volcanic belt. Western boundary of VKFA coincides with the principal system of Verkhoyansk overthrusts separating the Siberian Platform with Pre-Verkhoyansk fore-deep basin from the Verkhoyansk fold-thrust system. In the utmost SW, the structural boundary between the Siberian Platform and VKFA is represented by a series of tectonic thrusts and overthrusts composed by rock complexes of intraplate Sette-Daban rift.

Kolyma loop. Kolyma structural loop is the main VKFA geometrical tectonic element. Outcrops of the Precambrian and Paleozoic rocks form it. These complexes are separated from the Verkhoyansk fold-thrust system by the major Late Jurassic—Early Cretaceous overthrust (detachment thrust). In the south Kolyma loop consists of three blocks with continental crust (Omulevka, Pre-Kolyma and Omolon). These

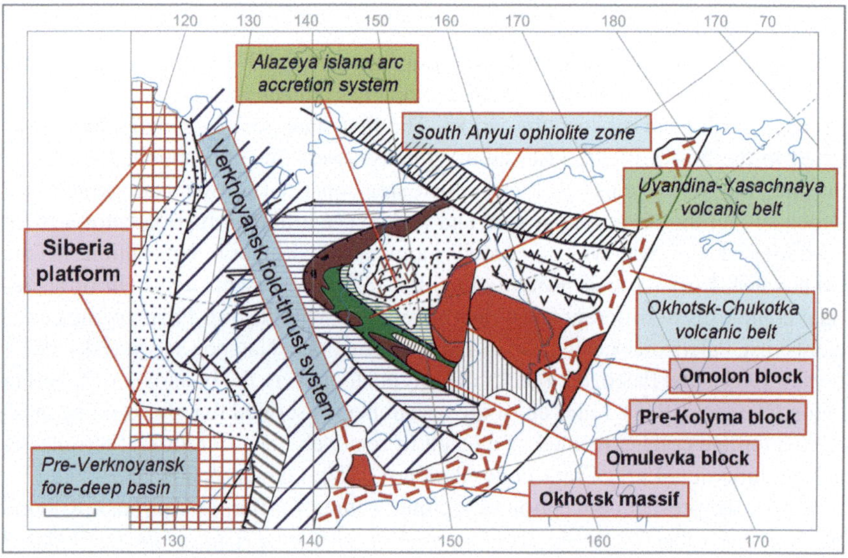

Fig. 8 Main tectonic units of the Verkhoyansk-Kolyma fold area. Numerals in the map: 1—Sette-Daban continental rift, 2—Kharaulakh continental rift, 3—Sugoy depression, 4—Yilin'-Tas zone and Zyryanovka fore-deep depression, 5—Oloy accretion island arc zone

blocks have the Precambrian (Paleoproterozoic) rocks in composition of their basement. Omolon block is mostly overlain by complexes of active continental margin (Kedon Fm., D_{2-3}).

Omulevka block is completely covered by the Paleozoic sedimentary and volcano-sedimentary complexes; and it is supposed that its basement has Paleoproterozoic age by resemblances of composition with the Pre-Kolyma block.

Blocks of the Paleozoic rocks (O_2–C_1) are the following: Omulevka (see above), Tas-Khayakhtakh, and Selennyakh-Polousny. Last two of them are situated in areas of ranges: Chersky to the northwest, Selennyakh Mts. and Polousny Range to the north. **Structural elements outward the Kolyma loop**.

The main of them are the Siberian Platform with basement uplift (Okhotsk Massif) and Paleozoic riftogeneous complexes. Okhotsk Massif is just one element of the Siberian Platform basement situated within VKFA. Sette-Daban and Kharaulakh Middle Paleozoic rifts probably were structures which in the Early Carboniferous gave the beginning of the Verkhoyansk paleobasin downwarping. A general scheme of relationship between continental rifting and formation of the Verkhoyansk paleobasin is shown in Fig. 9.

Pre-Verkhoyansk depression—(J_3v–K_1) is closely connected with formation of the Verkhoyansk foldthrust system and represented a fore-deep basin during overthrusting of the Verkhoyansk sedimentary complexes onto the Siberian Platform (Pushcharovsky 1960).

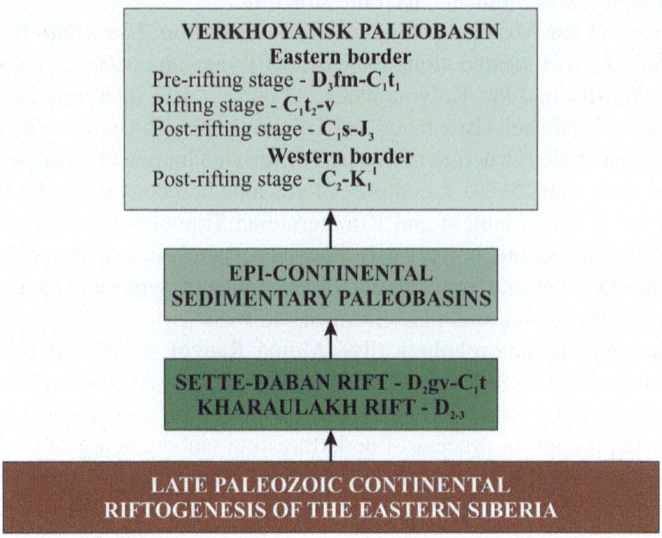

Fig. 9 Relationship between continental rifting in the eastern part of the Siberian Platform and formation of the Verkhoyansk paleobasin

Verkhoyansk fold-thrust system (and Verkhoyansk paleobasin) (C_1–T_2) occupies about one half of VKFA territory, has a complex configuration and internal structure. Structure of the Verkhoyansk system represents a complex series of thrusts and disrupted anticlines with sliced composition and clearly defined western vergence.

Yin'yali-Debin and Oldzhoi (Polousny) depressions (T_{2-3}–J_3). Yin'yali-Debin depression is immediately adjacent to the Chersky Range and was a residual part of the Verkhoyansk Basin in process of filling with sediments and their migration eastward (in present coordinates). Oldzhoi (Polousny) depression along the northern slopes of Selennyakh Mts. and Polousny Range was the new-born one with respect to the Verkhoyansk paleobasin: and from the Middle Triassic probably connected the South Anyui paleoocean with the main part of the Verkhoyansk Basin.

Structural elements inside the Kolyma loop. ***Alazeya accretional system (PZ_{2-3})*** represents a complex of tectonic slices and wedges formed by rocks of ensimatic island arc—basalts, andesitic basalts, tuffs, volcano-sedimentary rocks intruded by single small bodies of plagiogranite, diorite, and granodiorite. The Middle Devonian andesitic basalts are ancient outcropping rocks. But some researchers consider that Alazeya island arc system started to form at least in the middle Silurian.

Oloy accretional zone (PZ_3–J_3) distinguishes from the Alazeya one in the fact that in Oloy zone more young complexes are widespread; Triassic and Lower Jurassic greywacke are epi-accretional deposits overlaying tuff-terrigenous rocks with basalts and andesites. Until now, the structural relationship between Oloy and Alazeya zones is not quite clear (bad exposed rocks, poor paleontological and geochronological study of complexes, complicated tectonic structure).

Complexes of the Mesozoic stage of VKFA evolution. ***Uyandina-Yasachnaya volcanic belt (J_{2-3})*** is located along the Chersky Range and occupies adjacent area of Selennyakh Mts. and Pre-Kolyma block. At present, it is interpreted as a typical Andean-type volcanic belt (Bogdanov and Til'man 1992). The belt overlays the Late Paleozoic thrust-folded structure of this part of Kolyma loop with an angular unconformity (no more than 20–30°); complex of volcanic belt consists of the Bathonian conglomerates at the basement and Kimmeridgian-Tithonian volcanogenic series. Special studies on volcanic belt polarity allowed definitely decide the problem of its origin (Dylevsky 1994), i.e. rejuvenation of volcanogenic series and change of their composition to more acid took place from east to west.

Yilin'Tas zone (geomorphologically—Moma Range) is situated between the Chersky Range (Uyandina-Yasachnaya volcanic belt) and Alazeya uplift (Alazeya accretion system). According to its structural position, this zone formed at place of a fore-deep depression in process of underthrusting (subduction?) of thin Alazeya accretional crust under thick crust of the Siberian Craton boundary along the present Chersky Range. Yilin'Tas zone is filled by thick monotonous Middle Jurassic—Early Cretaceous polymictic and greywacke sediments synchronous to complexes of the Uyandina-Yasachnaya volcanic belt, which represented the Andean-type active continental margin.

Sugoy depression is situated between Omolon and Pre-Kolyma blocks. It is filled by the Triassic and Jurassic flysch deposits mostly of greywacke and volcanomictic composition. More ancient rocks in this depression are unknown. Fold structure of Sugoy depression is interesting by the fact that the most intense dislocations are registered in the northern part where Omolon and Pre-Kolyma blocks are touching, whereas the degree of deformation considerably reduces southward. Principal problem for Sugoy depression continues to be the nature of its basement. Complete absence of magmatic associations, great thickness of sediments, position between blocks with mature continental crust are the reason to suppose either melanocratic basement or oceanic crust of marginal sea type (Bogdanov and Til'man 1992).

4.2 Basic Principles of Tectonic Evolution

1. Chersky-Polousny bend of Kolyma loop is the primary structure arisen in the Late Neoproterozoic—Early Paleozoic as a result of considerable displacement of a part of the Siberian Craton along a transform-demarcation fault with formation of "Kolyma Gulf" of the Paleopacific Ocean.
2. In the Early and Middle Paleozoic, the Kolyma Gulf had two different types of continental margin: passive (Selennyakh-Polousny) and transform-demarcation (Chersky).
3. From the Middle Paleozoic (middle Silurian?) inside the Kolyma loop at the place of the Kolyma Gulf, continuous accretion of oceanic and island arc complexes took place.
4. At the end of Devonian parallel to the present NE boundary of the Siberian Platform, origination of rifting structures took place with filling of thick graben and olistostrome rock assemblages and continental-riftogenous subaqual volcanism. All the above was the reason of the Verkhoyansk pericratonic paleobasin formation, that existed from the Early Carboniferous to the end of Jurassic.
5. In Mesozoic, the common structure of VKFA was formed as a summation of the Pacific accretion, final collision of the Omolon and Pre-Kolyma blocks and closing with collision in South Anyui paleoocean. This stipulated the most intense injection of rock masses into the Kolyma loop corner with formation of thrust and fold-thrust structures inside the Kolyma loop and Verkhoyansk fold-thrust system.

4.3 Structure of Chersky-Polousny Corner of the Kolyma Loop

In Chersky-Polousny corner of the Kolyma loop, the disparity between strikes of the Paleozoic and Mesozoic structures is shown to the best advantage at Tas-Khayakhtakh Mts. (NW end of the Chersky Range) and Selennyakh Mts. (W end of the Polousny Range). Paleozoic structures form an angle 90°, whereas the Mesozoic structures smoothly round this right angle. Therefore, it can be supposed that the Paleozoic tectonic movements sharply differed from the Mesozoic ones.

Structure of Northern Tas-Khayakhtakh Mts. As we noted above, there is a visible disparity between the Paleozoic and Mesozoic structural attitudes here.

Internal structure of Tas-Khayakhtakh Mts. is formed by a thick tectonic overthrust of the Paleozoic section exfoliated by layerwise faults. Strike of the Paleozoic structures is NW–SE, and Mesozoic ones—NE–SW. Mesozoic autochthon and paraautochthon are composed by the East Verkhoyansk (Yin'yali-Debin) sandy-siltstone flysch complex (T_3), siltstone-mudstone flysch (J_{1-2}), transformed mostly into volcanomictic sandy flysch (J_3).

Complexes of Uyandina-Yasachnaya volcanic belt (J_2–J_{3v}) overlay with an angular unconformity the Late Paleozoic structure of Tas-Khayakhtakh Mts.

Structure of Selennyakh Mts. Structure of Selennyakh Mts. is described as a complex combination of the Late Paleozoic and Mesozoic thrusts, i.e. as a system of tectonic slices and plates (Tretiakov 1996). In the lower part of tectonic section, there are series of thrusts composed by the Lower Ordovician—Middle Devonian carbonate (mostly calcarenite) and clayey-carbonate sediments and Frasnian–Visean siliceous-volcanogenic rocks. Carbonate sections (O_1–D_2) of Tas-Khayakhtakh Mts. and Selennyakh Mts. have absolute resemblance and their biotas belong to a single province.

Above there are series of thrusts composed by green schist (metamorphosed from the Lower–Middle Paleozoic carbonate-terrigenous sediments). Earlier, these rocks were dated as the Pre-Cambrian and ascribed to the crystalline basement of Selennyakh Mts. The Siberian conodonts from Middle Ordovician to Middle Devonian have been recently found there in lenses and layers of limestone (Tarabukin 2006).

Above there are slices of amphibolite schist originated as green schist from carbonate-terrigenous turbidites. The uppermost elements of tectonic section are represented by gabbro-amphibolites and amphibolites overlain by metaultramafic and serpentinite mélange with ophiolite-clastic gravelite and olistostrome at the base of ophiolite tectonic plate (Kalgyn ophiolite complex of Uyandina allochthon) (Oxman et al. 1998).

Age of ophiolites approximately coincides with dating by different methods (Rb/Sr—560–600 Ma; K/ Ar—up to 720 Ma; $Ar^{39/40}$—640 + −16 Ma (Layer et al. 1993; Oxman 2000), i.e. this is the Late Neoproterozoic (Ediacaran = Vendian) or Ediacaran (Vendian)–Cambrian boundary. Several stages of metamorphism connected with tectonic deformations were determined for ophiolites: 419 ± 16 Ma (S_1/S_2); 370 Ma (D_2/ D_3); 312 Ma (C_1/C_2); 174 Ma (J_1/J_2) (Oxman et al. 1998). The first two dates indicate the intra-ophiolitic deformations and the last two dates are connected with two stages of formation of thrust structure of Selennyakh Mts.—obduction of ophiolites with Early–Middle Paleozoic rocks in the Early Carboniferous, and overthrusting the whole Paleozoic rock-thrust block onto the Triassic-Jurassic flysch of Polousny depression in the Late Jurassic—Early Cretaceous.

Neoautochthon for the Late Paleozoic structure of Selennyakh Mts. is represented by complexes of Uyandina-Yasachnaya volcanic belt (J_{2-3}) deformed during the Mesozoic thrusting (J_3/K_1).

Comparative description. So, Paleozoic structures of Selennyakh Mts. and Tas-Khayakhtakh Mts. are essentially different.

Paleozoic structure of Selennyakh Mts. has a certain resemblance with the structure at the boundary of the East European Platform in the Polar and partly South Urals. Lower Ordovician–Upper Devonian carbonate shelf tectonically dismembering section with the upper Lower Carboniferous siliceous-tuffite part can be compared with autochthonous Elets zone (Polar Urals). Obduction ophiolites and associated with them rocks occupy the uppermost position as the Polar Urals ophiolite blocks and Uyandina ophiolite allochthon by their structural position corresponds to Kraka ophiolite allochthon (South Urals).

Here, rock assemblage sections represent a typical passive continental margin with shelf, continental slope and foot of continental slope. Complexes of Uyandina ophiolite block were a part of marginal sea crust (Oxman 2000).

Late Paleozoic tectonic structure of Tas-Khayakhtakh Mts. looks differently. Tectonically double Lower-Middle Paleozoic shelf section divided by the Early Carboniferous riftogenous volcano-olistostrome complex is presented here. On the eastern slope of the Chersky Range, ophiolite complexes (Munilkan and Middle Indigirka) are bedded as tectonic lenses among the Middle Ordovician metamorphosed gravitites and chlorite-sericite schists and Early Carboniferous calcarenites with ophiolite-clastic olistoliths.

Thus, both tectonic elements of the Kolyma loop differ from each other not only by reconstruction of rock complexes (though their resemblances of shelf carbonate successions), but also by peculiarities of overthrust-slice structure.

At present, it is hard to determine where the overthusting was earlier. Most of data confirm that at Selennyakh Mts. thrust formation was after the Visean, whereas at Tas-Khayakhtakh Mts. It is not older than the Early Tournaisian.

4.4 Paleozoic Evolution of the Verkhoyansk-Kolyma Fold Area

Brief analysis of paleomagnetic data. At present, there are not so much reliable paleomagnetic data for VKFA. So, analysis of paleomagnetic data allows to draw a conclusion that in the Paleozoic all blocks with continental crust (Tas-Khayakhtakh, Omulevka, Pre-Kolyma and Omolon) were composite parts of the Siberian Platform or directly contacted with it (Neustroev et al. 1993).

Strikes of Paleozoic structures. Clear NW structural strikes are fixed within the Chersky Range (and at Kharaulakh Mts.) and further step-by-step changes into sublatitudinal and latitudinal strikes coincide with strikes of the Taimyr Paleozoic structures. First, the prolongation of structures from the Chersky Range to Kharaulakh

Mts. under the Upper Paleozoic—Mesozoic sediments was suggested in the 60-ies of last century (Rezanov 1968). This prolongation was substantiated by linear magnetic anomalies, orientation of the Mesozoic intrusions cutting fold structures in the Mesozoic sediments and strikes of the recent seismotectonic faults. A conclusion on the presence of linear gravity anomalies fixing the strikes of normal faults within the crystalline basement was made. So, the presence of a huge linear fold zone up to 1500 km long, only small part of which is opened at the Chersky Range, was assumed.

Main stages of tectonic evolution. The analysis of formations and structures of two boundaries of the Kolyma loop (Chersky and Selennyakh-Polousny), and also the Middle—Late Paleozoic accretion system allows to assume that in this part of the Siberian Craton as far back as in the Early Ordovician there was a gulf of the Paleopacific Ocean with two essentially various continental margins. Selennyakh-Polousny was a typical passive continental margin with carbonate shelf and carbonate (or calcarenite)-terrigenous slope complex. Chersky continental margin passed along a transform-demarcation fault which in the Paleopacific existed as an active oceanic transform fault throughout the Paleozoic. Further NW, there apparently was an intra-platform strike-slip discordance proceeding into the Taimyr collision zone; its role was limited to origination of the Kolyma gulf in the Paleopacific.

Existence of such Taimyr-Yana-Kolyma discordance proves to be true not only due to the "transparence" of the Paleozoic structures in Northern Verkhoyansk fold system but also due to various types of continental margins inside the "Kolyma Gulf". The important features are the face contacts of internal Kolyma structures and Alazeya accretion complexes to Chersky margin and presence of a series of transform-accretion tectonic wedges along the Chersky Range.

Time and mechanism of the "Kolyma oceanic gulf" origin. Most ancient internal Kolyma complex is represented by Kalgyn ophiolites (560–640 Ma) of Uyandina ophiolite allochthon in Selennyakh Mts. In the Early Ordovician, the differences in structure and sedimentary formations are observed, however, in the Middle Ordovician they are practically imperceptible. Therefore, it is quite possible to assume that the Kolyma Gulf formed by the Middle Ordovician, and intra-continental (Northern Verkhoyansk) part of transform-demarcation zone lost its strike-slip importance.

At least it is possible to offer three models of origin of the Kolyma Gulf.

1. Simple huge sinistral intra-continental strike-slip passing into a transform system of the Paleopacific. As a result of such strike-slip the displacement of the Indigirka-Kolyma fragment of Siberia was at least 1600–1800 km.
2. Sinistral displacement of the Indigirka-Kolyma block occurs to turn clockwise and angular opening of the Kolyma Gulf with a compensatory compression in Northern Verkhoyansk fold-thrust system and in Taimyr takes place. Opening of the Bay of Biscay in Northern Atlantic occurred in a similar way.
3. Formation of the Kolyma Gulf as a result of rifting and spreading opening of basin with advanced transform system. The similar structure of such type is known as the Guinea Bay of the Central Atlantic. Thus, formation of the Kolyma

Gulf connected with transform-demarcation zone and displacement of continental fragment block in the Vendian (Ediacaran) represents the most probable. This process is correlated with the age of collision metamorphism in Taimyr (570–600 Ma) and initial stages of the Iapetus Ocean opening (600–615 Ma). Intra-continental part of discordance fault by the Middle Ordovician had finally lost its strike-slip role but remained rather active throughout a long geological time. The movements along transform-demarcation zone occurred during the Paleozoic and Mesozoic.

4.5 *Stages of Tectonic Evolution*

Ordovician. Accumulation of carbonate sediments took place from the Early Ordovician at shelf frame of the Kolyma Gulf. Bathyal terrigenous and carbonate-terrigenous turbidites accumulated on the Selennyakh-Polousny continental slope. Along the discordant fault on the continental slope, carbonate-terrigenous gravitites (Tas-Khayakhtakh) or carbonate-fragmental gravitites and graptolite argillites interstratified with aphyric basalts were formed. Also in the Early Ordovician (Early Tremadoc), adjunction of Omulevka-Omolon (Proto-Omolon) block from Siberia with formation of a rift basin, in which deepwater clayey-carbonate sediments with subalkaline basalts of the Middle Ordovician accumulated, took place. Divided block almost completely (except for Northern Omulevka) was a land limited by a passive continental margin of the Paleopacific.

Silurian—Middle Devonian (Eifelian). Evolution of the main structural elements was inherited from the Ordovician. Omulevka-Omolon (Proto-Omolon) block remained lifted and was separated from Siberia by a narrow basin with terrigenous-carbonate sedimentation. Structural reorganization of the Kolyma Gulf began. In the Silurian, layering and first metamorphism of intra-Kolyma oceanic crust of the marginal sea and apparently origination of a primitive ensimatic (future Alazeya) island arc took place. Authentically established island arc tholeiite and calc-alkaline andesitic basalts in Alazeya accretion system are known in the Middle Devonian.

Along the transform zone, accumulation of slope sediments proceeded and was accompanied by effusion of spilites.

Sette-Daban and Kharaulakh continental rifts with accumulation of terrigenous deposits of graben facies occurred on the Siberian Platform in the Middle Devonian.

Middle–Late Devonian. Inside the Kolyma Gulf Alazeya, island arc system continued to develop, where volcanic rocks accompanied by greywacke accumulation of slopes and underwater terraces, deepwater trough, and formation of glaucophane schist zone are recognized.

Carbonate and slope carbonate-terrigenous sediments continued to accumulate (up to the Famenian) on the passive Selennyakh-Polousny margin shelf.

Active subduction of the Paleopacific crust under the Okhotsk (Omolon block) continental margin started. It was expressed in accumulation of facialchangeable basalt-trachyandesite-trachyrhiolite Kedon Formation of the Andean-type active continental margin. It is possible to reconstruct the rock order of Kedon Formation (D_{2-3}) on the Omolon block from a thick volcanic section through tuff section to rather thin tuff-terrigenous one in the NW of Omolon block. In the larger part of the Siberian Platform, accumulation of shelf carbonates continued and only in the Kolyma Gulf corner it is marked by a local Famenian uplift which preceded origination of the Late Paleozoic rift structure of the Verknoyansk Basin.

Sette-Daban rift reached its maximal evolution with typical continental olivine-basalt volcanism and Kharaulakh rift was practically amagmatic. There are no data whether these two rift systems were joined; however, this Sette-Daban—Kharaulakh rift system was the embryo of the Verkhoyansk epicontinental basin.

Early Carboniferous. First of all, this period is characterized by formation of the Alazeya accretion system and its obduction together with fragments of the Kolyma Gulf marginal sea crust onto the Selennyakh-Polousny passive continental margin. Along the transform-demarcation zone, there were intense movements accompanying plenty effusions of basalt volcanic rocks with simultaneous layering and metamorphism of slope sediments. At this time, protrusion of ophiolite fragments into the Ordovician green schists along the Chersky Range took place.

Intense Alazeya accretion caused local movements in the Kolyma loop corner, i.e. in the zone of transition from demarcation strike-slip to intra-continental discordance. Along it, an escarp formed with the subsequent formation of a thick olistostrome complex and eruptions of continental-rift basalts.

In the Early Carboniferous, disjunction of the uniform Omulevka-Omolon block into some separate fragments started. Such division was reflected in rock complex distribution, their facies and thicknesses, but in a lesser degree—in the style of sedimentation and volcanism.

In East Siberia, the Verkhoyansk Basin which inherited the continental rift system at its basement was finally formed. If western boundary of the basin is established by beach facies of sedimentary rocks, then the eastern boundary is authentically determined by intra-continental character of riftogeneous volcanism (Pospelov et al. 1995).

Late Carboniferous. Formation of overthrustfolded structure on periphery of the Kolyma Gulf was completely finished and transformed into an orographic uplift (Fig. 7). In Alazeya accretion system, post-accretion facially unstable terrigenous-volcanogenic rocks accumulated (andesite and dacite prevailed among volcanic rocks).

Active part of the Taimyr-Yana-Kolyma discordant strike-slip was reduced even more. Basalts continued to erupt along the active part of strike-slip. The Verkhoyansk Basin could extend at the expense of continental crust thinning at the basement of the Siberian Platform.

Permian. Epi-accretion complexes in Alazeya zone are represented by tuff-terrigenous deposits overlapping with unconformity the Alazeya accretion system intruded by the Late Paleozoic plagiogranite and diorite.

Orographic uplift continued to exist and separated the Alazeya epi-accretion basin from depression of the Verkhoyansk Basin. At that time, transform-demarcation strike-slip completely lost its structure-forming meaning remaining simply a transform fault in the Paleopacific. It was connected with the ending of the Paleozoic accretion and final formation of a rather thin accretional crust within the Paleo-Kolyma Gulf. **Conclusion**. Formation of mosaic folded areas in the Pacific sector of the Earth can be connected with structural heterogeneity of the continental margin of large paleocontinents, to which oceanic and island arc complexes were accreted. Such structural heterogeneity in NE of Asia was caused by a large cross transform-demarcation fault zone, which existed here from the Vendian (Ediacaran) to the end of Paleozoic. Namely this structural heterogeneity predetermined the difference of the Paleozoic of the Kolyma-style accretion from Okhotsk-style and, maybe, from Sikhote Alin style. Due to the Kolyma-style accretion, structural heterogeneity was filled by accreted complexes and gradually the Paleopacific margin was leveled. Subsequent accretion of nonlinear continental blocks with simultaneous change of direction of the tectonic movements was the main condition of mosaic folded areas formation.

5 Mesozoic Evolution of the Verkhoyansk-Kolyma Fold Area

Late Paleozoic tectonic structures of VKFA are substantially suppressed by the Mesozoic tectonic movements and reworking. Therefore, this area was traditionally considered as Mesozoides, though, as was shown, the main structural picture of this NE Asian part was formed in the beginning of Paleozoic.

The main Mesozoic structures and rock complexes within the Verkhoyansk-Kolyma area are the following:

1. **Verkhoyansk Basin (recent Verkhoyansk fold-thrust system)** is filled almost exclusively with sedimentary terrigenous complexes with age from Visean (Early Carboniferous) to Volgian (Late Jurassic), and in some places (mainly at the western periphery) by rocks of Valanginian-Hauterivian age. The main structural boundaries of the Verkhoyansk fold-thrust area are: in the west—system of frontal overthrusts along the Lena River, in the east—Tirekhtyakh thrust and its branch, along which Paleozoic of the Chersky Range is overthusted on the Triassic-Jurassic deposits. Adycha-Taryn shear zone, which as a non-ophiolite suture formed in process of the Verkhoyansk paleobasin closing, is situated along the eastern boundary of the Verkhoyansk fold-thrust area. More important structural zones of the Verkhoyansk fold-thrust area filled mainly by the Upper Triassic—Jurassic terrigenous sediments are: (1) its NE branch—Oldzhoi (Polousny) synclinorium, (2) Yin'yali-Debin synclinorium—along the Chersky Range.
2. **Uyandina-Yasachnaya volcanic belt (J_2–J_{3v})** formed by sedimentary, volcanic and volcano-sedimentary rocks of Bathonian-Volgian age originated mainly along the Chersky continental margin, also covered the adjacent parts of Polousny

(in area of Selennyakh Mts.) margin and both Omulevka and Pre-Kolyma blocks. Volcanic rocks are represented by differentiated basalt-andesite-dacite-rhiolite series with gradually growing upward explosive factor and distinctly expressed southwestward polarity (Dylevsky 1994).

Uyandina-Yasachnaya volcanic belt occurred as result of collision-subduction processes between the Late Paleozoic thin accretional consolidated crust and thick Archean-Paleoproterozoic continental crust of the Siberian Craton along its Chersky margin.

3. Sedimentary and volcano-sedimentary complexes of **Yilin'-Tas zone** and **Zyryanovka fore-deep depression**. They are represented by thick Upper Triassic—Upper Cretaceous tuffs, plenty tuffites and siltstones with total thickness together with the Cretaceous riftogenous plateau basalts (Moma rift) up to 9 km. These structures divide the Chersky Range and Alazeya Late Paleozoic island arc accretion system.
4. Weak deformed Triassic-Jurassic terrigenous (often with andesite tuffs) deposits of the sedimentary cover of Pre-Kolyma and Omolon blocks.

   Majority of magmatic assemblages in VKFA are Mesozoic. They are Uyandina-Yasachnaya volcanic belt, Main and Northern batholith granite belts, and Cretaceous riftogenous basalts of internal part of the Kolyma loop. They reflect both the geodynamic conditions of VKFA formation during the Mesozoic and the type and characteristics of the Earth's crust in this region.
5. **Main batholith granite belt** is stretched along the axial part and western slope of the Chersky Range, it is more than 1000 km long. It is represented by gradiorite-granite plutons structure (in its northern part—by granite-leucogranite association) with peak of formation at 143–138 Ma (Layer et al. 2001). By opinion of Yakutian scientists (Tectonics… 2001) this belt was generated as a result of subduction and collision of the "Kolyma-Omolon superterrane" with boundary of the Siberian Craton. Really this was underthrusting of thin accretion melanocratic crust under the Chersky Range which was the northeastern (in paleo- and modern coordinates) boundary of the Siberian Platform. Now, there are no sufficient data to consider this process as typical subduction according to the modern geodynamic concept. However, all geodynamic features testify the opportunity of such process: (1) formation of Yilin'-Tas depression as a fore-arc basin for the Andean-type Uyandina-Yasachnaya volcanic belt, gradual westward rejuvenation (up to 90 Ma) of granites of the Main batholith belt, sufficient depth of magma generation (26–29 km) in mature continental crust (Tectonics… 2001).
6. **Northern batholith granite belt** about 700 km long passes through the Selennyakh Mts. and Polousny Range and their northern slopes. The peak of pluton intrusions is in time interval 138–120 Ma (Layer et al. 2001), that testifies their formation later than that of the Main belt. By composition, granites of the Northern belt also differ from that of the Main belt: the latter are close to subduction granites (Layer et al. 2001), they are represented by a wide spectrum from granite-granodiorite up to quartz-diorite and monzodiorite. Thus, this part

of the Kolyma loop in the Early Cretaceous underthrusted (probably, obliquely-directed) under mature continental crust of Selennyakh-Polousny margin of the Siberian Craton.

Above we noted that Uyandina-Yasachnaya volcanic belt has its northern end at Selennyakh Mts. and is completely absent at the Polousny Range. Therefore, most probably underthrusting of crust of the internal part of the Kolyma loop occurred mainly under the Chersky margin of the Siberian Platform, and only partially—under the Selennyakh-Polousny margin.

We have collected the Pre-Cenozoic complexes of different tectonic zones of the Verkhoyansk-Kolyma folded area into a correlation scheme (Fig. 10). The latter reflects the relationship between rock assemblages and successions in adjacent tectonic units and their connection with the main tectonic events. Special legend (Fig. 11) shows more widespread types of geodynamic environments that caused the formation of different sedimentary, volcanogenic and intrusive rocks in process of the Earth's crust growth and consolidation.

Practically recent structure of VKFA formed as a result of composition of vector motions of two directions of tectonic movements from the Middle Jurassic to the early Cretaceous (Fig. 12).

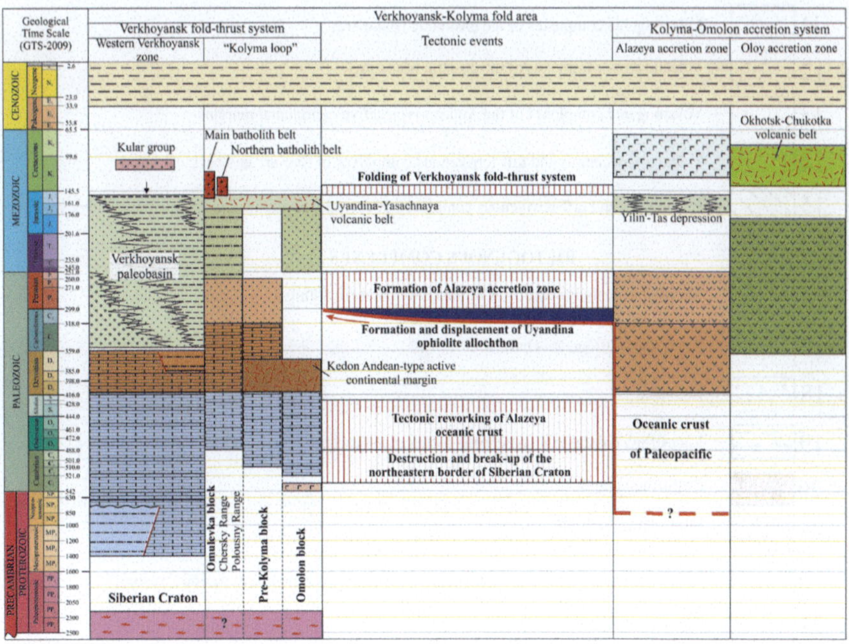

Fig. 10 Correlation scheme of relationship of rock assemblages and tectonic events in the Verkhoyansk-Kolyma fold area

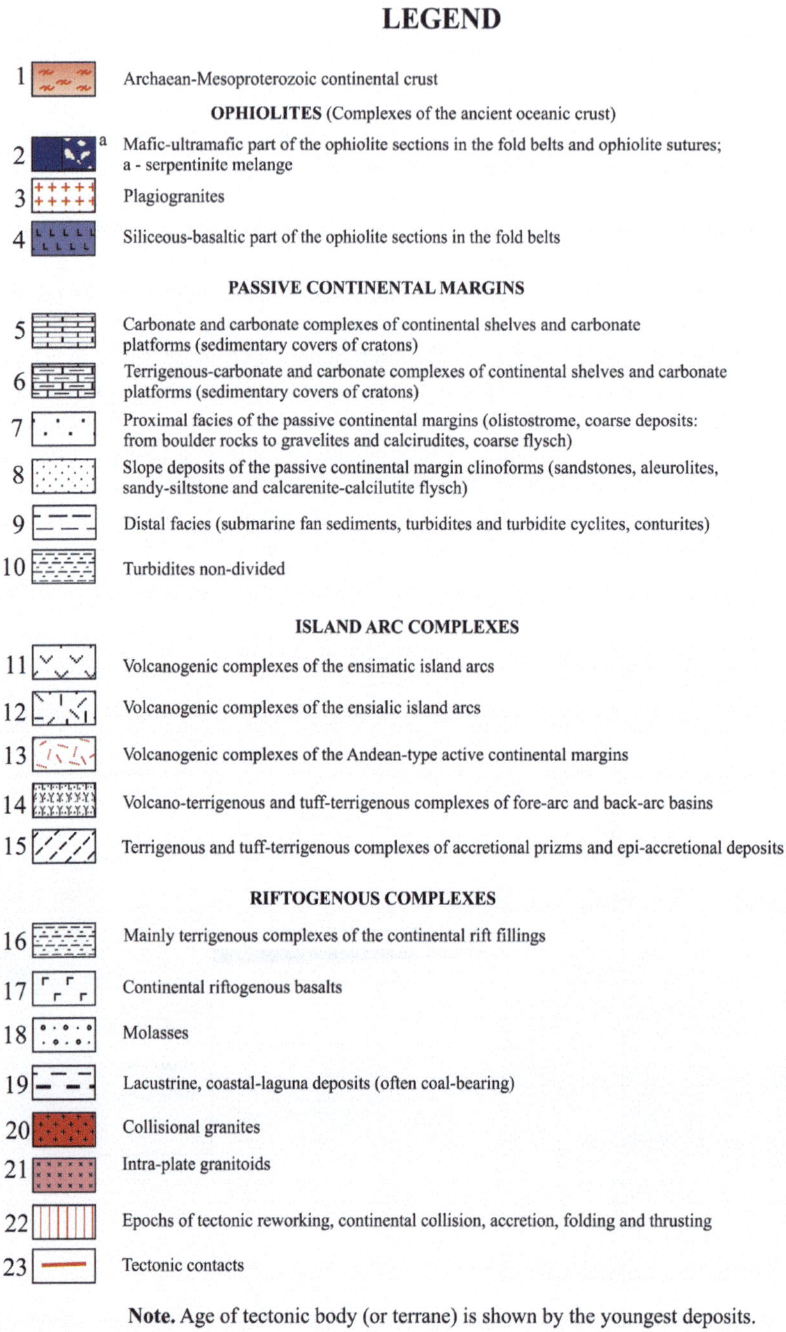

Fig. 11 Legend to the correlation scheme of relationship of rock assemblages and tectonic events in the Verkhoyansk-Kolyma fold area (Fig. 10)

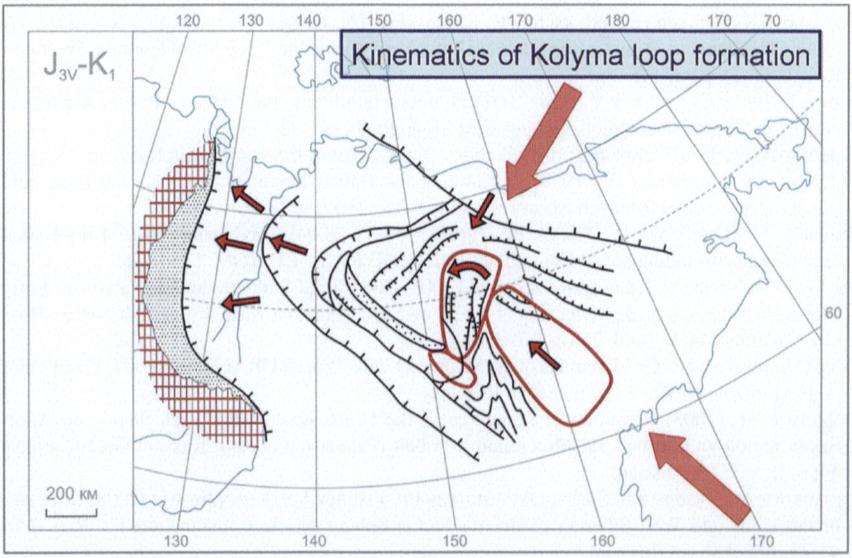

Fig. 12 Kinematics of the mosaic structure formation of the "Kolyma loop" in the Late Jurassic (Tithonian = Volgian)—Early Cretaceous. Finally the mosaic structure of NE Russia developed in the Early Cretaceous as a result of a combination of the Pacific accretion and collision connected with closing of the South Anyui Paleoocean

By features of structures and magmatic complexes, collision processes in the South-Anyui paleoocean predominated over accretion processes from the Meso-Pacific. This explains the main direction of underthrusting of Alazeya accretion crust southwestward (under Chersky margin of the Siberian Craton) and obliquely-directed and later on—under the Selennyakh-Polousny margin.

References

Alexandrov VK (1990) Nappe and overthrust structures of Pre-Baikalia. Nauka, Novosibirsk, p 103 (in Russian)

Bibikova EV, Karpenko SF, Sumin LV et al (1990) U-Pb, Sm-Nd and K-Ar ages of metamorphic and magmatic rocks of Pre-Olkhon (West Pre-Baikalia). In: Geology and geochronology of the Precambrian of the Siberian Platform and its setting. Nauk, Leningrad, pp 170–183 (in Russian)

Bochkarev VS, Brekhuntsov AM, Deshchenya NP (2003) Paleozoic and triassic evolution of West Siberia (data of comprehensive studies). Russian Geol Geophys 44(1–2):120–143 (in Russian)

Bochkarev VS, Krinochkin VG (1988) Precambrian and Paleozoic formations of West Siberia. In: Tectonics of platform areas. Nauka, Moscow, pp 80–104 (in Russian)

Bogdanov NA (1963) Tectonic evolution of the Kolyma Massif and Eastern Arctic in the Paleozoic. Acad. Sci. USSR Publishing House, Moscow, p 239 (in Russian)

Bogdanov NA, Til'man SM (1992) Tectonic and geodynamics of Northeast Asia (explanatory note to the tectonic map of Northeast Asia, scale 1:5 M). Institute of Lithosphere RAS, Moscow, p 54 (in Russian)

Bogdanova SV, Bingen B, Gorbatschev R, Kheraskova TN, Kozlov VI, Puchkov VN, Volozh YA (2008) The East European Craton (Baltica) before and during the assembly of Rodinia. Precambr Res 160(1–2), pp 23–45
Brown D, Juhlin C, Puchkov V (eds) (2002) Mountain building in the Uralides: Pangea to present. American Geophysical Union, Geophysical Monograph, pp 132–286
Bulgatov AN (1983) Tectonotype of Baikalides. Nauka, Novosibirsk, p 193 (in Russian)
Bulgatov AN, Gordienko IV (1999) Terranes of the Baikal mountain province and their gold mineralization. Geol Rudnych Mestorozhdeniy 41(3):230–240 (in Russian)
Bulgatov AN, Gordienko IV, Zaitsev PF, Turunkhaev VI (2004) Geodynamic map of the Baikal region and surrounding areas. Geological Institute SB RAS—CD ROM, Ulan-Ude
Buslov MM, Vatanabe T, Smirnova LV et al (2003) The role of faults in the late Paleozoic-Early Mesozoic tectonics and geodynamics of the Altay-Sayan and East-Kazakhstan folded areas. Russ Geol Geophys 44(1–2):49–75 (in Russian)
Deep Composition and Geodynamics of the Southern Urals (Project URALSEIS) (2001) Ed-in-chief A. F. Morozov. GERS, Tver, p 286 (in Russian)
Dobretsov NL (2003) Evolution of structures of the Urals, Kazakhstan, Tien Shan, and Altai-Sayan region within the Ural-Mongolian fold belt (Paleoasian ocean). Russian Geol Geophys 44(1–2):5–27 (in Russian)
Dyakonova AG, Ivanov KS, Surina OV, Astafyev PF, Vishnev VS, Konoplin AD (2008) Structure of the Urals and West-Siberian platform's tectonosphere on electromagnetic data. Rep RAS 423(5):685–688 (in Russian)
Dylevsky EF (1994) Zonation of Uyandina-Yasachnaya volcanic belt and its tectonic nature (Northeast Asia). Geotektonika 4:52–62 (in Russian)
Echtler HP, Stiller M, Steinhoff F, Krawczyk CM, Suleimanov A, Spiridonov V, Knapp JH, Menshikov Y, Alvarez-Marron J, Yunusov N (1996) Preserved collisional crustal architecture of the Southern Urals—Vibroseis CMP-profiling. Science 274:224–226
Fedorov YN, Ivanov KS, Sadykov MR, Pecherkin MF, Krinochkin VG, Zakharov SG, Krasnobaev AA, Erokhin YV (2004) Structure and petroleum potential of the pre-Jurassic complex in the Khanty-Mansi Autonomous District area: new approaches and methods. In: Ways for realization of the petroleum and ore potential of the Khanty-Mansi Autonomous District IzdatNaukaServis, Khanty-Mansiysk, vol 1, pp 79–90 (in Russian)
Fershtater GB, Krasnobaev AA, Bea F, Montero P, Borodina NS (2009) Intrusive magmatism of the earlier stages of evolution of the Uralian epi-oceanic orogen: U-Pb geochronology (LA ICPMS, NORDSIM, SRIMP-II), geochemistry, regular evolution. Geochemistry 2:130–150 (in Russian)
Gee DG, Pease V (eds) (2004) The neoproterozoic timanide orogen of the Eastern Baltica. Geological society, London, Memoirs, vol 30, p 248. Ivanov KS (1998) Main features of geological history (1.6–2.0 Ga) and composition of the Urals. Uralian Branch RAS, Ekaterinburg, p 252 (in Russian)
Gordienko IV (2006) Geodynamic evolution of Late Baikalides and Paleozoids in the folded periphery of the Siberian Craton. Russ Geol Geophys (Geologia i Geophyzika) 47(1):53–70 (in Russian)
Gordienko IV, Filimonov AV, Minina OR et al (2007) Dzhida island-arc system in the Paleoasian Ocean: structure and main stages of Vendian-Paleozoic geodynamic evolution. Russ Geol Geophys (Geologia i Geophyzika) 48(1):120–140 (in Russian)
Gordienko IV, Bulgatov AN, Ruzhentsev SV et al (2010) The late Riphean-Paleozoic history of the Uda-Vitim island arc system in the Transbaikalian sector of the Paleoasian Ocean. Russ Geol Geophys (Geologia i Geophyzika) 51(5):589–614 (in Russian)
Ivanov KS (1998) Main features of geological history (1.6–0.2 Ga) and the Urals structure. UB RAS, Yekaterinburg, p 252 (in Russian)
Ivanov KS, Erokhin YV (2010) Substantial composition and age of acid volcanics and granitoids of the pre-Jurassic basement of the Khanty-Mansi Autonomous District eastern square. Materials of the Russian scientific conference "Basement, structures of the West Siberian

Mesozoic-Cenozoic sedimentary basement framing, their geodynamic evolution and petroliferous problems". SIBNATS, Tyumen-Novosibirsk, pp 95–98 (in Russian)

Ivanov AI, Ryazanov GV (1992) Patom depression structural-kinematic analysis [in Russian]. Nauka, Novosibirsk, p 88 (in Russian)

Ivanov SN, Puchkov VN, Ivanov KS et al (1986) Formation of the Earth's crust in the Urals. Nauka, Moscow, p 246 (in Russian)

Ivanov KS, Smirnov VN, Erokhin YV (2000) Tectonics and magmatism of the collision stage (case study of the Middle Urals). Institute of Geology and Geochemistry of Uralian Branch RAS, Ekaterinburg, p 133 (in Russian)

Ivanov KS, Fedorov YN, Koroteev VA, Kormiltsev VV (2006) Uralides in the West Siberian basement structure. Gornye vedomosti 8(27):16–29 (in Russian)

Ivanov KS, Fedorov YN, Amon EO, Erokhin YV, Borozdina GN (2007) Age and composition of ophiolites from the basement of the West Siberian petroliferous megabasin. Rep RAS 413A(3):415–419

Ivanov KS, Koroteev VA, Pecherkin MF, Fedorov YN, Erokhin YV (2009) Western part of the West Siberian petroleum megabasin: geological history and structure of the basement. Russ Geol Geophys 50:365–379

Ivanov KS, Puchkov VN, Fyodorov YN, Erokhin YV, Pogromskaya OE (2013) Tectonics of the Urals and adjacent part of the West Siberian Platform basement: main features of geology and development. J Asian Earth Sci. Geol Evol Asia 72:12–24. https://doi.org/10.1016/j.jseaes.2013.02.029

Kashubin S, Juhlin C, Friberg M, Rybalka A, Petrov G, Kashubin A, Bliznetsov M, Steer D (2006) Crustal structure of the Middle Urals based on reflection seismic data. In: Gee D, Stephenson R (eds) European lithosphere dynamics. Geological Society, London, Memoirs, vol 32, pp 427–442

Kazansky AY, Metyolkin DV (2008) Fault deformation in the Paleozoic and Mesozoic history of the South and West Siberia structure formation on paleomagnetic data. Materials of Russian scientific conference "Basement, structures of the West Siberian Mesozoic-Cenozoic sedimentary basin framing, their geodynamic evolution, and petroliferous problems. SIBNATS, Tyumen-Novosibirsk, pp 98–101 (in Russian)

Klets AG, Kontorovich AE, Ivanov KS, Kazanenkov VA, Saraev SV, Fomin VA (2007) Geodynamic model for the pre-Jurassic as a basis for the petroleumgeological regionalization of the Precambrian-Lower Triassic structural stage of the West Siberian petroliferous province. In: Ways for realization of the petroleum and ore potential of the Khanty-Mansi Autonomous District. IzdatNaukaServis, Khanty-Mansiyk, vol 1, pp 79–90 (in Russian)

Kontorovich AE, Kontorovich VA, Filippov YF et al (2003) Pre-Yenisei petroliferous subprovince as a new perspective object of petroliferous search in Siberia. Geodynamic evolution of lithosphere of the Central Asian mobile belt (from ocean to continent). Conference Transactions. Publ. of the Institute of Geography of SB RAS, Irkutsk, pp 123–127 (in Russian)

Koroteev VA, Dianova TV, Kabanova LY (1979) Middle Paleozoic volcanism of the Eastern zone of the Urals. Nauka, Moscow, p 120 (in Russian)

Kosarev AM, Puchkov VN, Seravkin IB (2005) Petrologo-geochemical features of the Early Devonian-Eifelian island arc volcanites of the Magnitogorsk zone in the geodynamic context. Lithosphere 4:24–40 (in Russian)

Kosarev AM, Puchkov VN, Seravkin IB (2006) Petrologo-geochemical character of the Middle Devonian–Early Carboniferous island-arc and collisional volcanic rocks of the Magnitogorsk zone in the geodynamic context. Lithosphere 1:3–21 (in Russian)

Kuzmichev AB (2004) Tectonic history of the Tuva-Mongolian Massif: Early Baikalian, Late Baikalian and Early Caledonian stages. PROBEL-2000, Moscow, p 192 (in Russian)

Kuzmichev AB, Larionov AN (2011) The Sarkhoy series in the East Sayan: Neoproterozoic (~770–800 Ma) Andean-type volcanic belt. Geol Geophys 52:875–895 (in Russian)

Kuzmin MI, Parfenov LM (eds) (2001) Tectonics, Geodynamics, and Metallogeny of the Republic of Sakha (Yakutia), MAIK "Nauka\Interperiodika", Moscow, p 571 (in Russian)

Kuznetsov NB, Soboleva AA, Udoratina OV, Gertseva MV (2005) Pre-ordovician granitoides of the Timan-Uralian region and evolution of the Protouralides–Timanides, Geoprint, Syktyvkar, p 100 (in Russian)
Layer PW, Parfenov LM, Surnin AA, Timofeev VF (1993) The first $^{40}Ar/^{39}Ar$ age determinations of magmatic and metamorphic rocks of the Verkhoyansk-Kolyma Mesozoides. Dokl Acad Sci USSR 329(5):621–624 (in Russian)
Layer PW, Newberry R, Fujita K, Parfenov LM et al (2001) Tectonic setting of the plutonic belts of Yakutia, NE Russia, based on $^{40}Ar/^{39}Ar$ and trace element geochemistry. Geology 29:167–170
Medvedev AY, Al'mukhamedov AI, Reichow MK, Saunders AD, White RV, Kirda NP (2003) Absolute age of basalts from the pre-Jurassic basement of the West Siberian Plate (from $^{40}Ar/^{39}Ar$ data). Russ Geol Geophys 44(6):617–620
Mizens GA (1997) Upper Paleozoic flysch of the Western Urals. Institute of Geology and Geochemistry of Uralian Branch RAS, Ekaterinburg, p 229 (in Russian)
Morozov AF (ed) (2006) Structure and dynamics of lithosphere of Eastern Europe. Issue 2: Research results of Europrobe. Geokart, GEOS, Moscow (in Russian)
Narkissova VV (2005) Petrochemistry of the Late Ordovician-Early Devonian basaltoids of the Southern part of the Middle Urals (based on data on the Uralian superdeep borehole and near-hole area). Ph.D. thesis. Moscow State University, Moscow, p 27 (in Russian)
Natapov LM (1988) History of accretion structure of NE frames of the Siberian continent. Geology of seas and oceans: reports of Soviet geologists. Leningrad, pp 172–187 (in Russian)
Natapov LM, Stavsky AP (1985) Geodynamic model of NE USSR and its use for metallogenic analysis. Sovietskaya Geologia 3:70–77 (in Russian)
Neustroev AP, Parfenov LM, Rodionov BP (1993) Paleomagnetic data and nature of Tas-Khayakhtakh terrane of the Verkhoyansk-Kolyma area. Geol Geophys 8:25–37 (in Russian)
Oxman VS (2000) Tectonics of collision belt of the Chersky Range (NE Asia). GEOS, Moscow, p,236 (in Russian)
Oxman VS, Timofeev VF, Tretiakov FF (1998) Uyandina complex of polymetamorphic rocks: P-T-t evolution and geodymatic interpretation. Otechestvennaya Geologia 6:10–14 (in Russian)
Parfenov LM (1984) Continental margins and island arcs of NE Asia Mesozoides. Nauka, Novosibirsk, p189 (in Russian)
Peive AV, Ivanov SN, Necheukhin VM, Perfiliev AS, Puchkov VN 1976 (1977) Tectonic map of the Urals at a scale of 1:1,000,000 (and Explanatory Note to the Tectonic Map of the Urals). GUGK, Moscow
Popov VS, Kremenetskiy AA, Beliatskiy BV (2008) Pre-Ordovician Sm-Nd isotopic age of ultramafic rocks in ophiolite belts of the Urals: specified data. In: Structural-formational complexes and the problems of Pre-Cambrian geodynamics of Phanerozoic orogens. Transactions of the International Conference. Institute of Geology and Geochemistry of Uralian Branch RAS, Ekaterinburg, pp 100–103 (in Russian)
Pospelov II, Aristov VA, Kurenkov SA (1995) Upper Paleozoic olistostrome complex of the Tas-Khayakhtakh Range (Chersky mountain system). Dokl. RAS. V. 343(5):651–655 (in Russian)
Puchkov VN (1979) Bathyal complexes of the passive continental margins of geosynclines. Nauka, Moscow, p 146 (in Russian)
Puchkov VN (1997) Structure and Geodynamics of the Uralian Orogen. In: Burg J-P, Ford M (eds) Orogeny through time. Geological Society Special Publication, vol 121, pp 201–236
Puchkov VN (2000) Paleogeodynamics of the Southern and Middle Urals. Dauria, Ufa, p 145 (in Russian)
Puchkov VN (2003) Uralides and Timanides, their structural relations and the place in geological history of the Ural-Mongolian fold belt. Geol Geophys 1–2, 28–39 (in Russian)
Puchkov VN (2006) On the age of the Uralian ophiolites // In: Ophiolites: geology, petrology, metallogeny and geodynamics. Institute of Geology and Geochemistry of Uralian Branch RAS, Ekaterinburg, pp 121–129 (in Russian)
Puchkov VN (2007) Post-Variscan geodynamics of the Uralian region. Geodynamic evolution of lithosphere of the Central Asian mobile belt. From Ocean to Continent. 2. Irkutsk, pp 36–39

Puchkov VN (2009) The diachronous (step-wise) arc–continent collision in the Urals. Tectonophysics. https://doi.org/10.1016/j.tecto.2009.01.014
Pushcharovsky YM (1960) Pre-Verkhoyansk depression and Mesozoides of NE Asia. In: Tectonics of USSR, V. 5. Acad. Sci. Publishing House, Moscow, p 236 (in Russian)
Pystin AM, Pystina YI (2008) Metamorphism and granite-formation in the Proterozoic—Early Paleozoic history of formation of the Pre-Polar-Uralian segment of the earth crust. Lithosphere 6:25–38 (in Russian)
Rasulov AT (1982) Tectonics of the Early Mesozoic grabens of the eastern slope of the Urals. Sverdlovsk: Institute of Geology and Geochemistry, Uralian Scientific Center, Academy of Sciences of the USSR, p 43 (in Russian)
Reichov MK, Pringle MS, Al'Mukhamedov AI, Allen MB, Andreichev VL, Buslov MM, Davies CE, Fedoseev GS, Fitton JG, Inger S, Medvedev AY, Mitchell C, Puchkov VN, Safonova IY, Scott RA, Saunders AD (2009) The timing and extent of the eruption of the Siberian Traps large igneous province: implications for the end-Permian environmental crisis. Earth Planet Sci. Lett 277(1–2):9–20
Rezanov IA (1968) Features of structure and evolution of Mesozoides of NE USSR.Nauka, Moscow, p 224 (in Russian)
Rudenko VE, Stepanov DV, Vinogradova VP et al (2010) New geological and geochronological data of southwestern margin of the Angara-Vitim granitoid areal. Geologia i Mineragenia Zabaikalia, Chita, pp 151–160 (in Russian)
Ruzhentsev SV, Degtyarev KE (eds) (2005) Outlines on regional tectonics, vol 1. Nauka, Moscow, Southern Urals, p 249 (in Russian)
Sarayev SV, Khomenko AV, Baturina TP (2004) Vendian and Cambrian of the West Siberian southeast: stratigraphy, sedimentology, paleogeography. Geol Geophys Treat Petroliferous Deposits 1:7–18 (in Russian)
Savelieva GN (1987) Gabbro-ultrabasic complexes of ophiolites of the Urals and their analogues in the recent oceanic crust. Nauka, Moscow, p 245 (in Russian)
Savelieva GN, Shishkin MA, Larionov AN, Suslov PV, Berezhhaya NG (2006) Tectonomagmatic events of the Late Vendian in the mantle ophiolite complexes of the Polar Urals: data on U-Pb dating of zircons from chromites. In: Ophiolites: geology, petrology, metallogeny and geodynamics. Institute of Geology and Geochemistry of Uralian Branch RAS, Yekaterinburg, pp 160–164 (in Russian)
Sobolev ID, Avtoneev SV, Belkovskaya RP et al (1986) Tectonic map of the Urals at a scale of 1:1,000,000 (Explanatory note). Sverdlovsk (in Russian)
Stavsky AP (1984) Tectonics of the Kolyma-Indigirka area. Moscow: Geological Institute, Russian Academy of Sciences, USSR, Moscow, p 24 (in Russian)
Surkov VS, Smirnov LV (2008) Consolidated blocks of the Earth crust's in the West Siberian Plate's basement. Materials of Russian scientific conference "Basement, structures of the West Siberian Mesozoic-Cenozoic sedimentary basin framing, their geodynamic evolution, and petroliferous problems. SIBNATS, Tyumen-Novosibirsk, pp 207–210 (in Russian)
Surkov VS, Trofimuk AA (1986) Megacomplexes and deep structure of the Earth's crust of the West Siberian Plate. Nedra, Moscow, p 240 (in Russian)
Tarabukin VP (2006) Biostratigraphy and conodonts of the Ordovician of NE Asia. Yakutian Scientific Centre, Yakutsk, p 187 (in Russian)
Tessalina SG, Bourdon B, Gannoun A et al (2007) Complex proterozoic to paleozoic history of the upper mantle recorded in the urals lherzolite massifs by Re-Os and Sm-Nd systematics. Chem Geol 240(1–2):61–84
Tretiakov FF (1996) Folded structures of Selennyakh Mts. Geotectonika 4:43–57 (in Russian)
Tsygankov AA (2005) Magmatic evolution of Baikal-Muya volcanoplutonic belt during the Late Precambrian. Siberian Branch of Russian Academy of Sciences, Novosibirsk, p 306 (in Russian)
Volchenko YA, Ivanov KS, Koroteev VA, Ozhe T (2007) Structural-substantial evolution of the Platinum-bearing belt complexes of the Urals during the formation of the Uralian-type chromite-platinum ore deposits. Lithosphere 3:3–31 (part 1); Ibid. 4:73–101 (part 2) (in Russian)

Yazeva RG, Bochkarev VV (1998) Geology and geodynamics of the Southern Urals. Uralian Branch RAS, Ekaterinburg, p 203 (in Russian)

Yergaliev GK, Nikitin IF, Palets LM et al (1995) Vendian-Paleozoic evolution of Kazakhstan and Northern Tien Shan. Geol Kazakhstan 5–6:11–22

Yolkin EA, Krasnov VI, Bakharev NK, Belova EV, Dubatolov VN, Izokh NG, Klets AG, Kontorovich AE, Peregoedov LG, Sennikov NV, Timokhina IG, Khromykh VG (2001) Stratigraphy of petroliferous Basins of Siberia. The paleozoic of West Siberia. Izd, Novosibirsk. SO RAN, Filial "GEO", p 516 (in Russian)

Yolkin EA, Sennikov NV, Bakharev NK, Belyaev SY, Izokh NG, Kanigin AV, Klets AG, Kontorovich AE, Kontorovich VA, Obut OT, Saraev SV, Filippov YF (2008) Main features of the modern structure and formation history of the Precambrian-Paleozoic West-Siberian sedimentary basin. In: Materials of Russian scientific conference "Basement, structures of the West Siberian Mesozoic-Cenozoic sedimentary basement framing, their geodynamic evolution, and petroliferous problems. SIBNATS, Tyumen-Novosibirsk, pp 75–80 (in Russian)

Tectonic Domains of Central Asia

O. V. Petrov, I. I. Pospelov, T. N. Kheraskova, O. Tomurtogoo, Chen Bingwei, and Ren Liudong

Abstract The chapter offers insight into the geological structure and tectonic evolution of large geological structures within Central Asia; specifically—Tien Shan, Pamirs, Turan Plate, Mongolia and Northern China. The essays focus on the tectonic zoning of the regions illustrated by maps and schemes and describe main stages of formation of the continental crust and sedimentary cover.

1 Tien Shan

1.1 Northern (Caledonian) Tien Shan

Tien Shan Caledonides (Fig. 1) occupy the northern part of the Tien Shan mountain system. Kazakhstan—North Tien Shan epi-Grenville continental massif and Terskey thrust-folded zone contributed to their development. Consolidated crust of the latter has Early Paleozoic age and originated at the place of Terskey paleooceanic basin structurally closely associated with the Paleoasian Ocean. Terskey oceanic basin closed in the Late Ordovician.

O. V. Petrov (✉) · I. I. Pospelov
Russian Geological Research Institute (VSEGEI), 74 Sredny Prospect, St. Peterburg 199106, Russia
e-mail: vsgdir@vsegei.ru

T. N. Kheraskova
Geological Institute of Russian Academy of Sciences (GIN RAS), 7 Pyzhevsky Pereulok, Moscow 119017, Russia

O. Tomurtogoo
Mongolian Academy of Sciences, Prime Minister A. Amar street 1, Sukhbaatar district, Ulaanbaatar 14200, Mongolia

C. Bingwei · R. Liudong
Institute of Geology, Chinese Academy of Geological Sciences (CAGS), No. 26 Baiwanzhuang Road, Beijing 100037, P.R. China

O. V. Petrov and S. Dong (eds.), *Tectonics of Asia (Northern, Central and Eastern Asia)*, Springer Geology, https://doi.org/10.1007/978-3-030-62001-1_4

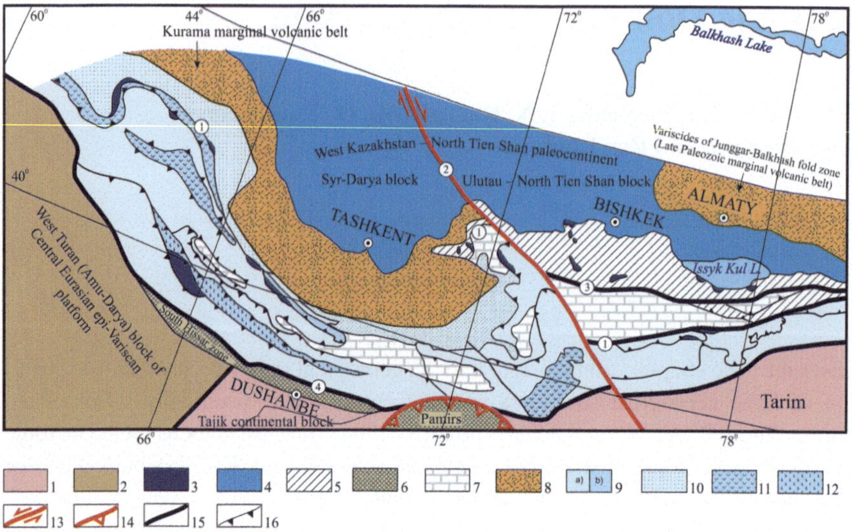

Fig. 1 Tectonic zoning of Tien Shan. 1—Achean–Paleoproterozoic blocks; 2—Epi-Variscan platforms; 3—ophiolites; 4–6—Tien Shan Caledonides: 4—West Kazakhstan—North Tien Shan terrane, 5—Terskey terrane, 5—Late Variscides of South Hissar and Northern Pamirs; 7–12—Tien Shan Variscides: 7—Turkestan-Alay epi-Grenvillian terrane, 8—Late Paleozoic marginal volcanic belts, 9—accretion prism complexes of the Late Paleozoic marginal volcanic belt: (a) sedimentary, (b) volcanogenic, 10—Late Paleozoic fore-arc depression, 11—blocks of island arc origin in the Late Paleozoic accretion prism, 12—blocks of oceanic basalts in the Late Paleozoic accretion prism; 13—Talas-Fergana strike-slip; 14—North Pamirs overthrust; 15—boundaries of the main tectonic zones and blocks; 16—thrusts and overthrusts. Numbers show: 1—Turkestan oceanic suture, 2—Talas-Fergana strike-slip, 3—Terskey suture, 4—South Hissar suture

Tien Shan Caledonides to the north form a united structural complex with the Central Kazakhstan's Caledonides. Dzhalair-Nayman ophiolite belt (the boundary between Southern Kazakhstan and Northern Tien Shan) formed in the Early Ordovician as a result of collision of Anrakhai (Southern Kazakhstan) and North Tien Shan micricontinents. U–Pb dating of plagiogranites from ophiolite complex gave 521 ± 2, 520 ± 4 and 513 ± 1 Ma, which indicates Middle Cambrian age of the oceanic crust. Cherts covering basalts contain the Upper Cambrian conodonts. Age of turbidites overlain by a combination of several tectonic thrusts of different-facial complexes is determined as Late Cambrian—Early Ordovician by brachiopods fossils and by detrital zircon dating (younger crystals have Tremadoc age—490 ± 1 Ma). Strong angular unconformity at the base of the Arenig deposits determines the Early Ordovician time of the main tectonic event (Alexeiev et al. 2011; Kroner et al. 2011).

In the extreme northeast, Caledonides are contiguous with Variscides of Junngar-Balkhash fold zone. To the south, Caledonides joint structures of South Tien Shan Variscides. In the eastern part of the area, boundary passes along the Terskey suture. To the west from Talas-Fergana strike-slip, the southern limit of Caledonides is southwestern boundary of Kazakhstan—North Tien Shan massif, its Syr-Darya block (with superimposed Kurama Middle—Late Paleozoic marginal volcanic belt) and Turkestan suture.

Structure of the Tien Shan Caledonides is not fully determined. This is explained both by its exclusive complexity and its partial destruction in times of intense geodynamic processes during the Late Paleozoic and Mesozoic-Cenozoic. The structure was also strongly changed by granite intrusions of different ages and numerous faults. The latter are subdivided into three types.

The first type of faults has sub-latitudinal ("Tethyan") strike in parallel with the Terskey suture (sub-latitudinal Nikolaev's lineament). Strikes of the superimposed orogenic depressions, filled up by the Middle—Upper Paleozoic deposits, coincide with strike of these faults. In addition, these faults displace the Ordovician granite intrusions and Mesozoic-Cenozoic sediments. All this allows supposing their Middle Paleozoic—Cenozoic age close to the age of "Nikolaev's lineament" (Burtman 1976).

The second type of faults is represented by steeply-dipping reverse—strike-slip faults, sometimes thrusts, with mainly northwest strikes in parallel with Talas-Fergana regional strike-slip fault and spread of the Lower Paleozoic deposits(?). The largest faults subdivide tectonic blocks; in each of them the Lower Paleozoic complexes differ by composition and geodynamic environments of formation. Most probably the second type faults have Early Paleozoic age and the first type faults cut them (Burtman 1976; Kheraskova et al. 1997).

Folded overthrusts (nappes) are assigned to the third type of faults. Their existence is revealed due to spatial combination of folded tectonic slices and nappes divided by the Early Ordovician olistostromes or serpentinite mélange.

1.1.1 Kazakhstan—North Tien Shan Epi-Grenville Continental Massif

Structure of Kazakhstan—North Tien Shan massif could be determined only by separated outcrops of uplifts appearing from under the Mesozoic-Cenozoic sedimentary cover or in roof sags of numerous Riphean—Early-Middle Paleozoic granitoid intrusions. Structure of the massif includes crystalline basement of Mesoproterozoic–Neoproterozoic age and Neoproterozoic–Paleozoic cover complexes subdivided into several structural stages.

In the Middle Tien Shan, Precambrian complexes underlie the terrigenous, terrigenous siliceous and terrigenous-carbonate Vendian (Ediacaran)—Lower Paleozoic sections. In the Northern Tien Shan, Precambrian and Lower Paleozoic complexes often have the tectonic relationships. Mostly complicated structure is typical of formations after high-pressure metamorphism and their exhumation into the upper-crust level. Such formations are known in the western part of the Kyrgyz Range where they compose several conjugated syn-, and antiforms of WNW strike. The Makbal antiform is the largest among them. The age of high-pressure metamorphism is 510–500 Ma (Alexeiev et al. 2011). Core of the Makbal antiform includes the Middle Riphean granitoids (1131 $\pm$ 4 Ma) (Degtyarev et al. 2012). It is suggested that the Makbal rocks were metamorphosed in the Middle Cambrian subduction zone dipping northward under the Northern Tien Shan microcontinent. Deformed granodiorite associated with the Makbal metamorphic rocks gave the age of 514 ± 5 Ma (Konopelko et al. 2012).

The most ancient basement complexes are represented by repeatedly metamorphic rocks appearing in the areas of Kirgiz, Kendyktas, Kastek, Zailiysky, Kungey and Terskey Ala-Too Ranges. Gneiss-migmatite complex revealed here has age of 2.2–1.6 Ga (U/Pb determination by zircons) (Kiselev et al. 1993). To the east, in the area of Kuyliu and Sarydzhaz Ranges, the most ancient rocks are represented by gneiss, amphibolites, marbles, and crystalline schist. Two generations of zircons were extracted from gneiss and crystalline schist; U/Pb isochronous 2.6 and 1.9 Ga ages was determined from them (Kiselev et al. 1993).

In the Chatkal Range (Syr Darya block), metamorphic complex is represented by migmatized ophiolites (Kassan complex). Ophiolitic association includes peridotite, pyroxenite, hornblendite, serpentinite, listvenite, gabbro-amphibolites, garnet amphibolites, and crystalline schist with garnet, tourmaline, staurolite. U/Pb isochronous 1925 ± 20 Ma age was determined by zircons from garnet-mica schist. Metaophiolites form Semizsay tectonic sheet (thickness up to 4000 m) is thrusted onto marbles and quartzite of unknown age. Ophiolites are covered with unconformity by metamorphosed flysch probably of Riphean age.

In addition, in the Kungey Ala-Too Range area, the Kochkor complex is mostly composed by granitegneiss lenses with which biotite-cordierite schist, marbles, and quartzite are associated. U/Pb age of zircons from metamorphic rocks is estimated as 1050 ± 20 Ma (Kiselev et al. 1987). These data and also new data received over the last years for Kazakhstan part (Degtyarev et al. 2008) Northern Tien Shan massif are the evidence of granitization of massif in the Grenville epoch of crust consolidation.

In the Kungey Ala-Too Range area, gneiss and migmatite are overlain by series of silicic volcanic rocks; their ages by Pb/Pb ratio from zircons are 840–740 Ma (Mikolaichuk 1998). The series of terrigenous-volcanogenic deposits (2000 m) in areas of Jatymtau, Naryntau, and East Akshiyrak Ranges has approximately similar stratigraphic position. Its lower part is composed with altered rhyolite and tuffs, intruded by micro-granite dykes and sills. Rhythmically intercalated terrigenous rocks with limestone layers and lava flows form the middle part. The succession top is composed by rhyolite and rhyolitic tuffs, the U/Pb age of which by zircons was determined as 830 ± 40 and 705 ± 10 Ma (Kiselev et al. 1993).

Upper Riphean and Lower Sinian deposits by their composition and texture were accumulated in limits of marginal volcanic belt of Tarim—Tien Shan continent (Kheraskova et al. 2010). Intense granitoid magmatism (in northern Tien Shan—837 Ma, Talas Range—825 Ma (Burtman 2006) is also genetically associated with this belt.

Superposed complexes overlie the above mentioned rocks with an angular unconformity and form pre-plate (uppermost Riphean—Vendian) and plate (Lower Paleozoic) cover complexes of Kazakhstan—North Tien Shan continental massif. Cover complexes were folded and frequently tectonically sliced by later deformations.

Uppermost Riphean—Vendian. Series of conglomerates, arcosic sandstones with Fe–Mn concretions, acid, intermediate, and basaltic lavas, tuffs and tuff turbidites correspond to this stratigraphic level in the south Kazakhstan—North Tien Shan continental massif (Bolshoi Karatau, Naryn-Too and Dgetym-Too Ranges). Volcanic rocks (up to 1000 m) contain a higher alkalic and antidromic eruption sequence—from rhyolite, rhyodacite, and trachyandesite in the lower part of succession to trachybasalts in upper part with the age of 690 ± 15 Ma (Korolev and Maksumova 1984; Sudorgin 1990).

Described above deposits are concordantly superposed by a series containing tilloides (3000 m). Tilloides (diamiktites containing fragmental rock material of ice stream) are separated by carbonate-terrigenous deposits. Formation of the latter deposits, by opinion of majority of researchers, took place in environment of continental riftogenesis. Everywhere the distribution of these complexes is clearly limited by lengthy co-sedimentation faults. In Talas zone, similar deposits contain fragments of ophiolitic assemblage (Geology and Metallogeny of Karatau 1986), which is evidence of neighborhood of an oceanic basin. It can be supposed that in the Vendian the crust of Kazakhstan—Northern Tien Shan massif suffered extension and continental rift genesis. That processes generated the opening of the Terskey oceanic basin. **Cambrian—Ordovician complexes** are represented by silicious-carbonate and carbonate sediments of continental shelf and terrigenous accumulations of continental slope. They are distributed in areas of Dgatymtau, Naryntau, East Akshiyryak, Kuyliu, and Saryjaz Ranges and in the eastern part of the Terskey Range (100–1000 m). Siliceous rocks are represented by phtanite and siliceous-clayey slates. Described above rocks contain the Middle—Late Cambrian trilobites, Late Cambrian conodonts, Early Ordovician and Llanvirnian graptolites and conodonts.

Middle—Upper Ordovician lies with stratigraphic unconformity above the Lower—Middle Ordovician and with a transgressive contact on more ancient rocks. Rhythmically interbedded polymictic sandstones, siltstones and shale (1500 m), sometimes with limestone layers, are predominant. These rocks contain brachiopods, nautiloidea and graptolites of the Middle Ordovician, Karadock and Ashgill (Klishevich and Sobolevskaya 1993; Stratified and intrusive formations of Kyrgyzstan 1982).

Described above deposits formed on the passive margin of Kazakhstan—North Tien Shan continental massif—on shelf and continental slope. Uppermost part of sections has the features typical for molasses and gives the evidence of collision processes in the Terskey zone situated to the south. The appearance of ophiolite and island arc volcanic fragments in terrigenous rocks (1000 m) also indicates the collision processes.

Lower Silurian overlies the Ordovician either concordantly or with erosion without angular unconformity, whereas Devonian—with an angular and azimuthal unconformity. Silurian is represented by molasses and fills the superimposed depressions. During the Devonian, the intense granitoid magmatism took place.

Tensity of fold deformations in the Vendian and Lower Paleozoic rocks is different. Quite a few folds are with gently sloping limbs, many gently sloping monoclines limited by the Late Paleozoic faults. Compressed overturned folds are presented somewhere. The strike of axes of large folds is in parallel to strike of the Terskey ophiolite suture. There are no evident angular unconformities in stratigraphic succession including the Vendian and Lower Paleozoic. Therefore, folded structure of the zone was created not earlier than in the Karadock.

1.1.2 Terskey Thrust-Fold Zone

Terskey thrust-fold zone was formed at the end of the Ordovician at place of the Terskey oceanic basin (Makarychev and Ges' 1981; Burtman 2006). Rocks of oceanic crust and island arcs are distributed as a wide zone extending through Central Tien Shan from the Chatkal Range in the west to the eastern end of the Terskey range to the east. These rocks are situated in tectonic relationships with the surrounding complexes belonging to Kazakhstan—North Tien Shan massif. In some outcrops ophiolites cover the Lower Paleozoic rocks as an allochthon. In another outcrops, ophiolites are limited by young faults and granite intrusions, so allochtonous bedding of ophiolites is only supposed.

It is possible to recognize the evolution of the Terskey thrust-fold zone from the pattern of relationships which were determined in the East Pre-Sonkul region. Here (in Karajorgo Range), assemblage of tectonic sheets, folded and destroyed by faults, includes the tectonic sheet composed by ophiolites (Kheraskova et al. 1997). Tuff-turbidites with layers of olistostromes form the lowermost part of visible geological section. Early Arenig conodonts are determined from limestone blocks of olistostrome (Mikolaichuk et al. 1997a). These rocks are overlain by the **first tectonic sheet** composed by limestone, tuff-silicite, and tephraturbidites of intermediate and acid composition with the Late Cambrian—Early Tremadock conodonts (1000 m). Upper part of this tectonic sheet succession including olistostrome is tectonically covered by the **second tectonic sheet**. The lower part of the latter is composed by limestone and thick layer of volcanogenic rocks (200 m) containing intercalated andesite, pillow andesitic basalts and acid tuffs. Upper part of the second sheet is composed by proximal tuff-turbidites with the Late Cambrian—Tremadock conodonts. Formation environments of the described above deposits corresponded to the slopes and foots of volcanoes probably situated in an oceanic island arc. The **third tectonic sheet** is formed by gabbro, gabbro-norite, and pyroxenite with swarms of parallel dykes (Mikolaichuk et al. 1997a). Study of zonal spinel has made it possible to draw a conclusion on the formation of rocks in spreading zone and their following alteration in subduction zone (Demina et al. 1995). Ophiolites are intruded by the Ordovician granites. Part of ophiolite complex (lying in allochtonous position) is composed by serpentinite mélange, basalts, and gialobasalts with chert layers. The Late Cambrian conodonts are determined from blocks of chert from mélange, and Arenig conodonts—from chert-overlain basalts (Kheraskova et al. 1997). Ophiolites are tectonically covered by an overthust of carbon-silicic slates with spicules of the Cambrian sponges.

Upper part of geological section of the Karajorgo Range is formed by two "Burenhai" tectonic sheets of the Precambrian rocks (fragments of Kazakhstan—North Tien Shan massif) (Kheraskova et al. 1997). **Lower "Burenhai" tectonic sheet** is composed by clayey-cherty slates, rhyolite, dacite, tuffs. Andesitic basalts and intraplate basalts with Riphean U/Pb age. **Upper "Burenhai" tectonic sheet** is composed by metamorphosed terrigenous deposits (1000 m) and overlying shelf limestone. Age of rocks from this sheet is unknown. Both tectonic sheets are intruded by granites for which Vendian age was determined (611–626 Ma—zircon Pb/Pb dating) (Mikolaichuk et al. 1997a).

Described assemblage of tectonic sheets of the overlying Karadgorgo Range formed during two stages. At the first stage (Sinian), the system of two "Burenhai" tectonic sheets was formed (at present—the uppermost). The most favorable conditions for origin of tectonic sheets were connected with formation of an accretional prism at oceanic margin or collision of sialic blocks in process of oceanic basin closing. As ophiolites are absent in composition of the "Burenhai" tectonic sheets, their formation was probably in an accretional prism. Therefore, it is possible to suggest that the Terskey oceanic basin had already existed by the beginning of the Vendian and accretion processes took place at its margin. In all likelihood, these processes were going at Naryn (i.e. southern) margin of the Terskey oceanic basin, as in geological section of margin of the Kazakhstan—North Tien Shan massif the "Burenhai" tectonic sheets lie higher than tectonic sheets composed by rocks of the Terskey oceanic crust.

The second stage of formation of tectonic sheets assemblage of the Karagorgo Range took place in the Early Ordovician. System of tectonic sheets including the Lower Arenig deposits is unconformably covered with tuff-conglomerates, tuff-sandstones, and finegrained tuffite with the Arenig–Llanvirnian graptolites (Kheraskova et al. 1997). These relationships are evidence of collision of the oceanic island arc with passive margin of Kazakhstan—North Tien Shan massif. This collision happened in the Arenig time. On the northern slope of the Chatkal Range (in Karaterek tectonic block), metamorphosed ophiolites and Ordovician deep-water volcano-sedimentary deposits are exposed. Pyroxenite, serpentinite, gabbroamphibolites, and actinolite schist (1000 m) and nonuniformly stratified series (more than 1000 m) are situated here with tectonic relationships. The last series are composed by distal terrigenous flysch with the Arenig—Lladeylo conodonts. It includes thick layers of pillow and tube vesicular basalts, and also tuffs, tuff-sandstones, with interlayers of sandstones, argillites, chert (Khristov et al. 1999). The island arc origin of this association is supposed.

Karaarcha tectonic sheet is situated in the west of the Kirgiz Range. It is composed by ophiolites which have tectonic contacts with the surrounding sedimentary and metamorphic rocks. Lower part of ophiolitic association is represented by pyroxenite, hornblendite, and layered cumulate complex composed by pyroxenite and gabbroid including olivine gabbro. For leucogabbro, Ar/Ar age by clinopyroxene (480 $\pm$ 4 Ma) was determined. Rocks are metamorphosed up to green schist degree. Thick series of tholeiitic basalts, alkaline basalts and andesitic basalts interlayered with silicites are situated structurally higher. In the upper part, there is a considerable amount of lava breccias and tuffs. Layers of siliceous rocks contain the Late Cambrian, Early and Middle Ordovician conodonts. Absolute Ar/Ar age of basalt determined by clinopyroxene is 460 $\pm$ 6 Ma (Travin et al. 2002). RREE spectra of calc-alkaline basalts show their formation above the subduction zone (Lomize et al. 1997).

Kenkol tectonic sheet clamped among the Early Paleozoic rocks at the southern slope of Kirgiz Range and extended to the northern slope of Talas Range is formed by serpentinite mélange lenses with gabbro and diabase blocks. The latter underlies a thick series of pillow toleiitic basalts and andesitic basalts, basaltic tuffs and jaspers (2500 m). Siliceous interlayers among volcanic rocks contain the Middle Cambrian algae, Late Cambrian conodonts, Early Ordovician and Llandeilo-Karadock radiolarians (Maksumova et al. 1988; Maksumova 1999). Petrochemical characteristics of lavas partly correspond to the oceanic basalts and partly to rocks of island arcs. Geochemical parameters of basalts (corresponding the N-MORB) indicate their formation in the spreading zone (Ges' and Makarychev 1985; Lomize et al. 1997).

Tectonic sheet about 800 m thick is deposited on the northern slope of the Jumgal Range among nappes composed by the Early Paleozoic volcano-sedimentary series. It is made up of serpentinite mélange with large blocks of pyroxenite, gabbro, sheeted-dyke complex of gabbro-diabase. Mélange sheet together with nappe system gently dips to the southwest. Structural study establishes the allochthon movement from south to north (Khristov and Chernyshuk 1987).

Further east, ophiolite belt extends to the central and eastern parts of the Terskey Range. Ultramafic and gabbroid rocks are situated here in tectonic slices and blocks. Thick complexes of basalts, often pillow ones, are also overbedded by volcano-sedimentary island arc deposits. Probably, the majority of rocks imputable to ophiolites in the eastern part of concerned the belt belonged to the basement of the Early Paleozoic oceanic island arc.

Volcanic part of island arc association is composed by pillow-lavas with interlayers of tuff-conglomerates, limestone, and chert with sponge spicules of the Early Paleozoic species (1000 m). By petrology and geochemical characteristics, basalts belong to low-Ti island arc tholeiites. Upper succession is composed by basalts, andesite, dacite, rhyolite and their tuffs with interlayers of tuff-sandstones and limestone containing brachiopods, gastropods, ostracoda and conodonts of the Botomian stage of Lower Cambrian (2000 m). These rocks are overlain by dacite and rhyolite tuffs, tuff-sandstones, tuff-aleurolite with the Late Cambrian conodonts in limestone beds (Mikolaichuk et al. 1997b).

Thus, the most part of the described ophiolite allochthons is associated with the island arc volcanic rocks and probably represent a complex of the oceanic island arcs basement. This is evidence that there was oceanic (ensimatic) volcanic Karadgorga island arc or, more probably, whole assemblage of such arcs in the Cambrian—early Ordovician in limits of the Terskey Basin. Island arc rocks are intruded by small massifs of the Early Ordovician M-type granites usually typical of island arcs. There age from zircons U/Pb and Pb/Pb is in interval 500–470 Ma. Granitoid intrusions are represented by quartz diorites, granodiorites, and younger quartz monzonite and quartz syenite (Mikolaichuk et al. 1997a; Ges' 1999).

Karadgorga oceanic island arc divided Terskey oceanic basin in two depressions—Naryn back-arc and Kensay fore-arc. Bottom sediments of the Kensay basin have remained in the Ichkeletau Range. Here, tectonic sheet includes motley mudstones, sandstones, and siltstones (with Arenig graptolites) containing interlayer of chert, limestone and turbidite horizons (1000 m).

In present structure, suture zone of the Terskey paleooceanic basin is represented by a system of the Late Paleozoic thrusts and strike-slips. Suture passes in western direction along the Terskey Range, rounds the Sonkul Lake depression, continues at the Moldotau Range, and thereupon extends in northwestern direction through the Susamyr and Eastern Talas Range into the Talas depression.

1.2 Southern (Variscan) Tien Shan

Southern Tien Shan Variscides represent a thrustfolded system which was formed on the southern margin of the Caledonian Kazakhstan—North Tien Shan paleocontinent during the Variscan stage of evolution in time of closing of the Turkestan paleooceanic basin.

Structure of the Southern Tien Shan Variscides has not been fully identified. This is explained not only by its complexity as its reworking in time of the intense geodynamic processes in the Late Paleozoic and Mesozoic-Cenozoic. Now it is the area of postplatform orogeny.

In the west, the Mesozoic-Cenozoic sedimentary cover of the Scythian-Turan plate of the Central Eurasian Platform covers the Variscan complexes. To the north, Variscides contact with the Kazakhstan—North Tien Shan continent overlain by Kurama marginal volcanic belt (Mossakovsky 1976; Kurchavov and Yarmoliuk 1984). The southern boundary of Variscides (from west to east) appeared to be along the West Turan (Amu-Darya) block of the Central Eurasian Platform, South Hissar ophiolite zone, Late Variscan Northern Pamirs—Kunlun—Qinling thrust-folded zone, and Tarim continental massif.

Formation of the thrust-fold structure took place from the late Early Carboniferous to the early Permian. Thrust-fold assemblage was probably formed by subduction of different complexes of the Turkestan oceanic basin under Kazakhstan—North Tien Shan continent and its Late Paleozoic Kurama Andean-type active continental margin. As a result, a structure, similar by the origin to accretional prism of the recent oceanic arcs, was formed (Biske 1996). The following indicators testify this point of view: appearance of ophiolitic and clayey mélanges and glaucophan-bearing rocks, eclogites along the Turkestan suture (for example, at Bykantau, Tamdytau and Nuratay Mts.); intercalation of steeply-dipping tectonic slices of the Early and Middle Paleozoic rocks by flysch-olistostrome complex of Late Carboniferous age; formation in frontal part of Kurama volcanic belt of lengthy Late Paleozoic depression, similar by structural position and rock complex filling (lower molasses) to fore-arc basins.

Flysch-olistostrome complex is widely distributed in the Bukan overthust. It has been studied in regions of Tamdytau, Bukantau, Northern Nuratau and others (Burtman 2006). Olistostrome includes large olistoliths and olistoplakas of organogenic limestone with fossils of different ages, and also sandstones, chert, alkaline basalts, rocks of all layers of oceanic crust, metamorphic schist. Olistoplakas of several kilometers in size are presented in Tamdytau Mts. They are formed by metamorphic schist, or by the Early Paleozoic trachybasalts, or by clastic rocks and tuffites (with the Ludlow graptolites) interbedded with basaltic and andesitic lava floods and sills. Lavas belong to calc-alkaline type of magma differentiation. Fragments of the Middle Paleozoic successions are also determined in composition of the tectonic slices.

South Tien Shan overthust assemblage is situated in allochthonous occurrence. Autochthon is represented by the Alay continental block which is a part of the Tarim continental massif. Terrigenous-carbonate sedimentary cover of the Alai block is of Silurian—Carboniferous age.

Overthust assemblage of Variscides, situated to the west from Talas-Fergana strike-slip, has in general antiform structure, complicated by lengthwise strike-slip axial folds and slice-dismembering, especially along the Talas–Fergana regional strike-slip. This overthust assemblage (Bukharin et al.1985; Biske 1996) includes ophiolitic complexes, and also the Ordovician–Silurian island arc and terrigenous series, Devonian—Lower Carboniferous silicic volcanic rocks of ensialic island arcs, metamorphosed in green-schist and amphibolite facies. In the Alay Range, one of slices consists of basalts with layers of chert, sandstones, and limestone with archaeocyatha and brachiopods of the Toyon stage of the Lower Cambrian. In another slice, outcropping basalts form mostly breccia, where the breccia layers intercalate with limestone with archaeocyatha of the Tommot stage of the Lower Cambrian (Biske 1987).

At Nuratau Mts., the Silurian deposits form isolated fragments of successions. Probably, thickness of the Silurian rocks exceeds 1000 m. Deposits with the Llandovery and Wenlok graptoliths are represented by argillites, carbon-siliceous-clayey slates, and sandstones. These deposits have a rhythmic texture and include some basaltic flows. The Lower Silurian deposits are represented by terrigenous and carbonate rocks with benthos fossils (Bukharin et al. 1985).

Devonian deposits lie unconformably on the underlying rocks. In the west of Northern Nuratau Mts. at the limbs of Shokhtau synform fold, the Early Paleozoic—Silurian deposits are covered with an angular unconformity and basal gravelites underlie a series of carbonate rocks with rich fauna fossils of corals, brachiopods, and foraminifers. Bottom boundary belongs to the Lokhkovian stage and upper boundary—to the Lower Moscovian substage. In general, this succession includes marks of numerous stratigraphic interruptions.

Overthust assemblage of the North Fergana region (Baubashata range) is divided into several structural units, the rocks of which were formed in different geotectonic environments (Burtman 2006). Baubashata unit occupies the lowermost position in the geological section, and Omtanchi, Kerey, and Shaidan units are above.

Structural unit of Baubashata outcrops at the cores of antiform folds. Succession consists of two stages divided by a tectonic contact. The lower stage is formed by terrigenous series (2000 m) with the Llandovery, Wenlok, and Ludlow graptolites and also by basalts, andesites, volcanic tuffs, and breccias. In the upper part of this stage, slates intercalate with limestone with the Ludlow–Pridoli brachiopods, corals, and trilobites. The upper stage with tectonic contact and structural unconformity covers the lower one; it is composed by carbonate and volcanogenic rocks (dacite, rhyolite, basalts) with corals and brachiopods of the Upper Ludlow, Pridoli, and all stages of the Lower Devonian. Middle Devonian is represented by subalkaline basalts and andesitic basalts. Sedimentation of carbonate deposits (3000 m) continued till Early Bashkirian age. Conglomerates and wild-flysch (400 m) with pebbles and blocks of limestone, containing the Bashkirian foraminifers compose the upper Baubashata stratigrahic succession.

Structural unit of Omtanchi forms several tectonic slices and nappes. The basis of visible section is composed by clayey-siliceous and siltstone slates with the Llandovery–Wenlock graptolites. Layers of polymyctic sandstones, conglomerates, dacite, and basalt flows are situated among slates. Petrochemical characteristics of lavas show their island arc origin (Biske 1996). The Upper Silurian succession is composed by clayey and clayey-siliceous slates and polymictic sandstones. Reef limestone with the Ludlow–Pridoli corals and brachiopods, basalts, andesites, and tuffs (1500 m) are bedded among them. Lower Devonian is represented by tuffs and slates with tentaculatas, Eifelian—by basalts, andesites, tuff breccias, tuff conglomerates, tuffs with layers of limestone including brachiopods and corals (500 m). Siliceous siltstone and layered limestone with conodonts and radiolarians occur up-section; this series (only 100–300 m thick) involves the time interval from Givetian to Serpukhovian ages. Conglomerates, sandstones, slates, cherts, and siliceous limestone (300 m) with the Late Serpukhovian—Early Bashkirian fossils lie on this series with erosion contact (stratigraphic unconformity). Series are represented by sedimentary mélange with blocks of the Visean limestone; rocks were partly transformed into mixtite, the matrix of which is composed of the Silurian slates.

Structural unit of Kerey was formed by weakly metamorphosed rocks of oceanic crust. Lower—Middle Devonian pillow basalts and picrite (thickness reaches 2000 m) are widespread. Chert layers and tectonic lenses of serpentinized ultramafic rocks are bedded among lavas. Volcanic rocks are covered by deep-water siliceous-carbonate sediments of Givetian—Serpukhovian age.

Basis of **structural unit of Shaidan** is formed by gabbro-ultramafic rock complex about 2.5 km thick (serpentinized harzburgite and dunite, serpentinite mélange with blocks of basaltic lavas, metamorphic schist, gabbro, rodingite). Pb/Pb age of zircons from dunite was determined as 532 ± 12 Ma. Structurally, verhlite and lherzolite are situated above; pyroxenite (100 m) and bended gabbro still above, alternating to gabbro-amphibolites (up to 800 m). U/Pb and Pb/Pb age of zircons from gabbro—395–475 Ma. Rocks have partly transformed into blastomylonites. The section is built up by sheeted-dyke complex of diabase and gabbro-diabase formed in two generations. By their composition, dyke rocks take up position between komatiites and oceanic tholeiitic basalts (Kurenkov et al. 2002). Tectonic nappe (1500 m) lies structurally up-section; the largest part of it is composed by green metamorphic schist, originated from oceanic tholeiitic basalts; glaucophane blue schist is present among green schist. Jasper interlayers contain the Silurian conodonts (Puchkov et al. 1987).

South Fergana tectonic zone occupies the northern slopes and foothills of the Turkestan and Alai Ranges and part of the Fergana Range. South Fergana overthust assemblage includes the following structural units: Isfairam, Abshir, Taldyk, and Shankol (Burtman 2006).

Two lowermost units of South Fergana assemblage (**Isfairam and Abshir**) are formed by deposits accumulated on the shelf, continental slope, and foot of Alai microcontinent. These rocks are slates and sandstones with the Llandovery, Wenlock, and Ludlow graptolites. The bottom of overlying carbonate series (thickness 2000–4000 m) in different successions is situated at different geochronological levels—from Ludlow to Middle Devonian. But the upper boundary has Early Moscovian age. The largest part of carbonate series is composed by shallow-water organogenic and organoclastic sediments. More deep-water clayey and clastic limestone with flint concretions, lenses, and layers (50–500 m) accumulated during the Serpukhovian—Early Moscovian time. Lower Moscovian limestone is overlain by flysch and olistostrome series (50–1000 m).

Two uppermost structural units of South Fergana tectonic zone (**Taldyk** and **Shankol**) are formed by nappes composed by oceanic crust complexes. **Structural unit of Kan** representing a fragment of suture zone of the Turkestan paleooceanic basin has a special position.

Southern Tien Shan ophiolite belt, stretching from northwest Uzbekistan to western regions of China, traces the suture of the Turkestan paleoocean, closed at the end of Carboniferous. In the western part, the age of volcanogenic-siliceous series of ophiolite association includes the interval from the Early Ordovician up to Early Carboniferous. Similar data about ophiolite age from the eastern regions are very poor; in this connection, the evolution of the Turkestan paleoocean between Kazakhstan and Tarim continental blocks remained open.

Structural unit of Taldyk (Kurenkov et al. 2002; Burtman 2006) is represented by metamorphosed ophiolites and overlying pelagic sediments. Structural unit of Taldyk is made up by structurally upper complexes of Alai microcontinent (structural unit of Abshir) and also composes some blocks in olistostromes. Rocks of layered, sheeted dyke and volcanic complexes belong to tholeiitic trend and were the derivates of picrite-basalt magmatic melt. The lower part of visible succession is formed by serpentinized rocks—harzburdite (predominates), peridotite, and lehrzolite (small amount). Layered complex (alternation of gabbros and peridotite) is typical for marginal suprasubduction basin or for island arc. Dykes are composed by picrite, tholeiitic and subalkaline basalts and hialoclastics. Widespread distribution of sprayed hialoclastics among the Devonian lavas evidences their formation at depth less than 1.5–2 km. Basalts of Sartale area erupted during interval from the Early Ordovician to Early Devonian (with large interruptions), of Khodzhaghoir and Kuroves areas—in Early and Middle Devonian, of Akbura area—in Middle Devonian, of Kirgiz-Ata area—in Early—Middle Devonian and Tournaisian. Greywacke sandstones intercalated with volcanic rocks in the upper succession contain debris of andesitic basalts, andesite, and dacites, which evidenced the existence of island arc.

Structural unit of Kan near the foot of Alai Range (Kurenkov 1983) is formed by serpentinite mélange. It includes serpentinites containing numerous blocks of ophiolite association rocks (pyroxenite, ophicalcite, gabbros, basalts and cherts), rhodingite, metamorphic rocks (altered to green schist and glaucophane schist) and also faintly metamorphized clastic, silicious, and carbonate rocks. For the most part, they were transformed into blastomylonite. Size of the largest blocks exceeds 1000 m. The Early–Middle Devonian and Famennian conodonts were determined from flint dynamoschist.

Serpentinite mélange of structural unite of Kan is covered with stratigraphic contact by serpentinite sandstones and another ophiolite-clastic sediments (gravelite, conglomerates, breccias) among which olistostromes are bedded. Serpentinite sandstones contain the Serpukhovian goniatites and foraminifers, and the Serpukhovian conodonts were determined from pelagic cherts among sandstones and ophioliteclastic breccias.

To the west of Talas-Fergana strike-slip, the Atbashi-Inyl'check oversthust zone is situated. The latter is formed by structural units of Kokkia, Chaturkul, Keltubeck, Atbashi, and Balykty, primary relationships between which were reworked by subsequent nappes and thrusts.

Shallow water carbonate deposits predominate in **structural unit of Kokkia**. Tectonic bottom boundary of structural unit migrates along thetectonic contact and is usually situated in the Lower Silurian—Middle Devonian part of succession. The base of unit (Ulan, Bordokoy, and Torugart ranges) is composed by sandstones and clayey slates with layers of basalts and limestone (1000 m), the latter with the Ludlow and Pridoli brachiopods and tabulata. Up-section, there is a thick carbonate series (up to 4000 m) with benthos fossils and foraminifers which are evidencing that the subface of series changes in limits from the Ludlow to Lochkovian and its roof—in limits from the Bashkirian to Lower Moscovian. Volcano-sedimentary rocks and basalts and andesitic basalts (100–1000 m) are bedded in the interval Middle—Upper Devonian among carbonates. Petrological and geochemical characteristics of basalts (Biske and Tabuns 1996) correspond to intraplate position of these volcanic displays, and their origin—to destruction of the continental crust. In the Lower Carboniferous part of succession, limestone includes layers of cherts. The whole succession is completed by carbonateterrigenous flysch with olistostromes. The youngest fossils of limestone olistoplacas have Bashkirian age, and flysch contains the Early Moscovian foraminifers.

Structural unit of Chaturkul is mostly composed of pelagic sediments—clayey and cherty slates with the Llandovery graptolites, flysch and carbonate-terrigenous sediments of Late Silurian age (1000 m). Overlying Devonian and Carboniferous sediments have a variable composition: carbonatecherty, volcano-cherty, and cherty types of successions are distinguished among them.

Structural unit of Keltubeck is formed by rocks of paleooceanic crust, partly metamorphosed up to green-schists. On the eastern slope of the Fergana Range, this unit represents a system of tectonic slices formed by mylonitized apo-harzburgite serpentinite, layered gabbro-amphibolite, basaltic rocks, and cherts. Limestone strata with the Early Devonian corals were found among lavas. To the west of northern slope of the Atbashi Range, unit represents the section of tectonic nappes and slices. Lower section is formed by serpentinite mélange with blocks of crystalline schist and gabbroid rocks, serpentinized pyroxenite and peridotite, gabbro and gabbro-amphibolite. Upsection, the above-mentioned complexes are succeeded by chert, tholeiitic basalts, ghialobasalts, and basaltic tuffs, metamorphosed to green-, and blue schist. Geochemical characteristics of basalts correspond to middle-ocean ridge basalts (Biske and Tabuns 1996). Chert and jasper of the siliceous-volcanogenic series of the ophiolite belt (Atbashi and Dzhangdzhir Ranges) contain conodonts found in 15 points (14—from chert, and 1—from jasper). They fix the following age levels D_1p–D_2, D_2gv, D_{2-3}, D_3f_1, D_3f_2, D_3fm_{1-2} (Alexeiev et al. 2007).

In area of the Dzhangdzhir Range, unit of Keltubeck is formed by tectonic nappes and slices, the total thickness of which reaches 3 km (northern slope of range). They are composed of ultramafic and volcanic rocks among which pillow-basalts are presented. Ultramafic rocks are represented as mélange with serpentinite matrix and blocks of serpentinized peridotite, gabbro, basalts, chert, green-schist, ophioliteclastic conglomerates, and breccias. Conodonts of the Middle and Late Devonian and radiolarians of the Tournaisian are determined in chert.

Structural unit of Atbashi is formed by crystalline schist, gneiss, garnet amphibolites, marbles, and glaucophane schist. At the first stage, metamorphism reached the epidote–amphibolite degree and then rocks were reworked by regressive metamorphism to green schist degree. Lens-like bodies of eclogite are situated among the metamorphosed rocks. Rb/Sr isochronous age of eclogite metamorphism determined by garnet, omphazite, phengite and rock has given 267 ± 5 Ma (Tangiry et al. 1995).

In Northern Nura-Tau, the Silurian rocks are intruded by large Koshrabad intrusion of granitesyenite, syenite and essexite with Rb/Sr isochronous age of 306 ± 4 Ma corresponding to the Late Carboniferous. These intrusive rocks belong to shoshonitelatite type (Dalimov et al. 1998). Thus, the Late Carboniferous granites fix the finishing time of the Variscan structure formation of the Southern Tien Shan.

In the easternmost Southern Tien Shan ophiolite tectonic belt (territory of China—Tonghuashan Mts.), plagiogranite is exposed at the top of ultramafic rocks. The concordant U/Pb age (406.5 ± 5.0 Ma) was obtained for plagiogranite by zircon dating as the Early Devonian (Huang et al. 2011).

2 Tectonics of the Pamirs

The Pamirs is a complex system of alternating blocks with continental crust mainly of the Archean-Paleoproterozoic age, divided by narrow zones of newly formed consolidated crust which has replaced paleooceanic structures of Paleotethys and Mesotethys. The general structure of the Pamirs is similar to transverse structure of the Qinghai-Tibetan Plateau in China, but more compressed. Besides, in the Pamirs there is no western continuation of the Yarlung Zangbo ophiolite zone (the Himalayas and southern edge of the Plateau)—the Shiok zone that is situated in India and Pakistan.

The northern restriction of the Pamirs is the Northern Pamirs deep overthrust (NP, Figs. 2 and 3) along which all Pamirs complexes, including newly formed

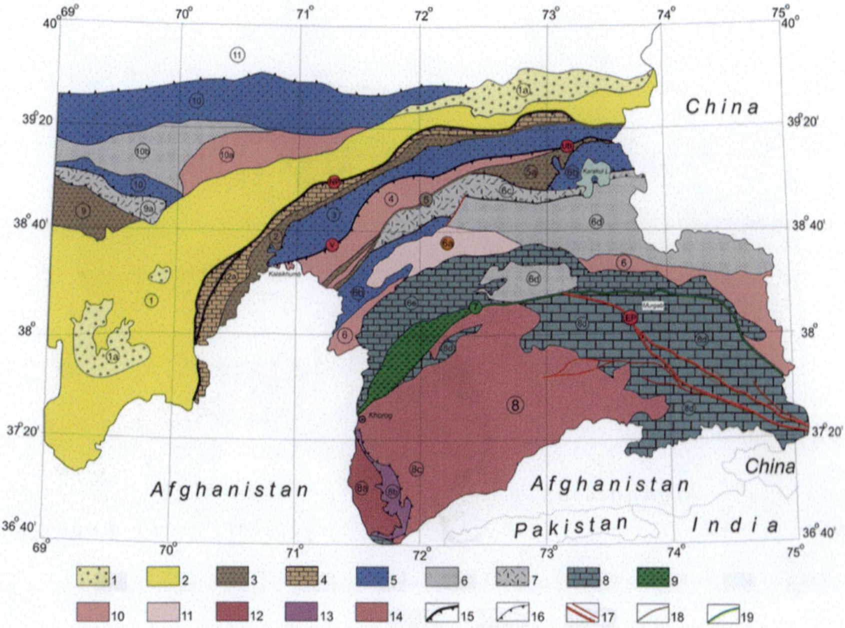

Fig. 2 Tectonic map of the Pamirs and Southern Tien Shan. 1—recent Quaternary molasses and loess; 2—Mesozoic-Cenozoic sedimentary cover of the Tajik depression; 3—oceanic and island arc complexes of the Paleotethysides (Northern Pamirs and South Hissar); 4—neoautochthon (C_2–P_3) of the Northern Pamirs Paleotethysides; 5—sedimentary cover (NP_3–D_3f) of Paleo-Gondwanaland; 6—complexes of the passive continental margin (C_1–P) of the Paleotethysides; 7—complexes of active Andean-type continental margin (C_2–P) of the Paleotethysides; 8 –sedimentary cover (C_1–Eo) of Paleo-Gondwanalnd; 9—metacomplexes (P–T) of Rushan-Pshart ophiolite zone; 10—continental blocks of Paleo-Gondwanaland (AR_3–PP_1) in the Southern Tien Shan, Northern and Central Pamirs; 11—Riphean (MP–NP_1) green schist sedimentary cover in the Central Pamirs; 12–14—Badakhshan block of Paleo-Gondwanaland: 12—Goran continental autochthonous block (AR_3), 13—Khorog ophiolite zone (AR_3–PP_1), 14—Shakhdara continental allochthonous block (AR_3–PP_1); 15—main overthrusts, 16—thrusts and overthrusts, 17—strike-slips, 18—ophiolite sutures of Paleotethysides, 19—ophiolite suture of Mesotethysides

consolidated crust of different age, are moved onto the deformed sedimentary cover of the Tajik depression (1, Figs. 2 and 3). The basement of the latter belongs to the Tajik-Tarim block with the Archean—Paleoproterozoic consolidated crust.

Nowadays, recent Quaternary molasse (1a, Fig. 2) continues to form on the deformed Mesozoic-Cenozoic cover of the Tajik depression.

In the Pamirs, following structures with consolidated crust of different age are identified from north to south:

- **Kalaikhumb-Sauksai ophiolite zone** with Carboniferous-Permian complexes and Late Permian—Early Triassic consolidated crust;
- **Kurgovat continental block** with the Paleoproterozoic crust;
- **Vanch-Karakul ophiolite zone** with the Early Triassic consolidated crust;

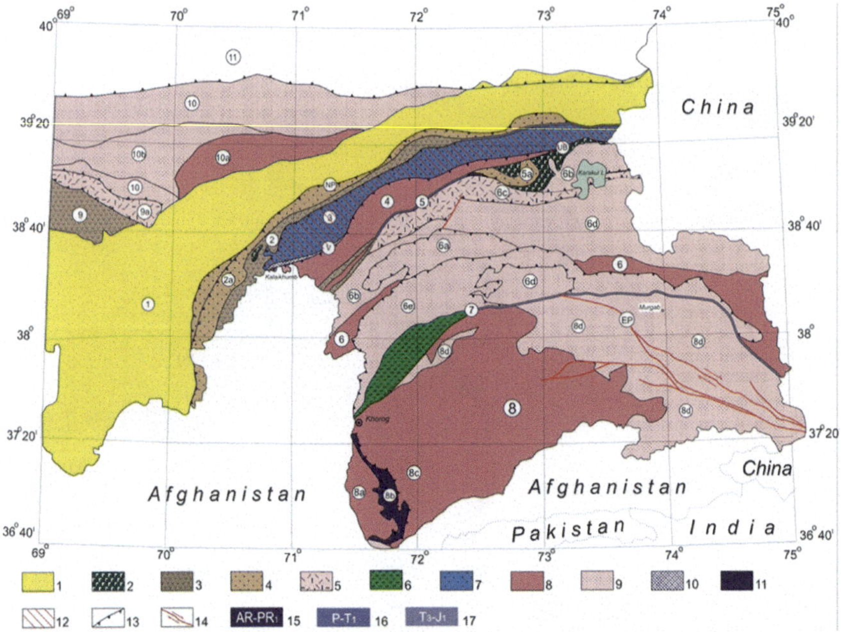

Fig. 3 Tectonic map of ophiolite zones and sutures of the Pamirs. 1—Mesozoic-Cenozoic sedimentary cover of the Tajik depression, 2—oceanic crust of the Northern Pamirs Paleotethysides, 3—island arcs complexes complexes of the Paleotethysides (Northern Pamirs and Southern Hissar), 4—sedimentary complexes (C_2–P_3) of the Northern Pamirs Paleotethysides, 5—complexes of active Andean-type continental margin, 6—metacomplexes (P–T) of Rushan-Pshart ophiolite zone, 7—allochthonous sedimentary cover (NP_3–D_3f) of the Northern Pamirs, 8—basement of Paleo-Gondwanaland blocks, 9—sedimentary cover of different ages on Gondwanaland blocks, 10—ultramafic part of ophiolites, 11—mafic–ultramafic section of ophiolites, 12—allochthon covered Paleotethyan complexes of the Northern Pamirs, 13—overthrusts and thrusts, 14—strike-slips; 15–17—ophiolite sutures: 15—Khorog ophiolite zone, 16—Northern Pamirs Paleotethyan sutures, 17—Rushan-Pshart suture

- **Central Pamirs continental block** with the Paleo-, Mesoproterozoic crust and deformed sedimentary covers of various ages;
- **Rushan-Pshart ophiolite zone** and suture with the crust generated up to the beginning of the Jurassic;
- **Badakhshan continental block** with the Neoarchean—Paleoproterozoic crust, covered with sediments of mainly Permian-Jurassic age in the south-eastern Pamirs.

2.1 *Northern Pamirs*

The general structure of the Northern Pamirs and its formations reflect all main stages of Paleotethys evolution, since its opening at the end of Devonian—beginning of Carboniferous to final continental collision in the early Triassic. Both ophiolite zones correlate with two adjacent Paleotethys basins, separated by a block with thick continental crust (Kurgovat microcontinent).

2.1.1 Kalaikhumb-Sauksai Ophiolite Zone

This zone clearly consists of two structural elements—Kalaikhumb autochthon and Viskharv allochthon. The autochthon represents formations and complexes of oceanic and island arc stages of Paleotethys development, and Viskharv allochthon includes formations of the platform (pre-riftogeneous) and continental riftogeneous stages preceding the Paleotethys opening.

Rock complexes of the **Kalaikhumb autochthon** (2, Figs. 2, 3 and 5) significantly differ along its strike and include the following elements: the section of paleooceanic crust with underlying serpentinite mélange; complexes of island arc system volcanic cones and intra-volcanic rifts; formations of the paleomarginal sea type. All these complexes occur at three chronostratigraphic levels, within limits of which several series of different composition facially replace each other (Fig. 4) (Pospelov 1987).

The lower chronostratigraphic level (C_1v_1–s_1) includes: (1) black spilite and tholeiitic pillow-lavas, (2) silt and clay-siliceous slates with basalt layers and diabase sills, (3) red-colored pillow-lavas and lava breccias with thin layers and lenses of limestones. First two series correlate well with the section of paleooceanic crust, and the third—with the basement of island arc volcanic uplift.

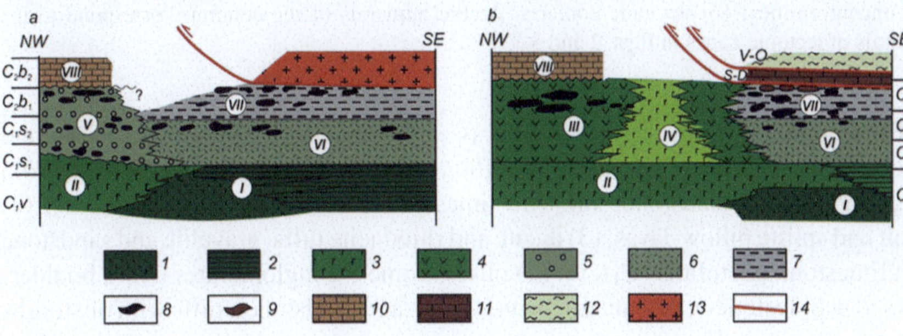

Fig. 4 Chronostratigraphic correlation of the lower carboniferous back-arc and island-arc complexes in the SW Darvaz Range (Northern Pamirs) (Pospelov 1987). 1—basalt, diabase, spilite; 2—siliceous tuffite, shales; 3—violet basalt and basaltic pillow-lavas and breccia; 4—andesitic basalt, andesite and tuffs; 5—conglomerates, gravelite, sandstones; 6—dacite, rhyolite and tuffs; 7—silt, siltstone; 8—olistoliths of Tournesian-Visean limestone; 9—olistoliths of Silurian-Devonian limestone; 10—Upper Carboniferous limestone of neoautochthone; 11—Silurian-Devonian limestone; 12—Vendian (NP_3)–Ordovician sandstone and green schist; 13—granite; 14—tectonic contacts. I—oceanic crust, II, III—island arc complexes, IV—intra-volcanic rift complex, V, VI—complexes of volcanic island slopes, VII—complexes of inter-island depression, VIII—neoautochthon (Middle Bashkirian and younger)

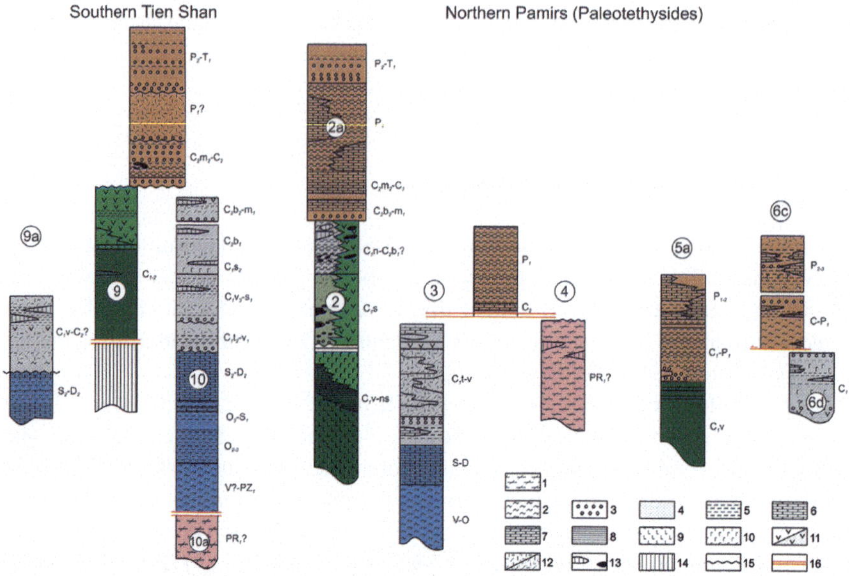

Fig. 5 Summary stratigraphic columns of the South Tien Shan and Northern Pamirs tectonic zones (Pospelov 1987). 1—crystal schist, amphibolites, gneiss and granite-gneiss; 2—phyllite, green schist, quartz-mica schist, quartzite; 3—conglomerates, pebble stones, boulder rocks; 4—gravelite, sandstone, siltstone; 5—mudstone, slate; 6—limestone, dolomite, marble; 7—sandy and clayey limestone; 8—chert, phthanite, siliceous tuffite; 9—diabase, tholeiitic basalt, spilite; 10—basaltic pillow-lavas and breccia; 11—andesitic basalt, andesite, tuffs; 12—dacite, rhiolite, tuffs; 13—limestone olistoliths (a—S-D, b—C_1); 14—sheeted dyke complex of diabase and gabbro-diabase; 15—unconformities; 16—tectonic contacts. Arabic numerals of the columns correspond to the numerals of tectonic zones in Figs. 2 and 3

The middle chronostratigraphic level (C_1s_{1-2}) includes: (1) andesitic basalt and andesite lavas and tuffs with layers of tuffite, volcanomictic sedimentary rocks from conglomerates to sandstone, silt with limestone olistostrome blocks (C_1t–v); (2) basalt and spilite pillow-lavas; (3) dacite and rhiodacite tuffs, gravelite and sandstone with limestone olistoliths (C_1t–v); (4) olistostrome—conglomerates (from boulder-rocks to pebblestones), volcanomictic gravelites and sandstones, tuffs with olistoliths limestone (C_1t–v). First two complexes correspond to volcanic rising of the island arc, and the others—to paleovolcanic slopes and depressions between them.

The upper chronostratigraphic level (C_1s_2–C_2b_1) unites two facies: (1) andesite and andesitic basalt tuffs, tuffites, volcanomictic sandstones and conglomerates with olistoliths of limestones (C_1t–v); (2) clayey-calcareous silt, siltstone, sandstone including olistoliths and olistoplacas of limestone of two ages (S-D and C_1t–v). The former forms deposits of volcano slopes and belongs to island arc uplift. The latter overlaps all rock complexes of the marginal sea structure being generated by overthrusting of the Viskharv allochthon northward.

Formations of oceanic stage of rather limited distribution are clearly identified in the Kalaikhumb autochthon. Formations of transitional (island arc) stages are widespread along the whole of the Northern Pamirs and allow detailed reconstruction of the ensimatic volcanic arc that had existed from the Late Visean to the Mid-Bashkirian. Numerous co-magmatic synchronous diorite and quartz diorite, granodiorite, plagiodiorite intrusives belong to island arc volcanic formations. As a rule, these intrusions belong to complexes of reconstructed central-type volcanic edifices.

All rock complexes in the northern part of the Kalaikhumb autochthon after formation of the trustfolded structure in the Mid-Bashkirian were overlaped by neoautochthon (C_2–P_3) (2a, Figs. 2, 3 and 5). Layered organogeneous limestone is recorded in the Carboniferous part of neoautochthon, and reef limestone, facially replaced northward by calcareous-terrigeneous sandstone and aleurolite of reef constructing slopes, in the Permian part.

The section of the **Viskharv allochthon** (3, Figs. 2, 3 and 5) consists of sandstone and sericite-chlorite schist (NP_3–O) which upward gives way to limestone and dolomite (S–D). These two series are considered as typical platform formations. They are characterized by consistency of sections and thicknesses (2500–3000 m) on wide areas, by the absence of volcanic rocks, velocity of sedimentation corresponding to platform regime, quartz and feldspar-quartz composition of terrigenous rocks. Similar sections are known in many areas of Central Asia: in the Southern Tien Shan in the Hissar and Zeravshan ridges, in the northern periphery of the Tajik depression, in Afghanistan and in the Central Pamirs (Figs. 2 and 5).

The Vendian (Ediacaran)—Devonian (NP_3–D) complex is covered with angular unconformity by slate, limestone, lavas of various composition and tuffs (C_1t–v). In a number of features, this rock assemblage is believed to be a riftogeneous (graben) formation: significant thickness (up to 400 m), lens-type alternation of sedimentary and volcanogenic rocks, contrast composition of effusives (alkaline basalt, andesite, rhyolite and trachyrhyolite). Numerous biotite and two-mica granites, leucocratic granite and aplite, which were co-magmatic to rhyolite and trachyrhyolite, intrude riftogeneous formation.

These formations originally formed on the continental crust of the Kurgovat block, and then in the middle(?) of the Bashkirian they were displaced from their basement and overthrust onto the complexes of oceanic and island arc stages of the Kalaikhumb autochthon. In turn, later the Viskharv allochthon was overthrusted by the Kurgovat block itself. After the process of stage-by-stage overthrusting, the formed thrust-folded structure was covered with carbonateterrigenous neoautochthon (C_2–P_1, see below).

2.1.2 Kurgovat Block

In western part, the Kurgovat block (4, Figs. 2, 3 and 5) is separated from the Kalaikhumb-Sauksai ophiolite zone by the Viskharv fault (V, Figs. 2 and 3) marked by lengthy and rather narrow dunite and peridotite protrusive bodies (V, Fig. 3). In eastern part its continuation is represented by the Uybulak fault (UB, Figs. 2 and 3).

The Kurgovat block is composed by metamorphic rocks: amphibole gneisses, crystal schists, amphibolites and marble (PR_1?), intruded by Proterozoic and Early

Carboniferous granite. The latter are directly connected with riftogeneous formation previous to Paleotethys opening. In recent structure, the Kurgovat block is overlapped by the slowly deformed sedimentary cover of limestone (C_2) and flyschoid sandstone with metamorphosed siltstone (P_1). Probably, originally this cover rested with sharp angular unconformity on only the Kurgovat block (Subdivisions... 1976), and later (during the Mesozoic), it has been stripped northward, overlapping the Kalaikhumb-Sauksai zone.

2.1.3 Vanch-Karakul Ophiolite Zone

Vanch-Karakul ophiolite zone (5, Figs. 2 and 3) is a structure that generated at the place of the southern branch of Paleotethys. In western and central parts, the zone is represented by narrow suture marked at the whole of strike by len-like bodies of metamorphosed garnet-bearing peridotite and pyroxenite, plagioclase peridotite, lherzolite, gabbronorite, serpentinite and serpentinite mélange. In the westernmost part, the suture is divides into two parallel branches separated in modern structure by lengthy and rather narrow (2 km) block of crystal schist, amphibole gneiss and marble (PR_1?). In the central part of zone, both sutures join in uniform zone, upthrust southward onto the northern border of the Central Pamirs. There, schistose and alterated diabase, basalt, spilite compose ophiolite zone, with participation of above-mentioned ultramafic rocks a swell as subalkaline basalt, their tuffs, quite often transformed into apovolcanic green schist.

In the east, the suture is exposed in the form of large ophiolite **Karakul allochthon** (5a, Figs. 2 and 3) replaced by overthrust gently sloping to the south in the Central Pamirs to Vendian(Ediacaran)-Ordovician terrigenous sedimentary cover or volcano-sedimentary section of the same age. In places, the gabbro-ultramafic complex is situated on the allochthon basis and is represented by successive section of the rocks outcropped in the western part of the allochthon—from peridotite and plagioclase peridotite, pyroxenite, lherzolite to layered gabbronorite. Monotonous thick section of diabase and basalt in places with pillow structure, with rare thin layers of siliceous tuffites (C_1v) is lying above the gabbro-ultramafic complex with tectonic contact at its subface (Fig. 6) (Pospelov 1987). Thick terrigenous section (C_1v_2–P_1) located above is composed by conglomerate, sandstone, siltstone and mudstone, less often limestone. Conglomerate-sandstone part of sedimentary section includes olistoliths of the Silurian-Devonian limestone, which is widespread over the area, including the Central Pamirs. The section is completed with shallow-water and reef limestone, various terrigenous rocks (P_{1-3}) of different facial mutual relations.

We consider the rocks of the Karakul allochotone basis as elements of Early Carboniferous oceanic crust, thrusted with fragments of the upper mantle according to tholeiitic composition of basaltdiabase complex, features of geological structure and structural position of mafic–ultramafic complex. Terrigenous-carbonate section accumulated on the oceanic crust was due to its sedimentological features either its cover (compensatory filling of paleooceanic depression), or passive continental

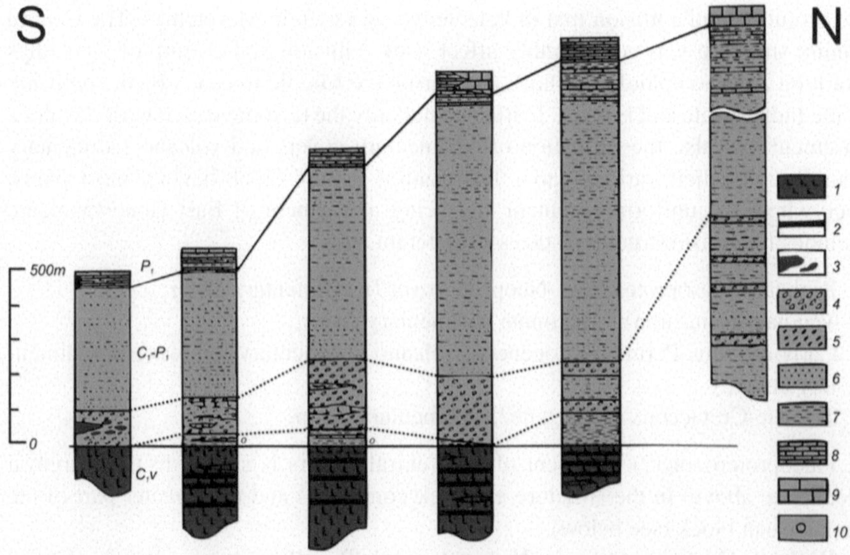

Fig. 6 Correlation of the Carboniferous—Lower Permian complexes in the Karakul ophiolite allochthon (Pospelov 1987). 1–2—Lower Carboniferous oceanic crust: 1—diabase, diabase porphyry, basalt, 2—chert, siliceous schist and tuffite; 3–9—sedimentary cover of oceanic crust: 3—olistoliths of Silurian-Devonian limestone, 4—conglomerate, 5—gravelite, 6—sandstones, siltstone, 7—mudstone, slate, 8—layered limestone, 9—reef limestones; 10—point of fossils in oceanic crust section

margin deposits. The change of proximal facies by distal flysch sediments, subarkosic composition of sandstones, and change in the thickness from south to north gives evidence of continental slope sedimentation (Fig. 6).

Karakul allochthon, as well as the adjoining part of the autochthonous Central Pamirs is intruded by quartz diorite, granodiorite, quartz monzonite and monzodiorite of Late Permian—Early Triassic age. The composition and time of formation of the complex is similar to more widespread North Kunlun batholith belt. This complex of collisional granite completed the formation of newly formed Triassic consolidated crust in the Northern Pamirs. Less common red continental molasse with pebbles and blocks of limestone, quartz diorite, granodiorite and granite (T_1) overlaps the collisional granite.

2.2 *Central Pamirs*

Essential influence of evolution of paleooceanic structures during the Late Paleozoic in the north and the Mesozoic in the south affected the general structure of the Central Pamirs (6, Fig. 2, 3). Located between the Northern Pamirs ophiolite zones and Rushan-Pshart ophiolite suture it regularly underwent structural reworking related to the evolution and collision first in Paleotethys and then in Mesotethys. The Central Pamirs structure was considerably affected by collision and closing of Neotethys (Yarlung Zangbo ophiolite suture), and during the Middle Eocene , by the collision of the Indian Plate and Eurasia. It affected not only the tectonic structure of the block basement, but also the formation of sedimentary covers and volcano-sedimentary complexes of their various parts. The Central Pamirs block having, most likely, everywhere the uniform basement and being a fragment of East Gondwanaland, includes four basic structural stages of different age:

- Riphean (Mesoproterozoic–Neoproterozoic?) sedimentary cover;
- Vendian (Ediacaran)—Devonian sedimentary cover;
- Carboniferous-Permian riftogeneous volcano-sedimentary complex and sedimentary cover;
- Triassic-Cretaceous (Paleogene?) sedimentary cover.

Paleoproterozoic(?) basement of the Central Pamirs is similar to the Kurgovat block (see above) in the structure and rock complexes and to a greater part of the Badakhshan block (see below).

Riphean (Mesoproterozoic–Neoproterozoic?) sedimentary cover (6a, Figs. 2 and 3) is conditionally determined from features of metamorphism, which are less pronounced than in the basement of the Central Pamirs, but more intensive than in slate and phyllite in the lower part of the Vendian (Ediacaran)—Devonian section (State Geological Map of USSR... 1983; Geological Map of the Tajik SSR... 1991). Till now, there are no data on the age of these rocks. Unlike metamorphic rocks of the basement, the Riphean(?) metamorphites are characterized by low degree of amphibolite facies and by associations mostly of green-schist facies, fast changes between various facies, and absence of migmatization. Quartzbiotite and quartz-sericite schists, quartzites, calcitic and dolomitic marbles, metasandstones, and conglomerates, sometimes metavolcanics (metabasalts) are typical of Riphean (Mesoproterozoic–Neoproterozoic?) sedimentary cover.

Vendian (Edicaran)—Devonian sedimentary cover (6b, Figs. 2 and 3) as was described earlier, is widely distributed from the Zeravshan Ridge in the north to the Rushan-Pshart suture in the south. It is characterized by consistency of facies and thicknesses that testify stable existence of the continental block before its disintegration (breaking-up) at the stage of Paleotethys formation. Sericite-chlorite and quartz-sericite schists combined with sandstone, siltstone and phyllite are common in the Vendian (Edicaran)—Ordovician part.

Carboniferous-Permian volcano-sedimentary complex (6c, Figs. 2 and 3). At least, three large lengthy outcrops of the Carboniferous-Permian rocks that have tectonic (allochthonous) relations are identified south of the Karakul allochthon. Two former complexes (C_1) are composed by schistose mafic effusive, diabase, metabasalt interstratifying with phyllite, sandstone and rare lenses of Visean limestone. In many places this complex is intruded by protrusions of serpentinized harzburgite and amphibolized pyroxenite. The third complex located north of the Karakul allochthon, in its lower part (C_1–P_2) is represented by various schists originated as a result of alteration from volcano-sedimentary rocks of mainly acid composition (quartz porphyry, felzite, their tuffs), interstratifying with gravelite, sandstone and limestone. Facially changeable series (P_3) of basaltoides, diabase, tuffs, volcanomictic sandstone and large bodies of reef limestone complete this section.

The structure formed by all the three types of sections represents a system of tectonic overthrusts and large folds with southern vergence (as in case of the Karakul allochthon). Rare serpentinite protrusions are situated along the fault planes. This complex is not adequately studied till the present. Therefore, our geodynamic characteristic of the complex is based only on the presence of bimodal volcanic rocks and elements of ultramafic–mafic section. In our opinion, these types of sections could be formed during the continental riftogeneous break up of the southern (Gondwanaland) margin of Paleotethys, where rifts did not transform in extensive oceanic basins. It is not improbable that there was a system of narrow ophiolite trough depressions filled with mafic volcanic rocks, and terrigenous sediments, and reef limestones simultaneously. The presence of subalkaline basalt and significant volumes of acid volcanic rocks testify in favour of continental riftogeneous origin, rather than oceanic one.

Carboniferous-Permian sedimentary cover (C– P) (6d, Figs. 2 and 3) has very wide distribution in northern part of the Central Pamirs. It is represented by monotonous section of metasandstone with rare layers of calcarenite, slate, phyllite and apobasaltic, apodiabasic metavolcanics. Its thickness is unknown because Carboniferous-Permian flysch forms a system of tectonic nappes, thrust slices and folds (mostly isoclinal) of southern vergence, conterminous with the vergence of Northern Pamirs structures.

Sedimentological features (domination of arkoses, low content of diabases), monotonous flyschoid types of texture evidence that this is an analog of typical bathyal slope sediments of passive continental margin. Its formation occurred along southern Gondwanaland margin of Paleotethys where slope sediments overlapped the shelf (NP_3–D) complex.

Triassic-Cretaceous (Paleogene?) sedimentary cover (6e, Figs. 2 and 3) begins with rhythmically alternating polymictic and quartz-feldspar sandstone, siltstone and carbonaceous mudstone (T_3–J_2). They are overlain by clay limestone, sandstone and slate ($J_{2\text{-}3}$). Conglomerate, sandstone and limestone (K_2) continue the growth of sedimentary section. First two series have some analogues in sedimentary cover of the Badakhshan block (see below). They can testify the accumulation of this sedimentary cover after the formation of the Rushan-Pshart ophiolite suture.

Paleogene(?) deposits complete the whole section. They lay with angular unconformity on Upper Cretaceous and more ancient rocks. Paleogene(?) deposits are either red-colored terrigenous sediments in eastern part of the Central Pamirs, or volcanic section of andesites, their tuffs, tuffites, sandstone, conglomerate and marl in western part.

All sediments of the Mesozoic cover take part in a system of tectonic sheets and slices of northward vergence, the rocks of underlying sedimentary complexes, and, first of all, Ordovician–Silurian terrigenous-carbonate sediments of Vendian (Edicaran)—Devonian section being also involved in a system of tectonic slices with recumbent folds. Thus, all sediments overlapping the Central Pamirs block form fold-thrust structures with opposite direction of vergence. This divergence reflects that processes of continental collision occurred differently in Paleotethys (Northern Pamirs) and Meso- and Neotethys (Rushan-Pshart and Shiok zones).

Central Pamirs includes several generations of granitoid magmatism. They are Proterozoic in Paleoproterozoic basement (similar to granite intrusions of the Kurgovat block); Early Triassic—connected with continental collision in the Northern Pamirs, Early Cretaceous—connected with collision in Mesotethys, and suprasubductional Early Paleogene(?)—connected with reduction and subduction in Neotethys.

2.2.1 Rushan-Pshart Ophiolite Zone

Rushan-Pshart suture (7, Figs. 2 and 3) separates the Central Pamirs and the Badakhshan block. In its western part the zone exists as a block with a width up to 20 km, and to the east it is transformed into a narrow suture zone covered with epicollisional sediments (K_2–N_1?).

All complexes of the Rushan-Pshart suture zone are profoundly altered and metamorphosed, that complicates determination of their formational characteristic and age. The oldest section in the Rushan-Pshart zone is the section of Lower—Upper Permian argillaceous and siliceous slates with thin layers of limestone and huge volumes of mafic effusive at discrete intervals of section (Pashkov and Shvolman 1979). Overlying Triassic(?) complexes consist of basalt, metasandstone, carbonaceous-argillaceous slate, marble's limestone. Younger sediments inside this zone are unknown. Permian–Triassic section is metamorphosed to the lower degree of amphibolite facies. Gabbro-ultramafic complex is unknown and only amphibolites along the fault, constraining the zone from the south, can be conditionally considered as upper mantle rocks (Pashkov and Shvolman 1979) (Fig. 7).

It is considered, that this section can reflect oceanic and island arc stages of the Mesotethys evolution, and this zone represents the newly formed Early Triassic consolidated crust (Tectonics of Northern Eurasia 1980).

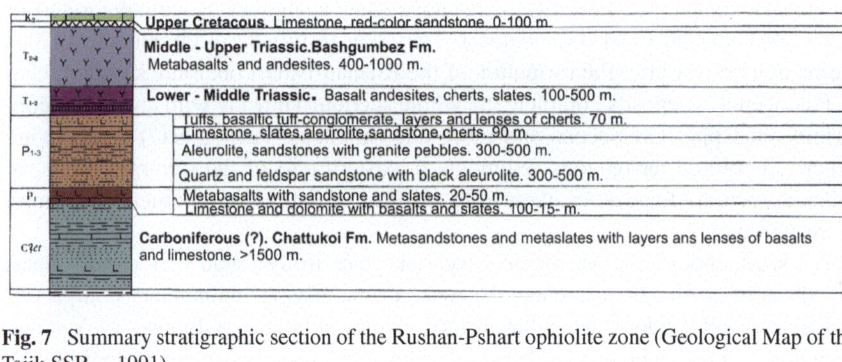

Fig. 7 Summary stratigraphic section of the Rushan-Pshart ophiolite zone (Geological Map of the Tajik SSR… 1991)

In the recent structure complexes of the Rushan-Pshart suture zone compose separate tectonic slices, in places thrusted northward onto the Central Pamirs block. The paleooceanic structure of Mesotethys existed in Late Paleozoic, probably during the Triassic; it was collisionally closed in the Late Triassic—Early Jurassic, and then in the Cenozoic, without formation of mountain system and continental molasses. I.e., similar to the Yushu-Jinshajiang ophiolite zone, the paleooceanic complexes were in general crushed (underthrusted) under colliding continental blocks. Most likely, this resulted in the formation of epicollisional molasse depression above the suture with accumulation of detrital clastic material transported from adjacent blocks—conglomerate, sandstone, mudstone (K_2), sandstone, gravelite, conglomerate (Pg? $+ N_1 + N_2$?).

2.3 *Southern Pamirs*

2.3.1 Badakhshan Block

Badakhshan block (8, Figs. 2 and 3) is located in the Southern Pamirs among Pamirs-Tibet blocks of Gondwanaland origin. In recent structure it is limited in the north by the Rushan-Pshart ophiolite zone and suture, and its southern border corresponds to the western continuation of the Yarlung Zangbo ophiolite zone—Shiok zone. In the west, the continuation of Badakhshan block is represented by the Kabul block in Afghanistan, and in the east—by the Karakorum block of the Western Tibet.

Archean—Paleoproterozoic basement of the Badakhshan block is exposed in its western part, and in eastern part it is overlapped by the Late Paleozoic—Mesozoic sedimentary cover.

The basement of the Badakhshan block consists of three main structural-formational complexes of Archean age (from west to east): Goran, Khorog and Shakhdara.

Goran complex (8a, Figs. 2 and 3) is composed by gneiss, quartzite, magnesite marble with lenses and layers of orthoand clinoamphibolite, eclogitelike rocks. Visible thickness is more than 2500 m. Age is determined as Archean with dating of 2.69 and 2.7 Ga (Subdivisions… 1976).

Khorog complex (8b, Figs. 2 and 3) with tectonic contact overlies the Goran complex. Gneiss, migmatite, amphibolite, eclogite with lenses and boudinage blocks of metagabbroids, metaultramafics and charnockites represent it (Fig. 8). Thickness along strike changes from 500 up to 1500 m. Absolute age determinations are 2.4, 2.46 and 2.54 Ga (Subdivisions… 1976).

Shakhdara complex (8c, Figs. 2 and 3) overlaps the Khorog complex with tectonic contact (metamorphosed mylonites) and is composed by gneiss, migmatite with marble horizons, amphibolite, and crystalline schist. Total thickness exceeds 6000 m. There are no definitions of absolute age. In opinion of some researchers, Shakhdara and Khorog complexes have certain similar features of structure and composition. A complex of primarily sedimentaryvolcanic rocks metamorphosed in

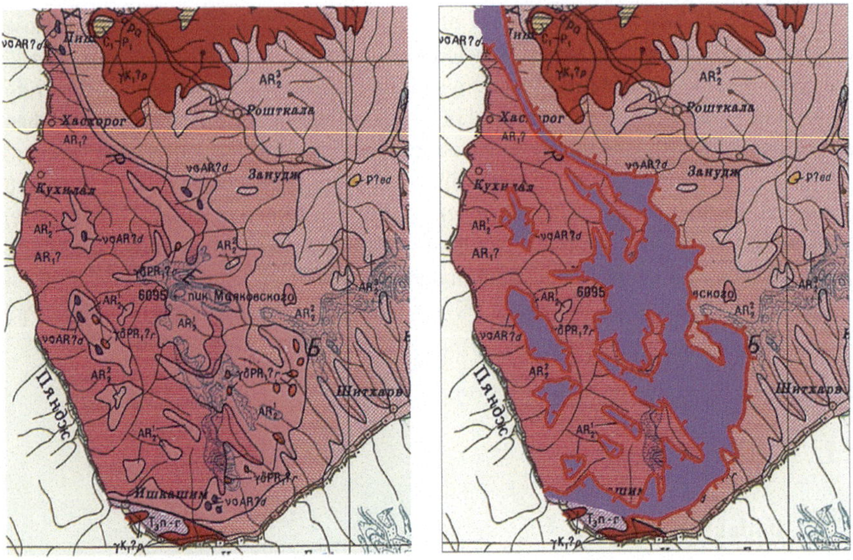

Fig. 8 Khorog ophiolite zone. Left—Geological map of the Badakhshan block, the SW Pamirs (State Geological Map of USSR… 1983, fragment); right—modern interpretation of structural position of the Khorog ophiolite zone (violet color—ophiolite zone, red lines—overthrust surfaces)

amphibolite facies lies above Shakhdara rocks. Their age is conditionally determined as Paleoproterozoic (PP_1?). The time of metamorphism is not determined. Several Riphean (MP–NP_2?) outcrops consist of green schist rocks, metasandstone, marble-type limestone and metabasalt. Khorog complex forms the tectonic plate (5–400 m) gently dipping eastward (Figs. 2, 3 and 8). Its rocks form a number of tectonic klippes above the Goran complex, or are exposed in separate tectonic windows from overlapping Shakhdara complex. The tectonic origin of the structure formed by the Khorog complex is a typical ophiolite zone (suture), connecting two blocks (probably, of the same age) with the continental type crust presumably during the Neoarchean—Paleoproterozoic. At present, the exact age of ophiolite suture and metamophism of its rocks is not determined. Archean ophiolites are represented by ultramafite, gabbro and gabbronorite. Thickness of ultramafic bodies varies from first meters to 30–80 m. They are dominated by metamorphosed garnet-bearing peridotite and pyroxenite, anorthosite peridotite, kortlandite, lherzolite and olivine gabbro-norite.

The first sedimentary cover later metamorphosed in amphibolite facies was accumulated after the collision of both continental blocks and formation of the Khorog ophiolite suture during the Paleoproterozoic.

Sedimentary cover of the Badakhshan block in the Southeastern Pamirs (8d, Figs. 2 and 3) in its lower part begins with low-thickness terrigenous rocks (C_1?–P_1). Above, it is composed by terrigenouscarbonate sediments with reef limestone (T_{1-2}) and thick mainly reef-carbonate bodies with horizons of terrigenous and volcano-sedimentary rocks (T_3–J_3) (Geological Map of the Tajik SSR… 1991).

Fold deformations in sedimentary cover are poorly manifested; mainly these are folds accompanying strike-slips with steep dipping axial surfaces. Eastern Pamirs dextral strike-slip is the main fault; it is accompanied by a number of smaller faults of varying amplitude and mainly of strike-slip kinematics (EP, Figs. 2 and 3). As a whole, the Eastern Pamirs dextral strike-slip is parallel to the Karakorum Fault in the West Tibet.

Intrusive complexes of the Badakhshan block base consist of Archean(?) (granites and plagiogranites) and Paleoproterozoic (granodiorites, quartz monzodiorites, granosyenite, charnockites, quartz diorites) rocks. The base and the sedimentary cover of the Badakhshan block are intruded by several phases of Early Cretaceous granitoids, among which biotitic and two-mica binary granites prevail, forming huge tabular intrusive bodies with an area up to 2700 km^2. Early Cretaceous granitic magmatism was connected with subductional and collisional processes that took place in time of Mesotethys closing and the formation of the Shiok ophiolite suture.

Thus, the Pamirs is a link between the Tibet and Kunlun in east and Afghanistan in west. Now geology and tectonics of the Qinghai-Tibetan Plateau in China is studied rather well. The regional study of geology in Afghanistan was ceased in 1980 (Geology and Mineral Resources of Afghanistan 2008). Now the comparison of geological structure and tectonic evolution of the Tethyan domain of Asia is possible by association of all geological information about East Afghanistan (Hindu Kush) and the Tibetan Plateau of China through the Pamirs which is the key for understanding of evolution of Tethyan paleooceans of the different age and growth of the Eurasian continent.

3 Turan Plate

Turan Plate envelops the territories of Turan lowland, Ustiurt plateau, Mangyshlack Peninsula, area of Aral Sea and adjacent areas of Turgai depression and Amu-Darya plain to the Fergana depression in the east (Fig. 9).

In the west, geological boundary of the Turan Plate passes in limits of the Caspian Sea aquatorium and has compound outlines (Atlas of the Lithology… 2002). Here, Turan Plate joints with the Scythian Plate forming united Scythian-Turan young epi-Hercynian platform. In tectonic map, western boundary of the Turan Plate is drawn by coast line of the Caspian Sea.

Turan Plate has been studied quite in detail by large-scale geological mapping, dense lines of regional and prospecting seismic profiles, by analysis of natural geophysical fields, by a huge volume of drilling works.

Like all plates, Turan Plate is formed by three structural stages:

- lower—crystalline folded basement;
- intermediate, or pre-plate (quasi-platform); its sediments are not included into the basement (although some researchers relate them to the basement);
- platform sedimentary cover.

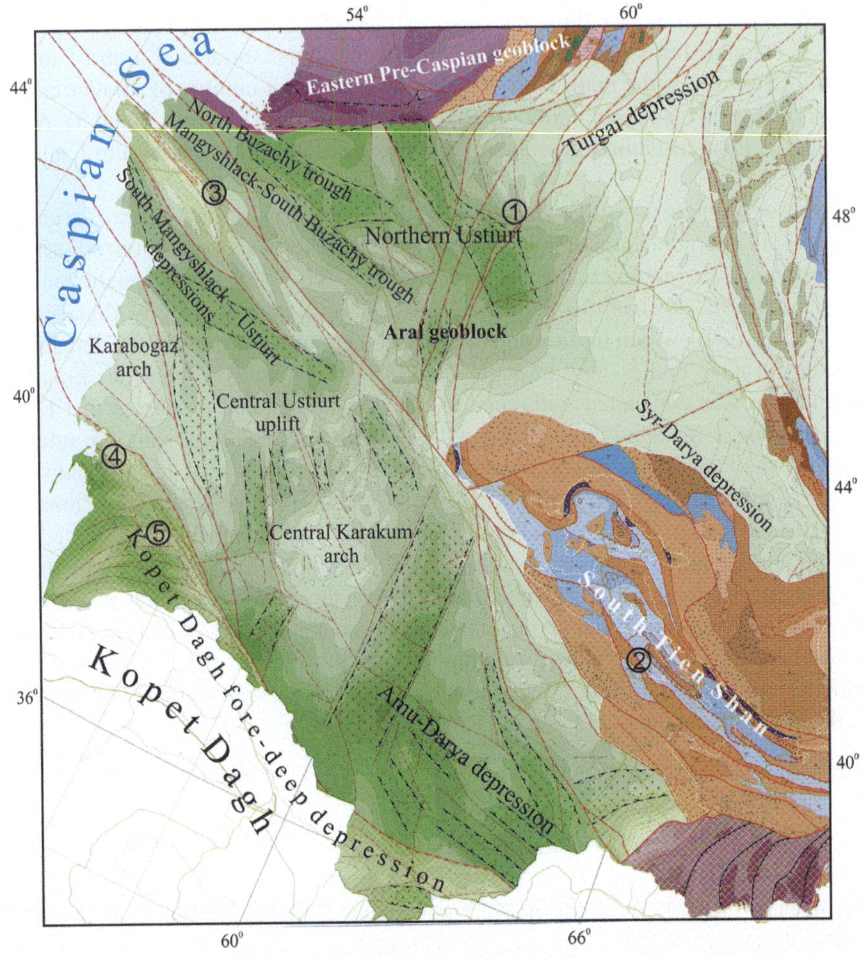

Fig. 9 Tectonic map of the Turan Plate (fragment of the Tectonic map of Northern-Central-Eastern Asia and Adjacent Areas, 1:2,500,000). 1—Akkulkov strike-slip, 2—Zeravshan fault, 3—Mangyshlack uplift and main fault system, 4—Bolshoy Balkhan, 5—Maly Balkhan

Turan Plate represents a recent intracontinental sedimentation basin inherited from the Early Mesozoic spacious depression in limits of which shelf basins of the northern continental margin of the Tethys Ocean (Para-Tethys) and superimposed on them Cenozoic basins of the fore-deep depressions of the Alpine collision fold belt were situated.

Modern conceptions on tectonic structure are reflected in published in the 70–80-ies tectonic maps of the USSR, Northern Eurasia and Europe compiled by large scientific teams (Tectonic map of Eurasia 1966; Tectonics of Northern Eurasia 1980; Tectonic map of Europe … 1979; Atlas of the Lithology… 2002). In all these maps, the northern part of the considered territory is related to the Scythian-Turan Plate in composition of young (epi-Variscan, or epi-Hercynian) Central Eurasian Platform.

Ancient massif of Northern Ustiurt is distinguished individually. Usually it is either included in the Ural-Mongolian belt or considered as block of the East European Platform. Uralian and situated to the west Donetsk Basin—South Emba faults are interpreted as a suture zone which marks collision margin of the East European paleocontinent and Paleozoic oceans: Uralian one in the east and Paleotethys in the south. Accordingly, the Scythian-Turan Plate basement is regarded as a collage of small continental blocks of ancient consolidation aggregated in various times to the East European Platform.

There are three main disputable problems of the plate deep geological structure: (1) age of basement of the Scythian-Turan Plate; (2) stratigraphic volume and rock complex composition of the Pre-Jurassic part of sedimentary cover succession; (3) genesis of contrastly expressed linear structures of the Scythian-Turan Plate.

In tectonic map, Turan Plate is shown as epi–Early Kimmerian plate, the sedimentary cover of which begins from the Permian deposits and enters into a composition of intermediate structural stage. Basement surface is a heterogeneous boundary of different ages. Mangyshlack—North Buzachy trough (with maximum depth of basement 8–13 km) and Tuarkyr (depth to 4 km) are distinguished in basement structure of the Turan Plate. Abrupt change along the Scythian-Sarmatian lineament influences not only the basement depth but also its velocity parameters. In area of maximum depths, the velocity of seismic waves reaches 6.5 km/s. In Mangyshlack sector, velocity has the average value of 5.7 km/s. It appears that velocities depend on the age of basement rocks. In Tuarkyr area, Paleozoic rocks enter into composition of basement. In Mangyshlack sector, the basement is covered by the Upper Carboniferous deposits.

Karabogaz arch is characterized by a sufficiently simple structure. In its limits, velocity parameters of consolidated crust are constant and reach 6–6.2 km/s characteristic for crust of the Pre-Paleozoic consolidation.

Drillings in Northern Karabogaz and in Peschanomyssky—Rakushechny uplifts revealed: (1) gneiss probably of Paleoproterozoic age; (2) Neoproterozoic crystalline and metamorphic schist; (3) granites with ages from 340 to 230 Ma.

Eastern Pre-Caspian—Turkmenistan area occupies the greater part of concerned structure. It consists of a system of isometric blocks divided by narrow trough-like depressions. Eastern Pre-Caspian, Ustiurt-Karakum, Amu-Darya, and Aral geoblocks are distinguished here. They are divided by South Emba (sublatitudinal) and Aral-Murgab (submeridional) troughs. In troughs, basement plunges down to depths of 12–14 km and has abnormally high velocities 6.3–6.5 km/s. Geoblocks are characterized by lower velocities in basement—6.0 to 6.2 km/s.

Age of basement rocks of the **Aral** and **Ustiurt-Karakum** geoblocks is determined by data of few drillings. Structures of the **Eastern Pre-Caspian** and **Amu-Darya** geoblocks could be interpreted by geological information from the adjacent Ural-Tau region (for the Eastern Pre-Caspian geoblock) and Hissar Range (for the Amu-Darya geoblock), where the top of basement approaches the Earth's surface.

Rocks no younger than the Riphean are present here in composition of the basement (Tectonics of Northern Eurasia 1980).

In **Ustiurt-Karakum** geoblock, rock complex of basement outcrops along the Central Ustiurt uplift zone and at Central Karakum arch. Crystalline and metamorphic schist found out by drillings are probably widely distributed in the geoblock basement. Their absolute age is 585 ± 25 Ma. Composition of basement also includes conglomerate-sandstone-slate series probably of Upper Proterozoic age and Silurian—Devonian granites (Karakum arch). Strongly metamorphosed carbonate-terrigenous rocks probably of Early Paleozoic age appear in section of basement near boundary with the Aral block (Knyazev et al. 1970; Kiriukhin 1974; Garetsky et al. 1971).

In **Aral** geoblock, basement is penetrated by drillings in limits of NW Pre-Aral area. Upper Proterozoic amphibolites, Upper Proterozoic crystalline and metamorphic schist, Lower Paleozoic effusiveslate deposits are established here.

Thus, analysis of geological data on stratigraphy of the Turam Plate basement makes it possible to distinguish the following types of blocks: of the Riphean (Baikalides), Paleozoic (Variscides), and Permian (Early Kimmerides) consolidation. Blocks of different ages grouped appropriately form: Ural—Kyzyl-Kum—Tien Shan Variscan belt, Caucasus—Kopet Dagh Late Kimmerian belt and spacious Scythian-Turan Baikalian area.

Boundary of Baikalian blocks of the Scythian-Turan area with Variscan blocks of the Ural—Tien Shan belt marked in the north by the Main Ural Fault displaced to the east by Akkulkov strike-slip (NW Pre-Aral area) and further passes along the Aral fault stretched to Sultanuiz-Dag. Further, this boundary turns to the east and passes along the Zeravshan fault into South Tien Shan. Aral block occupies a different position and probably has Late Caledonian (Early Devonian) age of consolidation.

In **South Emba area** Pre-Jirassic succession of the sedimentary cover begins from volcanogenicterrigenous Lower Paleozoic series (thickness up to 4000 m) overlain by the Devonian—Early Carboniferous terrigenous deposits (2000–4000 m). The latter in the north are covered by terrigenous (to 1000 m) and in the south by carbonate series (to 2500 m) of Visean—Early Permian age. In the south of area, succession includes a long-term discontinuity; and the uppermost Upper Permian—Lower Triassic (in some areas Jurassic) red-color terrigenous deposits overlie the Upper Visean—Early Permian carbonate rocks. In northern part of **South Emba area**, the stratigraphic succession is complete. Here, overlie Lower Permian terrigenous series underlie the Kungurian saliferous deposit; the latter in turn is covered by the Upper Permian—Triassic red-color terrigenous rocks. Upper Permian—Triassic terrigenous section is 0–2000 m thick, and saliferous—0 to 1500 m. In the Pre-Jurassic part of sedimentary cover succession, two structural–erosion surfaces are distinguished. One of them (Pre-Devonian) is observed everywhere, but the second one (Pre-Visean), only in the southern part of South Emba area.

3.1 Pre-Jurassic Intermediate Complex of Sedimentary Cover

In **Ural—Kyzyl-Kum** area, the Pre-Jurassic complex is distributed fragmentary filling separated graben-like depressions. Usually the succession begins from the Carboniferous mainly carbonate and terrigenous rocks and finishes by the Upper Permian—Triassic terrigenous and volcanogenic-terrigenous series. The latter overlies on underlying deposits with a sharp angular unconformity which divides the section into two structural stages: lower—folded and upper—weakly deformed. Thickness of deposits of each stage reaches 2000–5000 m.

In **Western Turan (Ustiurt—Amu-Darya)** area in the Pre-Jurassic part of sedimentary cover, the most ancient deposits fixed by drillings are the Upper Devonian terrigenous-carbonate rocks. Here, boreholes penetrate mainly Lower–Upper Carboniferous carbonate and clayey-carbonate deposits, Upper Carboniferous—Lower Permian effusive and volcanogenic-sedimentary series and also Upper Permian—Triassic redand multi-color terrigenous, sometimes volcanogenic-terrigenous deposits (Bakirov et al. 1970; Garetsky et al. 1971; Knyazev et al. 1967; Knyazev et al. 1970; Kunin 1974; Shlesinger 1974). Two structural stages are distinguished by deformation characteristics and bedding conditions: lower—Paleozoic deformed practically everywhere, and upper—Upper Permian, weakly deformed. Structural boundary between these stages is well determined by the reflected wave method as a regionally distributed surface of structuralerosion unconformity (Volozh et al. 1981). Maximum thicknesses are determined both for the lower (to 4000 m) and upper (to 5000 m) stages. Thickness changing of both stages correlates with the basement relief. Zones of maximum thickness are connected with down warping of basement, and of minimum thickness—with its uplifts.

Thus, sedimentary cover of the Turan Plate and inter-mountain depressions of Central Asia consists of the Devonian—Triassic pre-plate (intermediate) complex and Jurassic (Middle Jurassic)—Quaternary plate complex, and also Upper Oligocene—Quaternary orogenic complex.

Distribution in the Turan Plate of the Pre-Upper Permian deposits of intermediate complex, throughout a long time of their formation (Devonian—Early Permian), is limited by areas of the Early Baikalian and Caledonian consolidation of basement and its down warping (Ustiurt, Chu-Sarysu and Turgay depressions). Upper Permian—Triassic intermediate deposits distributed more widely are situated on the Variscan basement and separated by depressions. Orogenic rock assemblages predominate over paraplatform assemblages and volcanic rocks in composition of the intermediate (pre-plate) complex. These assemblages are characterized by a sharp mutability of composition and thickness and discontinuity in their sedimentation. Terrigenous deposits prevail.

Carbonate rocks are typical for the Upper Devonian—Lower Carboniferous paraplatform rock assemblage; they are presented in subordinate amount in the Lower Permian and Middle Triassic. Devonian, Upper Carboniferous, Permian and Lower-Middle Triassic deposits have red color of rocks. Devonian and Permian sections contain evaporates.

Thickness of separated complexes reaches the first km. However, total thickness of the intermediate (pre-plate) complex rarely exceeds 5000–6000 m because of intermittent distribution and wide presence of unconformities.

3.2 Plate Complex

Plate complex composes the main volume of sedimentary cover of the Turan Plate. Its deposits also take part in filling of inter-mountain depressions and downwarpings of epi-platform orogenic areas in the east of Central Asia. Formation of plate complex began in the Jurassic. Solid sedimentary cover had been formed in western and southern parts of the Turan Plate to the end of Middle Jurassic. In its northeast part (Syr-Darya Chu-Sarysu depressions), formation of the solid sedimentary cover started in the second half of Early–Late Cretaceous. Upper Cretaceous and Paleogene deposits are mostly distributed in time when spacious areas of recent uplifts of Kyzyl-Kum and Tien Shan were involved in processes of downwarping and sedimentation. Starting from the Late Oligocene, the area of plate complex sedimentation progressively decreased due to the growth of Tien Shan, Northern Pamirs and Kopet Dagh mountain ranges, formation of molasses at their peripheries and in depressions. Middle Pliocene—Quaternary deposits are minimally distributed in the plate complex section. That time, the largest territory of the Turan Plate was involved in processes of uplift and erosions.

Marine and lagoon sediments of epi-continental basin predominate in the plate complex composition. Carbonates, carbonate-terrigenous, and terrigenous series, changing each other either in successions or in areas, are distributed approximately in equal degrees. Maximum of carbonate sedimentation fell within the Late Jurassic, Neokomian, and Late Cretaceous. Epochs of carbonate sedimentation also correspond to that of saliferous accumulation. Salt deposits are widely distributed among the Upper Jurassic and Neokomian rocks.

Continental sedimentation predominated in the Early Jurassic and in the first half of Middle Jurassic when coal-bearing terrigenous series were formed. Hauterivian—Lower Barremian multi-color series of alluvial genesis is widely distributed. In the eastern part of Turan Plate, the Upper Jurassic and Early Cretaceous continental basin (subaeral) and alluvium red-color rocks are distributed (They can be determined as sedimentary wedges of an orogenic complex of the Late Mesozoides of Central Asia). Jurassic, Cretaceous, and Paleogene (without Upper Oligocene) series of plate complex are usually similar by their composition to more thick complexes of Kopet Dagh, Bolshoy and Maly Balkhan. Upper Oligocene—Quaternary series are represented by orogenic molasses situated around the epi-platform orogen of Tien Shan and Northern Pamirs and also deformed sedimentary cover complex in Kopet Dag. Everywhere they represent the molasses and only in limits of recent Turan Plate they are included into the plate complex.

Thickness of plate complex reaches 6000–7000 m (Amu-Darya syncline), 5000 m (Ustiurt syncline and South Mangyshlack—Ustiurt system of depressions), 2000–3000 m (Syr-Darya and Chu-Sarysu synclines), and 3000–4000 m (Turkmenistan anticline). About one third of thickness of plate complex in the deepest Amu-Darya, Ustiurt synclines and South Mangyshlack—Ustiurt system of depressions belongs to the Jurassic deposits.

In depressions of the east of Central Asia, the orogenic complex is represented by continental aerial molasses. The latter is characterized by a sharp change of thickness and composition depending on the remoteness from sources of detrital drift, by discontinuity of sedimentation, presence of numerous disconformities and erosion cuttings. At the edges of depressions, coarse-detrital deposits prevail, but in the central part of depressions they associate with lacustrine thin-detrital and saliferous deposits. Usually, the thickness of orogenic complex reaches here 5000–6000 m and more.

In Western Turkmenistan depression, sandy-clayey deposits of intra-continental basin predominate and are characterized by changing of salinity. Along the Pre-Kopet Dag fore-deep depression from west to east and from below upward the succession, basin deposits are gradually replaced by alluvium, and at Kopet Dagh foothills—by coarse-detrital alluvium and proluvium ones. Simultaneously, deposit thicknesses decrease to 3500–4000 m and numerous discontinuities and unconformities appear the in succession. Author greatly appreciates Dr. M. Antipov from the Geological Institute of the Russian Academy of Sciences for his help in writing this chapter.

4 Major Tectonic Features of Northern China and Adjacent Areas

Northern China and adjacent areas refer to the Chinese part north of the Kunlun-Qinling-Dabie tectonic belt and related regions of the adjacent countries. For the conciseness of the text, we preliminarily give tectonic partition according to the formation time of the crustal consolidation and plate complexes (Fig. 10).

4.1 Structures with Crustal Consolidation at the Initial Stage of Mesoproterozoic

4.1.1 North China (Sino-Korean) Platform

Northern China area north of the Qilian-Qinling-Dabie tectonic belt, including the Bohai Bay and Yellow Sea regions and the northern part of the Korean Peninsula, as a whole takes the shape of a triangle. Bounded by deep faults with neighboring tectonic elements, the Sino-Korean platform is one of the oldest platforms in the

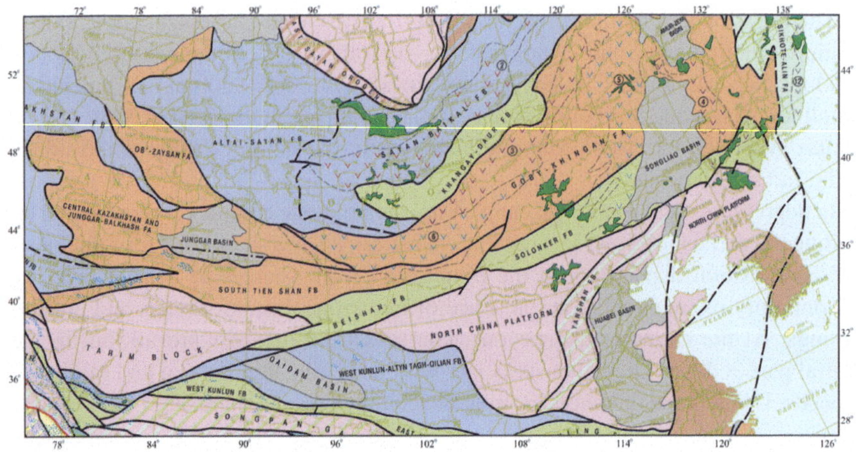

Fig. 10 Tectonic partition of North China and adjacent areas (fragment of the Map of Tectonic Zoning, Fig. 1). FA—fold area, FB—fold belt

world (Fig. 11). Its basement experienced several stages of cratonization, they are mainly three stages of Paleoarchean (3800—3000 Ma), Neoarchean (3000–2400 Ma) and Paleoproterozoic (2400–1800 Ma). As to its tectonic evolution, the Chinese part of the platform (the North China Platform) is emphasized here.

(1) ***Paleoarchean stage*** is actually the stage of Qianxi cycle and older period. In the Anshan region, Liaoning Province, the U–Pb zircon ages of 3804 and 3600 Ma

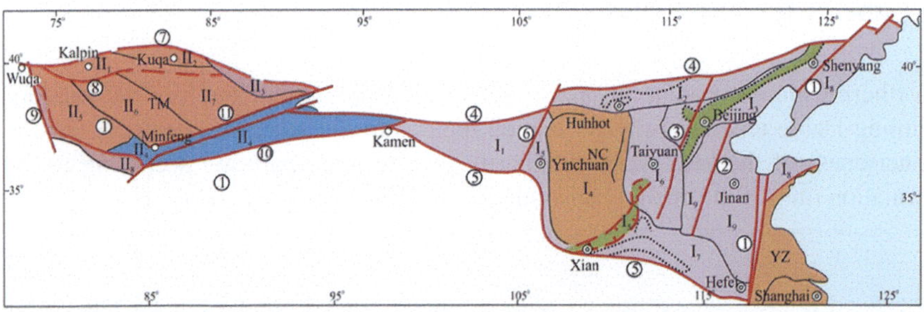

Fig. 11 Sketch map of the Tarim block and North China Platform (Cheng et al. 1994). I_1—Alxa block; I_2—Northern marginal uplift of the North China Platform; I_3—Yanliao aulacogen; I_4—Ordos Basin; I_5—marginal graben of the Ordos Basin; I_6—Shanxi uplift; I_7—southern marginal aulacogen of the North China Platform; I_8—Jiaoliao uplift, I_9—Luhuai fault uplift (Meso-Cenozoic North China–Hehuai Basin); II_1—Kalpin uplift; II_2—Kuqa depression; II_3—Kuruk Tag uplift; II_4—Altun-Dunhuang uplift; II_5—Southwestern Tarim depression; II_6—Central Tarim uplift; II_7—Northeastern Tarim depression; II_8—Tikanlik uplift. Faults: 1—Tancheng-Lujiang fault, 2—Cangzhou fault; 3—Taihangshan fault; 4—northern marginal fault of the North China Platform; 5—northern marginal fault of the Qilian-Qinling; 6—Langshan–Zhun Bayan fault, 7—southern marginal fault of the Tien Shan; 8—Bachu fault; 9—Talas-Fergana strike-slip fault; 10—Altyn Tagh strike-slip fault; 11—Minfeng fault

were obtained from granite (Liu et al. 1991), and the Chentaigou granite that intruded the Chentaigou supracrustal rocks gives the zircon U–Pb ages of 3300–3100 Ma (Song et al. 1994). In the eastern Hebei Province, fuschite-bearing sillimanite-quartzite of the Caozhuang complex gives the detrital zircon ages of 3550–3850 Ma (Wu et al. 1991; Wang et al. 1994). While the tonalite that emplaced in the complex has the zircon U–Pb age of 3200–3300 Ma (Wu et al. 1991). The data above show that the supracrustal rocks older than 3.8 Ga exist in the North China Platform and multiphase magmatism activity occurred in the place. Shen Qihan et al. in 1995 discerned three major tectono-magmatic events of circa 3800, 3000, and 3200 Ma in the platform.

Biotite leptynite of the supracrustal rocks at Shuichang, eastern Hebei Province has the concordant U–Pb zircon age of 3047 Ma, while the K-feldspar granite intruding the rock gives the U–Pb ages of 2960 and 2980 Ma (Wang et al. 1994). In Anshan region, the eastern and western granites and the Gongchangling granite have the U–Pb zircon ages of 2990 and 3000 Ma (Wu et al. 1991). Therefore, it can be deduced that the formation time of the nucleus of the North China Platform was in the period from 3.8 to 3.0 Ga.

(2) ***Neoarchean stage*** is actually the so-called Fuping cycle. The Anshan Group of the stage is a series of metamorphosed banded iron ore formation (BIF). The lower part are the hypersthene-feldspar granulite, gneiss, granulite interbedded with magnetite quartzite and are characterized by biotite-, and hornblende-rich rocks. The upper part is dominated by the granulitite and schists and is interbedded with multilayers of magnetite quartzite. In the protolith formation we can distinguish the ultramafic–mafic-intermediate-acid volcanic rocks (mafic rocks are dominant) interleaved with pyroclastics and magnetite quartzites. Other correlated metamorphic sequences to the Anshan Group are the Qingyuan Group Complex in the northern Liaoning Province, the Jiapigou Group in the southern Jilin Province, Zunhua Group in the eastern Hebei Province, Taishan Group in the western Shandong Province, the Fuping Group in the Taihangshan Mountains and the Dengfeng and Taihua Groups in the Henan and Shaanxi Provinces. All the groups give the isotopic ages between 2800 and 2600 Ma.

Wutai Group is mainly distributed in the Wutai and Hengshan Mountains and in the Fuping region of Hebei Province. The group is in unconformable contacts both with the lower Fuping Group and the overlying Hutuo Group of Paleoproterozoic. The Wutai Group consists of biotite-, and hornblende-rich granulites and schists, magnetite quartzite and minor carbonate rocks, and locally strongly migmatitized. In the Wutai Group and correlative sequence the isotopic ages range between 1900 and 2400 Ma, with the maximum not exceeding 2600 Ma, so the time of the Wutai Group is still controversial. Considering that the widespread occurrence of potassic granites in the North China Platform concentrates in the period of 2500–2350 Ma (Wu et al. 1991), it can de reasoned that the Fuping cycle (possibly including part of the Wutai Group) was the critical cratonization stage of the North China Craton and the North China crystalline basement was essentially formed after the stage.

(3) ***Paleoproterozoic stage*** or the Zhongtiao cycle stage. The Paleoproterozoic of the North China craton is characterized by the Hutuo Group in the Wutaishan Mountains, the Zhongtiao Group in the Zhongtiaoshan Mountains, the Liaohr Group in the eastern Liaoning Province, and the Fenzishan Group in the eastern Shandong Province. Both the top and the bottom of the sequence are in unconformable contacts with the overlying and underlying strata, respectively, and the isotopic ages show the range of 2300–1800 Ma. The full-scale and ultimate cratonization of the North China Craton took place at this stage and finally the folded basement of the craton formed.

Stable cover sequences of the craton were developed from the Mesoproterozoic to Middle Triassic. But the evolution of the craton had been affected by the adjacent two mobile belts in the south and north. Accompanying the development of the adjacent southern and northern oceans, the margin of the craton began to extend and the epeiric aulacogens and the marginal mobile zones accompanying orogens were formed. Since the Mesozoic, the craton has been influenced by the activity of both the Tethyan and Pacific tectonic domains and the E–W striking tectono-magmatic belt in the southern margin and the N–E striking tectono-magmatic belt in the eastern margin superposed on the craton. According to the tectonic features of different parts of the craton, nine second-class tectonic units in all can be distinguished in the craton (Fig. 11).

I_1—*Alxa block* is situated in the westernmost of the North China Platform and is neighboring to the Qilian Mountains in the south and to the Beishan.

Mountains in the north, respectively. The block consists of the crystalline basement of Archean and Paleoproterozoic, and low-grade metamorphosed folded basement of clastic-carbonate-volcanic rocks. The Paleozoic sequence is absent and the Mesozoic is of terrestrial basin sediments. Calc-alkaline granites of the Meso-Neoproterozoic, which bear some features of both the Tarim and North China platforms, are surrounding the block.

I_2—*Northern marginal uplift* of the North China Platform is dominated by the Archean metamorphic rocks. The uplifting state existed since the Paleoproterozoic. The Langshan-Zaltay-Bayan-Obo aulacogen of the Meso-Neoproterozoic is in the northwestern part of the uplifting region. With developed E-W striking structures, the uplifting zones have experienced multiphase magmatic activity, such as the Early and Late Paleozoic, and also Early and Late Mesozoic, with stronger Late Paleozoic and Late Mesozoic activity.

I_3—*Yan-Liao rift* of the Mesoproterozoic was depressed centering at Jixian-Lingyuan area, where a thick sequence of Changcheng System (including potassic mafic-intermediate volcanic rocks) and Jixian system deposited. Circa 1000 Ma, the region rose as a land and then subsided and deposited the Qingbaokou System of Neoproterozoic. At 850 Ma, the whole region uplifted as a land. The area had undergone tectonic movements of the Phanerozoic; the Yanshanian movement of Late Mesozoic was the strongest.

I_4—*Ordos basin* is located in the western part of the North China Platform. It is an inherited basin and has hardly experienced deformation. Stable sediments formed since the Cambrian, especially the Mesozoic strata are completely developed. With the total thickness over 6000 m, all the strata are of fluvial-lacustrine and marsh sediments.

I_5—*Marginal grabens of the Ordos basin*, such as the N-S striking Yinchuan graben and S-type Fenwei graben are formed as a result of the Cenozoic extension in the western and southern margins of the Ordos basin. The Fenwei graben is mostly developed along valleys and runs through the Shanxi Uplift and occurs as echelon depressions in tile northeast, which intermittently extends to Datong and Huailai counties, where are accompanied by volcanic activity of the Paleogene and Pleistocene and earthquakes have been frequently manifested in recent years. In the west, the Yinchuan graben can be affected by the uplifting of the Qinghai-Tibetan Plateau. The compression from the plateau results in some thrusts.

I_6—*Shanxi uplift* is situated to the east of the Ordos Basin, and crystalline and folded basements of the Archean and Paleoproterozoic are widely distributed. The cover sequence is dominated by the Paleozoic, especially by the Carboniferous—Permian, and well developed; they are the important coal measures of the region. Uplifting occurred since the Mesozoic.

I_7—*Rift* at the southern margin of the North China Platform approximately takes the NWW orientation. The middle and upper Xiong-er Group of the Early Mesoproterozoic are composed by mafic to acid volcanic rocks, potassic basalt and rhyolite of alkaline volcanic rock series, suggesting the extensional setting of the continent. The rift closed and folded at the end of Mesoproterozoic and finally consolidated in the Jinning tectonic stage between 1000 and 800 Ma.

I_8—*Jiao-Liao uplift* presents in the northeastern part of the platform. The Archean and Paleoproterozoic are widely exposed in eastern Liaoning Province. The Mesoproterozoic is absent in most part of the region. The Neoproterozoic is represented by flysch formations and the Paleozoic is composed by the stable cover sequence of marine and marine-terrestrial facies. Mesozoic intrusions are widely distributed. In the Changbaishan Mountains, the Quaternary basalt flows are splendid.

I_9—*Lu-Huai uplift* is located in the eastern part of the platform. The basement is of Archean and Paleoproterozoic age, with the Archean cropping out at northern and southern ends of the uplift and the Mesoproterozoic is absent. While the Paleozoic is widely distributed and fault depression basins were developed since the Mesozoic to Cenozoic. Structures are dominated by the extensional faults of NNE and NWW orientation and turned to compression in the late stage.

4.1.2 Tarim Block

This block is nearly completely covered with the Mesozoic and Cenozoic basin sediments. Only in Keping, Kuruk Tag, southern Tarim, Tikanlik and Altun (*Altun* is the name of mountain massif, but *Altyn Tagh* is used for geological terms) uplifts of the Precambrian metamorphic basement and the overlying stable cover sequence are exposed (Fig. 12).

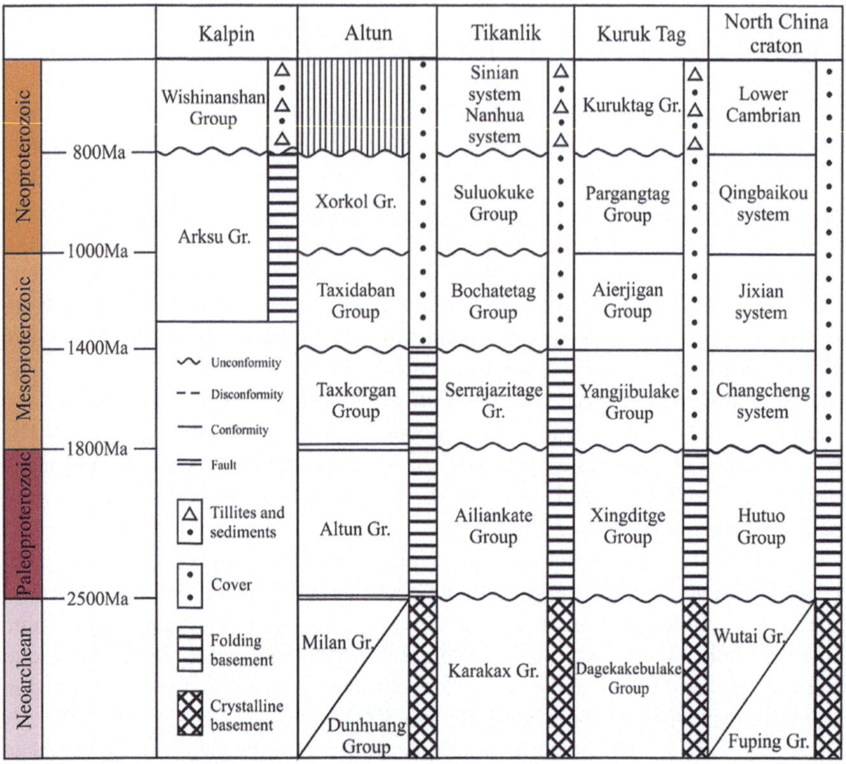

Fig. 12 Stratigraphical subdivision of the basement of the Tarim block and its correlation to that of the North China block

As the influence of adjacent orogenesis, the surrounding above mentioned uplifts were evolved in various ways and each has its own Precambrian basement and the corresponding cover sequence. Compared with major blocks, the uplifts were more mobile and were deformed in more complicated way.

*Kuruk Tag Uplift (**II**$_3$)*

The oldest sequence exposed in the area is Dagelakebulake Group and is dominated by stripped migmatite and plagio-biotite-hornblende gneiss. As the age data of 3262 ± 129 Ma (Hu et al. 1992) and 2582 Ma (Gao et al. 1993), it can be considered that the group is mainly of Neoarchean and the lower part may be older. The Xingditage Group of Paleoproterozoic is dominantly distributed along the southern slope of the Kuruk Tag Ridge. Uncomfortably overlaid by the Yangjibulake Group of Mesoproterozoic, the Xingditage Group consists of schists, with the bottom magnetite-bearing quartzite, lower part clastic rocks and the upper clastic rock interbedded with marble, suggesting the terrestrial clastic-carbonate rock formation. Migmatite granite, granodiorite, amphibolite, and quartz diorite have the U–Pb zircon ages of 1972 Ma, 1920 Ma, 2300–2100 Ma and 1800 Ma, respectively (Cheng et al. 1994).

The overlying strata of Mesoproterozoic to Early Neoproterozoic are a series of stable sequence with slightly deformation and metamorphism and can be taken as the earliest cover over the basement which is further subclassified into the lower crystalline basement and the upper folding basement (Fig. 12).

*Tikanlik Uplift (**II_8**)*

Situated on the southern margin of the Tarim block and adjacent to the Western Kunlun fold belt, the metamorphic basement of the Tikanlik uplift consists of the Neoarchean Karakax Group, Paleoproterozoic Ailiankate Group, and Early Mesoproterozoic Serrajazitage Group. The Karakax Group is a series of middle-high grade metamorphic rocks of migmatites and gneisses. The Ailiankate Group is a series of lower-middle grade schists of greenschist facies. Though there is no direct contact between the two Groups, they both are overlain with unconformities by the Serrajazitage Group which includes schist, slate, and phyllite of slightly deformed and metamorphosed rocks, with their protoliths of marine mafic-acid volcanic rocks, pyroclastics, and dolomite. The keratophyre of the upper part gives the Rb–Sr whole-rock age of 1764 Ma (Jiang 1996). Therefore, the Neoarchean Karakax Group may be the crystalline basement, while the Paleoproterozoic Ailiankate Group and Early Mesoproterozoic Serrajazitage Group are the cover sequence of the block (Fig. 12).

*Altun-Dunhuang uplift (**II_4**)*

The Milan Group on the northern slope of the Altun is composed of leptynite, gneiss, and hypersthene granulite. The U–Pb age of the group is 2462 Ma (Cheng et al. 1994). The Dunhuang Group in the Dunhuang area consists of granulite and migmatite. Gneiss and migmatite at both eastern and western sides of the Hongliuxia Canyon give the hornblende Ar–Ar age of 2936 Ma (Li 1992) and here it is tentatively taken as Neoarchean. The Altyn Tagh Group is focusly distributed in the western segment of the Altun Mountains and is with fault contact to the underlying Milan Group and overlying Baxkorgan Group. The Altyn Tagh Group is a series of upper greenschist—lower amphibolite facies metamorphic rocks, such as amphibolite schist, biotite-garnet schist, marble and quartzite and strongly migmatized rocks. The time of the group is generally considered as Paleoproterozoic. The Baxkorgan Group is a series of metaclastic rocks interbedded with mafic metavolcanic rocks of lower schist facies. Marble and dolomite are intensively foliated and wrinkles of 4″ penetrative foliation are strongly developed. The Baxkorgan Group with unconformity overlies the Daxidaban Group of gentle occurrence. The Taxidaban Group is composed of weakly deformed and metamorphosed clastic rocks and intermediate-mafic volcanic rocks, dolomite, and marble. According to the stromatolite and micropaleoplants it is determined that the Group is of Late Mesoproterozoic.

Therefore, the underlying Baxkorgan Group can be deduced as the Early Mesoproterozoic. The Taxidaban Group and overlying Early Neoproterozoic strata have the features of a platform cover sequence.

Kalpin Uplift (***II₁***)

The oldest rocks presented here belong to the Aksu Group which consists of schists interleaved with amphibolite schist, magnetite-quartzite, and siliceous rocks. The outcrops can be extended northeastwards to the southern slope of the Khan-Tengri peak, southern Tien Shan Mountains, Heying Mount and Horo area. Typical blue schist has been observed in the southwestern slope. The Aksu Group has the metamorphism age of 962 ± 12 Ma and the galena vein gives the Pb–Pb time of 1000 ± 31 Ma (Cheng et al. 1994). Phengite in blue schist has the muscovite K–Ar age of 720 Ma (Xiao and Tang 1991). Thus this Group can be determined as the Meso-Neoproterozoic. The Aksu Group has undergone strong deformation and metamorphism and the protoliths include mafic lavas, siliceous rock and flysch, that is a thick oceanic sequence. Though metapyrolite is absent, the sequence is a typical "eugeosyncline" formation, suggesting early extension of the continent and occurrence of the initial ocean basin. The Aksu Group is uncomfortably overlain by the Wushinanshan Group of Late Neoproterozoic when the whole Tarim block was stable and clastic rocks interbedded with tillites, carbonate rocks, siliceous rocks, and volcanics deposited, and tillite is most typical in the area.

The Tarim block is more complex in structure than the North China Platform. The major difference lies in that both the basement and cover of the Tarim block are younger than that of the North China Platform, and tillite is present in the cover, and there is no substantial gap between the Meso-Neoproterozoic and the Phanerozoic. On the Kuruk Tag uplift at the northern margin of the Tarim block, the stable cover sequence began to accumulate from the initial period of Mesoproterozoic, which is similar to that of the North China Platform, while the sedimentation on the Tikanlik and Altun uplifts at the southern margin of the Tarim block primarily occurred in the Late Mesoproterozoic. All sediments are older than that of the Yangtze Platform where the cover sequence approximately formed since the Late Neoproterozoic, when tillite was deposited.

4.2 Structures Related with Consolidated Crust up to the End of Neoproterozoic

4.2.1 Yangtze Platform

The platform occupies the territory that is situated southward of the Qinling-Dabie orogenic belt, eastward of the Songpan-Garze belt, including the eastern Yunnan Province in the west to Jiangsu Province in the east, that is, nearly whole reaches of the Yangtze River basin. The range can be extended to the southern Yellow Sea in the east, and further east to the southern part of the Korean Peninsula.

Platform basement consists of the Neoarchean, Paleo-Mesoproterozoic, and Early Neoproterozoic metamorphic rocks. The passive margin rocks of mid-Neoproterozoic to mid-Triassic (Li 2001) and some terrestrial basin sediments since

Late Triassic were deposited uncomfortably over the basement. Sporadically exposed rocks of Neoarchean to Paleoproterozoic are dominantly of upper greenschist to amphibolite facies, and locally granulite facies metaplutonic and supracrustal rocks. Metaplutonic rocks have the geochemical feature of TTG (Gao et al. 1993), and supracrustal rocks include theoleiite, calc-alkaline series volcanic rocks, carbonate and clastic rocks of continental margin. According to the geological and geochronological data (Xing et al. 1993; Gan et al. 1996), rocks of Neoarchean to Paleoproterozoic may be more abundant in the depth. In the Mesoproterozoic, mafic to ultramafic plutons are observed to intrude in the Sibao Group in the northern Guangxi Autonomous region. Meso-Neoproterozoic mobile marginal formations were developed in southern Anhui—northern Jiangxi Provinces and Kangdian region in Sichuan Province. Ophiolite or oceanic residues have been reported in the Dahongshan Mountains, Hubei Province, southern Anhui and northeastern Jiangxi Provinces, northern Guangxi Autonomous region and in Longmenshan Mountains. Dykes of the Neoproterozoic were found in southern Anhui Province.

According to some data, the basement of the Yangtze Platform formed after two movements of 1000 and 800 Ma (Ren et al. 1999). These two movements are manifested by the angular unconformity between the underlying Kunyang Group and the overlying Chengjiang Sandstone of Nanhua System in eastern Yunnan Province. It is clear that the bottom of the Nanhua System is about 800 Ma and the upper limit of Kunyang Group is 1000 Ma, in other words, the Jinning movement lasted for about 200 million years. Actually, the Jinning movement is the result of superposition of two movements, that is, the Wuling movement and the Xuefeng movement. This map includes only a small part of the northern Yangtze Platform, that is the lower Yangtze Platform marginal depression south of the QinlingDabie orogenic belt and the northern Jiangsu Basin. The former platform marginal depression was formed in the period between Paleozoic and Triassic and experienced multistage (Indosinian and Yanshanian) tectono-magmatism since the Mesozoic and resulted in complex platform cover folds. The northern Jiangsu Basin bounded by the Tancheng-Lujiang Fault in the west with the North China platform and the Dabie orogenic belt extends to the northern Yellow Sea in the east and is the terrestrial fault basin since the Cretaceous.

4.2.2 Bureya-Jiamusi-Xingkai Block

The tectonic domain includes the Bureya-Jiamusi block, Xingkai block (in the literature) and the Zhangguangcai Hills eastward of the Songliao Basin. Striking NS, the block is situated in eastern Heilongjiang Province and adjacent Sikhote-Alin in the east and Khing'an fold belt and Songliao Basin in the west (Figs. 10 and 16).

The Precambrian metamorphic rocks of the basement are mainly of Mashan and Heilongjiang Groups of the Neoarchean to Paleoproterozoic. The Heilongjiang Group actually consists of continent marginal rocks and residues of paleooceanic basin of the Late Mesoproterozoic to Paleozoic, and the Group underwent intense reworking in two tectono-thermal events in the Late Paleozoic and Early–Middle

Jurassic. The granites once considered as Precambrian are mostly Late Paleozoic. In the Zhangguangcai Hills the volcano-sedimentary rocks and the Paleozoic granites constitute the active continent-margin formations. Thus, in eastern Heilongjiang Province the Luobei–Yilan, Zhangguangcai Range, Mudanjiang–Hulin collision belts of the Paleozoic can be discerned and some blocks like Hegang, Jiamusi, Mashan and Xingkai bounded by the collisional belts. We can say that the Bureya–Jiamusi–Xingkai block actually comprises of the complex mosaics of some Precambrian blocks and Paleozoic collision belts.

4.3 Structures Related with Consolidated Crust up to Middle Devonian

4.3.1 Kunlun-Qilian-Qinling Composite Orogenic Belt

Situated to the south of the Tarim–North China blocks, the southern margin of the composite belt is neighboring the Bayan-Har—Songpan-Garze tectonic belts in the western segment and the Yangtze Platform in the eastern segment (Fig. 13). The composite belt is a major orogenic belt traversing central China and thrust over the Precambrian blocks mentioned above and is as a whole manifested as the fan-like structure styles which were reworked to different degrees in later movements. The geological bodies comprising the belt include the Neoproterozoic or even older metamorphic basements similar to the adjacent Precambrian blocks, ophiolites and oceanic residues of the Neoproterozoic and Paleozoic, continental margin rocks and syn-collision melanges of the Neoproterozoic, Paleozoic to Triassic, and locally developed post-collision volcanic rocks and granitoids (Xiao et al. 2003; Dong et al.

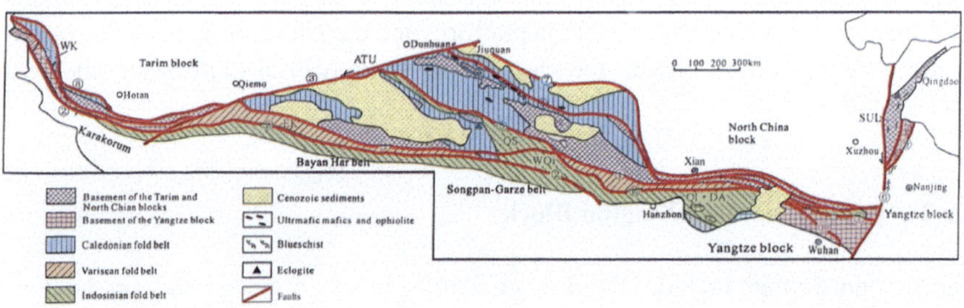

Fig. 13 Sketch map of the Kunlun-Qilian-Qinling composite orogenic belt. WK—Western Kunlun; ATU—Altun Shan; EK—Eastern Kunlun; NQD—Northern margin of Qaidam; WQi—Western Qinling; QS—Nanshan, Qinghai; QL-DA—Qinling–Dabie; SUL—Sulu. 1—Kunlun-Qinling-Dabie central fault (central suture); 2—Kunlun-Qinling southern margin fault; 3—Altyn Tagh strike-slip fault; 4—Northern margin fault at Qilian (suture); 5—Southern margin fault at Qilian; 6—Tancheng-Lujiang strike-slip fault; 7—Foreland fault of Qilianshan; 8—Northern margin fault at western Kunlun; 9—Jiasha-Xiangshui fault; 10—Lintan-Shanyang fault

2003; Lu et al. 1999; Zhang et al. 1999; Zhu et al. 2000; Deng et al. 1995; Liu et al. 2003; Wang 2002; Li et al. 2000; Yuan et al. 1994; Wu et al. 1998; Chen et al. 1996).

Composite belt is the distinguished Central China fault zone from the Kunlun Mountains in the west to Qingling Mountains in the middle and Dabie Mountains in the east (Fig. 13.4). The northern part of the composite belt is uplifted and exposed the Neoproterozoic and older than Mesoproterozoic crystalline and folded basement similar to that of the Tarim and North China blocks, while the southern part of the composite belt may be underlain by the pre-Late Neoproterozoic metamorphic basement correlative to that of the Yangtze Platform. These suggest that the Central China fault belt initially developed much early, at least from the Mesoproterozoic. Ophiolites and concomitant mafic eruptives with deep-water sediments from Late Neoproterozoic to Early Paleozoic were sporadically exposed (Xiao et al. 2003; Gao et al. 1990; Jiang et al. 1992). The later reworked Altyn Tagh and Tancheng-Lujiang transcurrent faults cut the central belt and its adjacent folded zones to three segments, the western Kunlun tectonic belt in the west, the eastern Kunlun-Qinling-Dabie tectonic belt in the middle, and the Sulu tectonic belt in the east. To the north of the central orogenic belt, the Altay, Qilian, and northern Qaidam, and also the northern Kunlun orogenic belt, Qinling-Dabie belt are of Caledonian origin, while the southern Kunlun orogenic belt situated south of the central belt includes the Variscan and Indosiniaan folded belts.

High to ultrahigh pressure metamorphic rocks of the Early Paleozoic have been observed in the Altun Mountains, Qilian, Northern Qaidam basin and Dabieshan regions. High pressure metamorphic rocks of the Triassic are present in the southern part of the Sulu, Dabieshan, and Eastern Qinling Mountains (Li 2004a).

Structure features of the individual tectonic belt of the composite orogenic belt will be described briefly as follows (Fig. 13).

Qilian tectonic belt

Transitional Corridor belt is of passive continental margin feature in the Paleoproterozoic and the outcrops of the Precambrian metamorphic basement are rarely present. From Cambrian to Ordovician the area was dominated by the earlier terrestrial clastic rocks and later carbonates rocks, which were interbedded with the volcanic basalt-dacite rocks. The southern part of the area is uncomfortably overlain by the foreland basin molasse of the Middle Devonian.

Northern Qilian fold belt is a complex mélange zone of paleooceanic and continental residues. Intensively developed faults of the zone lead to the distribution pattern of alternating horsts and grabens. The uplifted domain is actually composed by the rocks of the consolidated Precambrian metamorphic basement, while the deeply eroded depression area shows the strongly folding, sheared mantle rocks, pillow basalt and radiolarian siliceous rocks, which constitute two stages of ophiolite suites of the Mid-Late Cambrian and Early-Middle Ordovician and a well-developed HP–UHP metamorphic zone. Huge thickness of the Silurian flysch formation is mainly distributed in the northern part of the northern Qilian fold belt. Upwards sequence is of deep-water fan facies to hypabyssal slope and coastal facies and the sediments were transferred from the south, suggesting the filling of the residual sea basins. The Devonian flysch uncomfortably overlies the Lower Paleozoic. Above, the molasse represents the Upper Paleozoic-Triassic sequence composed of the stable sediments which are similar to that of the North China Platform.

The central Qilian uplifting belt is represented by a widespread crystalline basement of the Paleoproterozoic and folded basement of the Meso-Neoproterozoic. The Paleozoic is sparsely exposed and the Middle Cambrian and Lower Ordovician are mainly of stable shallow to lake littoral facies clastic and carbonate rocks, locally interleaved with basalt and andesite and uncomfortably overlie the metamorphic basement of the Precambrian. The cover sequence of Carboniferous, Permian and Triassic are present in large part of the central Qilian belt.

Southern Qilian fold belt has a continuous succession of the Ordovician to Silurian which uncomfortably overlies the Cambrian. The Ordovician consists of volcanic rocks of bi-modal feature, greywacke, and carbonate. The Silurian is mainly of flysch which was derived from the central Qilian and two margins of the Qaidam block. The strata demonstrate strongly isoclinal folds and were intruded by abundant intermediate-acid magmatic rocks. The Upper Paleozoic and Triassic are the stable deposits similar to that of the south China block. In the Laji Mount area (east of the Qinghai Lake), the Early Paleozoic is similar to that of the northern Qilian belt and is comprised of multiple suites of intermediate to mafic volcanic rocks and deep-water sediments. The Laji Mount zone may be an intra-continental aulacogen from east to west penetrating between the central and southern Qilian paleoblocks. The aulacogen might expand simultaneously to the northern Qilian paleoocean.

The Qilian tectonic belt as a whole is well developed in the intrusions of the Early Paleozoic, among which the major intrusive bodies are mainly distributed along the central and southern Qilian Mountains. In the northern Qilian, over 20 minor plagiogranite bodies are concomitantly exposed and the intrusions mainly of diorite-granodiorite, I-type granite series feature, are present in the transitional Corridor and central Qilian belt, while in the central-southern Qilian Mountains the large area of S-type granite series is characterized by leucocratic monzogranite.

Altyn Tagh tectonic belt

Situated between the Tarim and Qaidam blocks, the Altyn Tagh belt is actually extended along the Altyn Tagh faults where the metamorphic basements of the Paleo-Mesoproterozoic and interleaving granulite facies metamorphic rocks and stable platform cover sequence of Neoproterozoic and initial Paleozoic are widely distributed. From the Early Ordovician, mobile sedimentary rock association of "eugeosyncline feature" began to form along the Altun belt. The mafic–ultramafic rocks, intermediate-mafic volcanic and siliceous rocks were contemporaneously developed. At the end of Middle Ordovician, the "eugeosyncline" closed and the rocks were folded. The Altyn Tagh belt began to form at least before the Paleozoic and manifested strong strike slip in the Mesozoic-Cenozoic: dextral strike-slip in the Mesozoic and sinistral strike-slip in the Cenozoic.

Kunlun tectonic belt

This belt is located between the Tarim and Qaidam blocks in the north and in the south, respectively, by the southern boundary fault of Kunlun Mountains it contacts with the adjacent Bayan Har tectonic belt. The Kunlun tectonic belt is dissected by the Altun fault into the eastern and western segments, i.e., the eastern Kunlun tectonic belt and western Kunlun tectonic belt, respectively. The Central Kunlun fault belt corresponds approximately with the Kudi-Subexi zone in the western Kunlun and Qingshuiquan zone in the eastern Kunlun Mountains. Along the fault, ophiolites and associated oceanic sediments and eruptives and related accretionary wedges of the Late Neoproterozoic to Early Paleozoic occur sporadically. Here the granitoids of the Ordovician to Silurian are of calc-alkaline feature; their geochemical properties suggest the polarity that the paleooceanic crust subducted southwards (Wang et al. 1990).

Taking the Central fault as the boundary, the metamorphic basement of the northern Kunlun developed similar to that of Tarim and Qaidam blocks, while the metamorphic basement of the southern Kunlun was not consolidated at the initial stage of the Neoproterozoic (similar to that of the Yangtze Platform). In the Late Paleozoic, the Central fault belt (Paleozoic suture) was reworked and locally small oceanic basins opened and the sediments of paleooceanic features were formed there. The Middle–Late Paleozoic granites have the active continental margin calc-alkaline geochemical characteristics, suggesting the northward subduction, i.e., the tendency to northern Kunlun and margins of the Tarim and Qaidam blocks (Wang et al. 1990).

Pamir syntax

Pamir syntax is the western end of the western Kunlun tectonic belt. In the literature, the Pamir is generally subdivided into northern, central, and southern Pamir (see chapter "Tectonics of Pamirs" of this Explanatory Note). The central fault of the western Kunlun may extend to the northern marginal fault of Northern Pamir which correlates with the southern zone of the western Kunlun belt, while the Central Pamir approximately corresponds with the Karakorum tectonic belt (Fig. 14).

The Northern and Central Pamirs have the stratigraphic, structural, and magmatic properties similar to that of the Western Kunlun and Karakorum tectonic belts, respectively, but show a stronger effect of reworking. Especially the compression from south to north in the Cenozoic resulted in the Pamir syntax and arc structures by shearing and thrusts. At two sides of the arcs, plenty of slippery faults occur. The old faults like in Central and Western Kunlun, south marginal of Western Kunlun and Karakorum were reset near the Pamir and were further evolved into strike-slip faults.

Fold belt in the Northern Qaidam Basin

This NW striking fold belt is located between the Qilian tectonic belt in the north and the Qaidam block in the south, and includes the Serteng, Luliang Shan, and southern Xitieshan (Fig. 13).

Here we can find the Paleo-Mesoproterozoic basement over which in short Early Neoproterozoic period the stable sediments were formed, and then the ophiolite

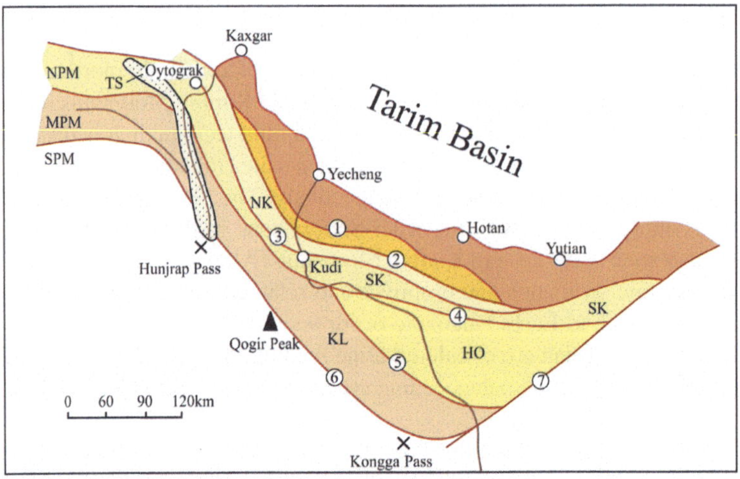

Fig. 14 Sketch tectonic map of western Kunlun Mountains (revised after Jiang et al. 1992). Tectonic units: NK—Northern Kunlun; SK—Southern Kunlun; HO—Hoh Xil; KL—Karakorum; Pamirs (NPM—Northern Pamirs; MPM—Middle (Central) Pamirs; SPM—Southern Pamirs); TS—Taxkorgan basin. Major faults: 1—Foreland fault of the Western Kunlun; 2—Northern Kunlun fault; 3—Central Kunlun fault; 4—Southern margin fault of Kunlun (Kengxiwar); 5—Quanshuigou fault; 6—Karakorum fault; 7—Altyn Tagh strike-slip fault

association of Late Neoproterozoic occurred, including basalt pillow-lavas, dyke swarms, and gabbro. The whole rocks have the isotopic ages of Rb/Sr (768 Ma) and Sm/Nd (780 Ma) and are consistent with the protolith age (750–800 Ma) (Yang et al. 2004) of eclogite of the fold belt in the Northern Qaidam Basin.

In Early the Ordovician, mafic, ultramafic, and intermediate-mafic volcanic rocks and minor siliceous rocks were formed; the intermediate-mafic volcanic rocks have the feature of island arc volcanic rocks. All rocks underwent the slight greenschist facies metamorphism. At the end of Late Ordovician, the sequence was folded and metamorphosed. UHP-eclogite can be observed in some amphibolite and chlorite schist. The cover strata of alternating DevonianCarboniferous ocean-continental facies uncomfortably overlie the underlying old sequence.

Qinling-Dabie orogenic belt

Qinling-Dabie orogenic belt is sandwiched between the North China and Yangtze platforms, including the western Qinling fold belt in the west and the Sulu fold belt in the east.

In light of the central fault (central suture), the main body of the orogenic belt can be subdivided into southern and northern belts (Fig. 13).

Northern belt is actually the prolongation of the Qilian tectonic belt. The northern, central, and southern belts of the Qilian extend eastwards and were converged into the northern Qinling fold belt. The difference consists in that in both northern and southern sides of the Qinling belt the Neoproterozoic ophiolite and related volcanic rocks are developed. As the Caledonian granites of subduction-collision types (450–380 Ma) are present in both northern Qinling and southern margin of the North China Platform, the Yangtze Platform is thus postulated to have been subducted under the North China Platform and the continental collision took place from south to north.

Southern belt is the southern Qinling fold belt that is located inside the central fault belt and can be further subdivided into the northern Variscan and the southern Indosinian fold belts on the boundary of Lintan-Shanyang fault. To the west of Fengxian County, western segment of the northern Variscan, huge thickness of flysch sequence of the LowerMiddle Devonian were uncomfortably overlain by the Upper Devonian molasse. While in the Zhashui County, eastern segment of the northern Variscan, in the Late Carboniferous "geosyncline" deposits still continued to accumulate; those were not transformed into fold belt until the Permian, as the southern part was folded only in the Indosinian. Some ages of 996–767 Ma of quartz diorite and granites were obtained along the Douling-Tongbai-Dabie metamorphic complexes (Li et al. 1991), suggesting the area in Meso-Neoproterozoic had been the active northern margin of the Yangtze Platform. In the Early Paleozoic, the southern Qinling belt was changed into passive margin. From the Devonian to Permian, nearly all sediments were the carbonate rocks, while all deposits in the Triassic were flysch formation and they were distributed in Fengxian County and further westward. ***Western segment of the Qinling-Dabie orogenic belt*** is the western Qinling fold belt which is bounded from the eastern Kunlun belt by the NW striking Kengxiwar fault. Besides, the Meso-Neoproterozoic basement rocks and some Paleozoic strata occasionally present, the belt is almost covered by the Triassic flysch sequence, which is connected to the Songpan-Ganze flysch basin in the south, while it extended northwestward through the Nanshan Mount, Qinghai Province, then subducted beneath the passageway between the southern Qilian belt and northern Qaidam basin block (Fig. 13).

Eastern segment of the Qinling-Dabie orogenic belt is the Sulu fold belt and is situated east of the Tancheng-Lujian fault. It extends eastwards through the northern Yellow Sea, then correlated to the Gyeonggi block of the Korean Peninsula. Similar to the metamorphic complexes of the Tongbai-Dabie belt, here the Mesoproterozoic Zhangbaling and Haizhou Groups which are mainly of medium–high pressure green-schist facies and blue-schist—whiteschist can be found. Neoarchean-Paleoproterozoic Donghai and Jiaonan Groups are predominantly distributed in the Shandong Province, and the rocks occur as widespread eclogites, granulites, and granitic gneisses. Since the Late Mesozoic, the fold belt was reworked by the crustal extension activity and the metamorphic rocks of the root part were exposed to the surface, then intruded by the magmatic rocks since Cretaceous or uncomfortably overlain by the terrestrial eruptives.

4.3.2 Altay-Zaisan Fold Region

Altay region inside the territory of China is only a small part of the whole southern Altay region and is stretched westwards in Kazakhstan and is called Rudny Altay. Its eastern stretching to Mongolia is called Gobi Altay while its Russian part is generally known as Gorny Altay, Northern Altay.

Pre-Sinian metamorphic rocks are present in the Chinese Altay. Micropaleoplant fossils in the slightly metamorphosed strata and the Sm/Nd isochron age of gneiss-migmatite granite (1400–1500 Ma) suggest Paleo-Mesoproterozoic age of the sequence which was called Xemirxek Group. Some diabase and gabbro intrusions in the Group give the Nd model age (TDM) of 945–977 Ma and the Sm/Nd isochron age of 974 $\pm$ 63.4 Ma, implying the event of mafic magma intrusions in the Early Neoproterozoic (Fang et al. 2002). In the period of Early Paleoproterozoic to Late Neoproterozoic, the terrestrial deposits predominantly of flysch feature were in unconformable contacts with both the underlying Xemirxek Group and the overlying strata of the Late Ordovician to Silurian,—with the lower part of coarse-clastic molasse and the upper layer of fineclastic flysch interleaved with acid volcanic rocks. In the Devonian to Early Carboniferous, two types of formation were developed: in the north—intermediate-mafic to intermediate-mafic volcanic lavas, pyroclastic rocks and flysch, local spilite-keratophyre, in the south—some components of ophiolite, such as mafic, ultramafic volcanic rocks and radiolarian siliceous rocks. As regard to the intermediate-acid magma activity, there were major I-type and minor S-type granites in the Early Paleoproterozoic, on the contrary, in the Late Paleoproterozoic, S-type and minor I-type granites. In the Late Neoproterozoic, the A-type granites occurred.

In the south, the Altay orogenic belt is neighboring the Zaysan–Ertix tectonic belt along the Ertix fault which extends to the east as "Mid-Mongolian Tectonic line"; its further stretching is the Huma–Hailar fault in northeast China. Taking this critical line as the boundary, the northern part is the Central Mongol-Ergun tectonic belt and the southern is Beishan—Inner Mongol—Xilinhot tectonic belt (Figs. 15 and 16).

4.3.3 Ergun Fold Belt

Ergun fold belt is the eastern part of Central Mongol-Ergun tectonic belt and crustal rocks are similar to that of the Altay orogenic belt. The oldest metamorphic basement consists of various schists, amphibolite, granulites, leucogranulite, granitic gneiss, marble, and migmatites. The overlying Sinian (Ediacaran) to Lower Cambrian is represented by epicontinental clastic rocks, calc-alkaline series volcanic rocks, and sediments that bear fossils of Spongea Archaeocyatha. The Ordovician and Silurian are carbonate and clastic rocks interbedded with eruptives and conformably overlain by flysch and volcanic layers of the Late Paleozoic. Part of the ophiolite mélange sequence is intermittently exposed along the Huma–Hailar fault. The formation time of ophiolite is still under debate, Meso-Neoproterozoic or Paleozoic? If taking into consideration the radiolarian siliceous rocks, it is reasonable to deduce that the majority is of Devonian. In the fold belt, granite and migmatite are well developed and the time periods focus on the Middle-Late Paleozoic and minor Early Paleozoic. Proterozoic granites occur as batholith, while in the Mesozoic the syenogranite and alkaline granite were formed along the faults.

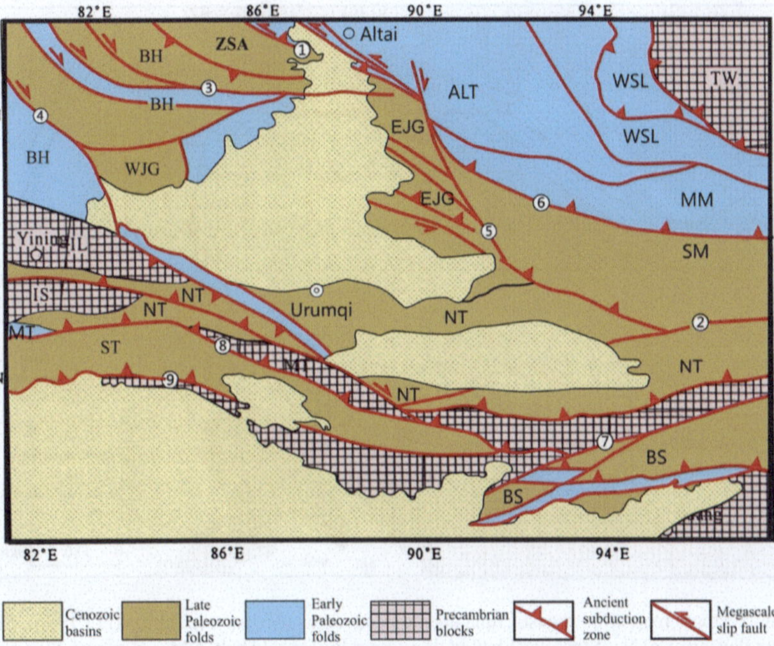

Fig. 15 Sketch tectonic map of northwestern China and adjacent areas. TW—Tuva block; WSL—western Lake Saysan region; ALT—Altay; ZSA—Zaysan—Ertix (Irtysh) zone; BH—Barkaxgar (Balkhash); WJG—Western Junggar; EJG—Eastern Junggar; IL—Ili block; IS—Issyk Kul block; NT—Northern Tien Shan; MT—Middle Tien Shan; ST—Southern Tien Shan; BS—Beishan; SM–Southern Mongolia; MM—Middle (Central) Mongolia. 1—Ertix (Irtysh) fault; 2—Bogda fault; 3—Chinghis strike-slip fault; 4—Central Kazakhstan fault; 5—Koktokay fault; 6—Mid Mongolian Tectonic line; 7—Beishan shear zone; 8—northern margin fault of Southern Tien Shan; 9—southern margin fault of Tien Shan

4.4 Structures Related with Consolidated Crust up to Late Carboniferous—Early Permian

4.4.1 Barkaxgar-Junggar (Junggar-Balkhash) Fold Series

In the Paleoasian Ocean domain, the Junggar-Balkhash fold series is a broad and complex fold-thrust series which neighbors the Altay tectonic belt along the Ertix (Irtysh) fault in the north and faces the Tien Shan orogenic belt to the south. The western segment is situated in Kazakhstan and is predominantly of Caledonian folding. The eastern segment in northern Xinjiang is of Variscan folding. The further eastern extension is up to the Koktokay strike-slip fault and contacts the central and southern tectonic belts. The fold series is also characterized by the involving of multiple basement blocks and miniblocks of the pre-Sinian (pre-Ediacaran). The major blocks are Kokchetav, Yili, Issyk-Kul, Junggar(?) blocks, among them the NW striking Yili and Issyk-Kul blocks were further extended SE and were subducted

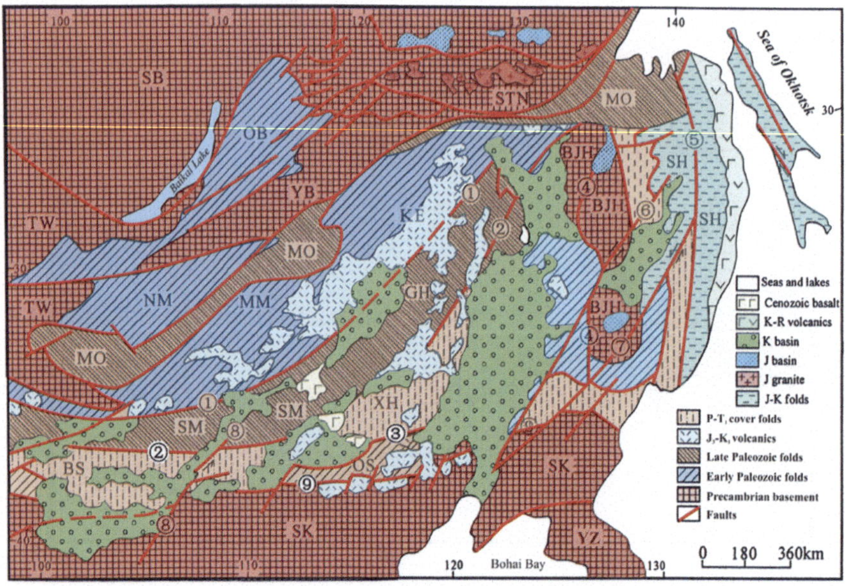

Fig. 16 Tectonic map of northeast China and adjacent areas (revised according to Li 1996). SB—Siberian Platform; STN—Stanovoy block; OB—Outer Baikal fold belt; NM—Northern Mongolian fold belt; YB—Yablonov block; TW—Tuva block; MO—Mongol-Okhotsk fold belt; MM-KE—Central Mongolian–Ergun orogenic belt; SM-GH—South Mongolian– Khinggan orogenic belt; BS-XH—Beishan-Xilinhot orogenic belt; BJH—Bureya-Jiamusi-Xingkai block; OS—Ondor Sum fold belt; JY-SL—Jieya-Songliao Basin; SH—Sikhote-Alin orogenic belt; SK—North China (Sino-Korean) continent block. 1—Huma-Hailar fault; 2—Hegenshan-Erenhot fault; 3—Xar Moron fault; 4—Mudanjiang fault; 5—Middle Sikhote fault; 6—Yilan-Yitong fault; 7—Mishan-Dunhua fault; 8—Langshan fault; 9—Northern margin fault of the North China Platform

beneath the Northern Tien Shan Mountains (Fig. 15). In the Chinese scope, the fold series can be subdivided into Ertix (Irtysh), Western and Eastern Junggar fold belts.

Ertix (Irtysh) fold belt

Ordovician rocks are the oldest ones in the belt and occur in fault blocks of jasper rock and mafic lavas, but ophiolite of metaperidotite or mafic cumulates is not found. The Middle-Lower Devonian are predominantly of flysch formation and the bimodal volcanics presented in the upper part are overlain uncomfortably by the Upper Devonian—Lower Carboniferous succession, with the bottom of molasse and predominantly of pyroclastic and terrestrial clastic rocks. Separated zircons from the dynamometamorphic garnet-biotite schist near the Ertix fault had given the maximum value of 2349 Ma and minimum of 353 Ma, apparent ages 957–2339 Ma (Hu et al. 2001), suggesting the metamorphic rocks may be an involving of residual slices of the pre-Sinian basement.

As to the magmatism actions, besides the calcalkaline series of volcanic rocks of the Devonian-Carboniferous, the I-type granite of Middle Variscan and S-type alkaline granites of Late Variscan are also present in the belt. The strata older than the Early Carboniferous manifested strong folding, while those since the Late Carboniferous were gentle and open folds typical of the cover sequence.

Western Junggar fold belt

Up to now, no strata older than Ordovician have been found in the belt. But the leucocratic plagiogranite dyke in the Tangbale ophiolite melange in the south gave the titanite Pb/Pb ages of 508 Ma, i.e., Late Cambrian (Xiao and Tang 1991). The Lower Ordovician consists of slate and schist, and protoliths have the feature of terrestrial clastics. The Middle Ordovician includes altered pillow-lava and radiolarian siliceous rocks. The Lower Silurian is a thick flysch and uncomfortably overlies the Tangbale ophiolite melange. At Maila, the Middle Silurian has a similar oceanic and ophiolite section, such as the pillow lava and radiolarian siliceous rocks, being overlain by flysch formation of the Upper Silurian. The Middle Devonian has molasse in the lower part and flysch-like formation interleaved with pyroclastic rocks in the upper part. At Dalabute (= *Darbut* in numerous articles) in the central zone, the Middle Devonian contains spilitic pillow-lava and radiolarian siliceous rocks, the Upper Devonian is flysch. The Carboniferous is dominantly the terrestrial clastic rocks, with local intercalation of intermediate-acid volcanic rocks and overlain with unconformity by subaeral intermediate-acid volcanic rocks of the Lower Permian (Fig. 17, the column V).

As mentioned above, termination of the marine sedimentation during the period from the end of Late Carboniferous to beginning of Early Permian indicates the finishing of the oceanic basin existence in the Western Junggar fold belt. In the Tangbale ophiolites, the intermediate-mafic–ultramafic cumulates were not well developed and peridotite was absent, only thin layers of gabbro cumulate and minor diabase, dacite, and rhyolitic dacite dykes were exposed (Xiao and Tang 1991), suggesting that ophiolites seem to have formed in the setting of narrow oceanic basin after a slow spreading at the initial stages. Typical blue-schists were discovered at both the eastern and southern sides of the Tangbale ophiolites. Meanwhile, glaucophane has been noticed in the clastic rocks of the Lower Silurian, which effectively implies that the subduction of the ocean basin began before the Early Silurian. Oceanic ophiolite of the Late Paleozoic may represent the backor fore arc setting.

In Western Junggar, not only S-type granites of the Early-Middle Late Paleozoic, but also post-collisional alkaline granites of the Late Paleozoic have been widely distributed.

Eastern Junggar fold belt

Two NWW striking ophiolite belts are distinguished, the northern Nar Mad (another name—*Almantai*) ophiolite belt and the southern Kelameili ophiolite belt (Figs. 15 and 17, VI). Scattering geological bodies of the Early Paleozoic show that the tectonic pattern of oceanic basin evolution, typified by the Nar Mad ophiolite, had existed in the Early Paleozoic in Eastern Junggar. The Nar Mad ophiolite includes harzburgite, dunite, cumulate gabbro, diabase, pillow-lavas and radiolarian siliceous rocks. The time of the ophiolite origin was once considered as Devonian, but the later age

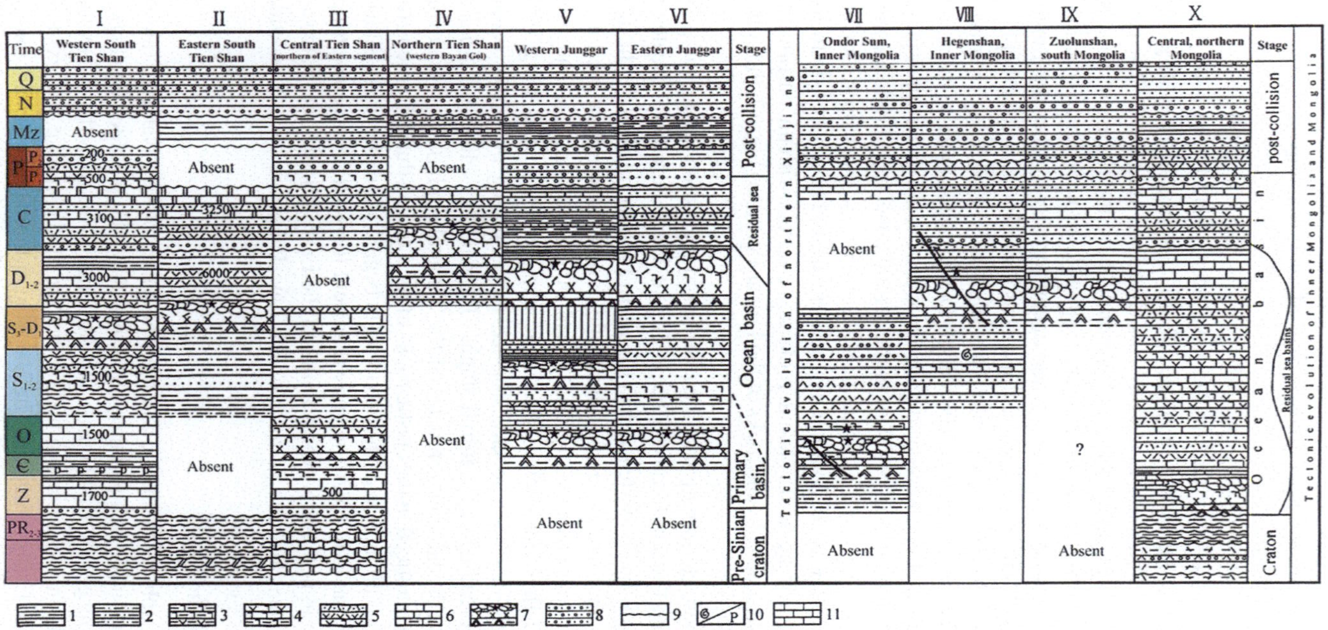

Fig. 17 Comprehensive geological correlation table of Tien Shan—Eastern and Western Junggar—Inner Mongolia—Mongolia (revised after (Xiao and Tang 1991), only 6 is slightly changed). 1—blue schists; 2—quartz-sericite-chlorite schists; 3—chlorite-plagioclase amphibolite; 4—intermediate-mafic volcanic rocks and schists; 5—intermediate-mafic lavas and tuffs; 6—marble and crystalline limestone; 7—ophiolites; radiolarian siliceous rocks; 8—conglomerate, sandstone-conglomerate; 9—unconformity; 10—fossils and P-bearing layers; 11—limestone

determination gave the results of Sm/Nd age 561 Ma, 479 Ma and Rb/Sr age of 392 Ma (Li 2004b; Jin et al. 2001), suggesting possible the Cambrian. The Early Paleozoic terrestrial clastic rocks and carbonate rocks are Northward of the ophiolite belt. They were formed as a sequence of the passive continental margin. While in the south features of the geological bodies and their distribution demonstrate the marginal properties of arcs and basins. The Nar Mad ocean basin closed in the Silurian and the collisional orogenic belt formed.

Discovery of the Devonian—Early Carboniferous radiolarians in the jasper rock in the Kelameili ophiolte of Late Paleozoic and the fossils of Tuvaella fauna from the Silurian strata at both the northern and the southern sides implies that the paleooceanic basin represented by the ophiolite complex formed after the Silurian. During the Late Silurian, the paleocontinental blocks at both sides of the ophiolite suture were accreted to the Siberia Platform. The presence of the molasse coarse-clastic rock 1600 m thick in the lower part of the Upper Carboniferous at southern Kelameili Mount shows the time of closing of the oceanic basin.

Composite granite belt at the northern Kelameili ophiolite belt comprises the Late Carboniferous—Permian granodiorite, biotite granite, binary-mica granite and granosyenite; their chemical composition varies from calc-alkaline to alkaline. While in the southern Kelameili ophiolite belt, only some small stocks of alkali hornblende syenite and syenite were intruded in the contemporaneous volcano-sedimentary rocks.

4.4.2 Tien Shan Orogenic Belt

Chinese part of Tien Shan orogenic belt is situated between the Junggar fold belt and Tarim block and on the whole can be subdivided into Northern Tien Shan, Central Tien Shan, and Southern Tien Shan tectonic belt.

Northern Tien Shan tectonic belt

In southern part of the Northern Tien Shan tectonic belt, sparsely exposed oceanic crustal residues of the Late Paleozoic, including dolerite and diabase dykes intruding the pillow-lavas, radiolarian siliceous rocks and flysch formation (Fig. 15, IV), are distributed along the Bayan Gol and Manas zone. In the northern part of the Northern Tien Shan tectonic belt, difference lies between two sides of the Turpan–Hami Basin: south of the basin is occupied by volcanic-sedimentary rocks and intermediate-acid intrusions of the Devonian to Early Carboniferous, which have the features of island arc, while north of the basin—by pyroclastic–terrestrial clastic rocks of the Devonian to Carboniferous, which have the back-arc basin-affiliated feature (Li 2004b). Northern Tien Shan tectonic belt is of the Variscan (Early Devonian) (Wang et al. 1990) and manifested predominantly by intermediate-mafic volcanic and pyroclastic rocks and was slightly folded. In the main cycle (end of the Early Carboniferous) rifting with spreading caused formation of an oceanic basin with appropriated oceanic crust complexes. At the middle of Late Carboniferous, rock assemblages of paleooceanic basin were folded, and at the end of Late Carboniferous the back-arc basin was closed. In the Permian, terrestrial volcanic rocks and molasse formation prevailed.

Southern Tien Shan tectonic belt

Belt bounds with the Central Tien Shan belt along the northern marginal fault (its western extension out of China once was called the Nikolayev line) in the north, and neighbors the Tarim Basin along the southern marginal fault in the south (Fig. 15). Actually, the pre-Early Silurian structure features of the Southern Tien Shan were identical to that of the northern Tarim platform. In the pre-Sinian (pre-Ediacaran) stage, the structure was of mobile type and the Aksu Group schists and cover sequence were extended into the Southern Tien Shan Mountains. From middle of Silurian to Early Devonian, ophiolite complex and overlying siliceous rocks and volcano-flysch deposits were formed (Fig. 17, I, II). In the Middle Devonian, rocks were represented by carbonate and clastic sediments interbedded by intermediate-mafic to intermediate-acid volcanic rocks, suggesting the existence of arc basin stage. Lower Carboniferous molasse with unconformity overlays the Devonian. At northern Halik Mount, the HP and UHP eclogite and blueschists of Early Carboniferous have been reported (Xiong et al. 2006).

Central Tien Shan tectonic belt

Belt is sandwiched between the Northern and Southern Tien Shan Mountains and can be called Central Tien Shan up warping belt. New research reveals that the conventional old metamorphic basement is actually of Paleozoic through the discovery of fossils. The real old basement preserved as slices or blocks occurring like a chain of islands in the Central Tien Shan Mountains. The residual basement is of Paleoproterozoic, including schist, gneiss, and migmatite. While the stable cover sequence of Meso-Neoproterozoic is dominated by terrestrial clastic and siliceous-magnesia carbonate rocks. Over the block, the Lower Paleozoic was the second cover sequence and partially transformed into the surrounding mobile setting (paleooceanic of Caledonian). In the Late Paleozoic, the Central Tien Shan belt was basically uplifted or changed as the island-arc zone at the Northern Tien Shan ocean basin. In Permian, the Southern, Central, and Northern Tien Shan belts were amalgamated in the same named unit, and together with the Junggar and Tarim blocks they became part of the Paleoasian Continent.

4.4.3 South Mongol—Khingan Tectonic Belt

Belt is located between the Huma–Hailar fault and Hegenshan–Erenhot fault; the Central Mongol-Ergun tectonic belt is situated to the north and the Beishan-Xilinhot tectonic belt—to the south (Fig. 16). This belt is composed by the sedimentary formations of carbonate, fine clastic rocks, flysch, and pyroclastic rocks of the Ordovician to Early Carboniferous (including minor Sinian (Ediacaran)-Cambrian) and Precambrian crystalline basement blocks which were mostly intruded by the Paleozoic granites and overlain by the Mesozoic volcanic layers. The Sinian (Ediacaran)-Cambrian sequence of the belt has the characteristics of terrestrial clastic sediments, while the Ordovician to Lower Carboniferous sequence was of nearby-arc basin volcano-sedimentary formations. It is specially pointed out that the Silurian in the belt is rich in fossils of Tuvaella fauna, suggesting the belt was once connected to the Junggar block (nowadays northwest China), and the belt had affinity to the Siberian Platform. Along the line of Erenhot to Hegenshan Mount in the southern Mongol-Khingan tectonic belt, the largest Paleozoic ophiolite block in northeast China is exposed, the time of the ophiolite is generally considered as Devonian. Inside the belt there is a great amount of volcanic rocks of Late Jurassic to Early Cretaceous, which are commonly known as "the Great Khingan volcanic rocks". In fact, distribution of the volcanics is not limited in the Great Khingan Range. They can be observed as overlying almost in all orogenic belts south of the Mongol-Okhotsk orogenic belt. Further more, the structural strain of the region is NE, but not NNE as conventionally convinced.

Beishan-Xilinhot tectonic belt is sandwiched between the North China Platform in the south and South Mongol—Khingan tectonic belt in the north. In the western segment, the Beishan fold belt could be connected to the Tien Shan orogenic belt before the cutting of the shearing of the Beishan shear zone (Figs. 15 and 16). In light of the tectonic features, the belt can be subdivided into Beishan fold belt, Ondor Sum fold belt, and Xilinhot fold belt.

Beishan fold belt

Belt is located in the western part of the Beishan-Xilinhot tectonic belt. Metamorphic basement of the Precambrian is widely distributed, including the Archean and Paleoproterozoic gneiss and schist. The Mesoproterozoic and Sinian (Ediacaran) are of stable cover sequence. Both the basement and cover are quite similar to that of the blocks in Dunhuang area. The Cambrian consists of the bottom carbonaceous shale and overlying sandstone and slate. The Middle Ordovician is a suite of dismembered association of oceanic crustal residue slices, like jasper rocks, pillow-lavas, cumulate gabbro, diabase dykes, plagiogranite, and ultramafic rocks. The Upper Ordovician is the sandstone-slate alternation interbedded with marble and intermediate-mafic volcanic rocks. Silurian is represented by rock association of andesite, dacite, rhyolite, clastic rock, and reef limestone, demonstrating the volcanic arc and backarc basin setting. The association corresponds with the small Shibanjing—Xiaohuangshan oceanic basin in the north, which was subducted southwards. The unconformity between the Devonian and underlying sequence shows that the region changed into the intraplate stage which is especially obvious from the superposed fault-limited basins of Carboniferous to Permian, being accompanied by a great amount of leucocratic granite series dominated by monzonite and granosyenite.

Ondor Sum fold belt

This belt is situated at the northern margin of the North China Platform, and is neighboring with the Xilinhot fold belt along the Xar Moron fault zone in the north (Fig. 16). It is a Caledonian fold belt and was strongly reworked in the Variscan and mainly composed of geological bodies of the Paleozoic. The molasse of Late Silurian represents the unconformity with the underlying strata and the time of the discontinuity constrains the orogenic movement of the belt. Discerned oceanic crustal ophiolite and accompanying blueschist in the fold belt are called the "Ondor Sum ophiolite belt" which formed in the Early Cambrian. Another ophiolite suite called "Kedanshan ophiolite belt" is situated to the north and separated from "Ondor Sum ophiolite belt" by the Cenozoic basin, and is also accompanied by blueschist formed in the Ordovician, which means it is younger than the first one.

Xilinhot fold belt

Xilinhot fold belt is situated to the east of the Beishan fold belt, and sandwiched between the southern South Mongol—Khingan tectonic belt and the Ondor Sum fold belt. The principal geological bodies of the belt include the volcano-sedimentary rocks of the Late Carboniferous to Early Permian and ophiolites of possible Carboniferous. Ophiolites are sporadically distributed in the Engger Ulu (its possible western extension to the Hongshishan Mount ophiolite at northern Beishan Mts.), Solon Mount (possibly including the Solonker belt in southern Mongolia), and Houtou Mount, Inner Mongolia. The host rocks of ophiolites are of Late Carboniferous to early most Early Permian. One marine transgression event in the middle of Early Permian formed the Zhesi Group and correlated sequence overlapped the older geological body. Above the unconformity, the strata of Permian to Lower Triassic changed gradually from marine to continental facies, suggesting the final closing of the Paleoasian Ocean. But intense folding occurred in the Triassic. There is some controversy on this point. Some researchers (Ren et al. 1999) consider the belt, especially that represented by the Linxi belt, as Indosinian fold belt. In this map compilation, we take it as the Late Variscan fold, while the Indosinian fold is expressed in the map as cover folding in later cycles.

4.5 Structures Related with Consolidated Crust up to Middle Triassic

4.5.1 Yushu-Jinshajiang Fold Belt

Belt is situated approximately in the axial part of the Qinghai-Tibetan Plateau and its western extension may enter into northern Karakorum. Eastern part of the belt occurs at ophiolite belt along the Jinshajiang River, which is generally called Yushu-Jinshajiang ophiolite belt. The Bayan Har—Songpan-Garze tectonic belt, which has the platform type carbonate formation of the Ordovician to Lower Permian is to the east of the ophiolite belt. The formation is continuous in sequence, stable in lithofacies and prolific in fossils. While a suite of mobile sediments of the Lower Paleozoic to Lower Permian is to the west of the ophiolite. Great amount of volcanic rocks, formed in the Middle Devonian and Early Permian, was characterized by mafic volcanics, spilite-keratophyre, radiolarian siliceous rocks, and minor carbonate rocks. The intermittently exposed ultramafic rocks and the Early Permian mafic lavas with radiolarian siliceous rocks together constitute the ophiolite suite. The tectonic movement between the Early and Late Permian, position of metamorphic zones and the island arc granites in the western margin demonstrate the NE to SW subduction of the paleooceanic crust.

Similar profiles have been found near the Xijir Ulan Lake (Zhang et al. 1994), where the Upper Permian to Lower Triassic strata with unconformity overlie the radiolarian siliceous rocks and lavas of the Lower Permian. The radiolarian siliceous rocks, lavas, and ultramafic rocks constitute the ophiolite suite. To the west of the ophiolite, great amount of intermediate-mafic to intermediate-acid volcanic rocks and collisional granites of Triassic to earlier Late Triassic were intruded, superposed the island arcs and formed the continental margin volcanic arc belt. The terrestrial deposits and coal measures of Late Triassic (Norian stage) indicate the final closing of the paleooceanic basin. The Changdu Caledonian fold belt where the flyschoid formations of the Ordovician to Silurian are overlaid with unconformity by the Lower-Middle Devonian deformed flysch sequence is to the west of the Yushu-Jinshanjiang fold belt. Above the unconformity the stable cover sequence is of the whole Upper Paleozoic. In the Mesozoic, the terrestrial basins prevailed.

4.5.2 Bayan Har—Songpan-Garze Tectonic Belt

Belt represents a broad area between the southern Kunlun margin fault in the north and Yushu-Jinshajiang fault in the south. Its main body is located south of the eastern Kunlun tectonic belt and spins out to the west, i.e., south of the western Kunlun Mountains. The whole area is almost covered by the Triassic flysch formations. Research reveals that many blocks composed by the Precambrian metamorphic rocks were buried under a thick flysch sequence. Rare exposed small blocks of the crystalline basement show that they were formed in the Late Paleoproterozoic to Late Neoproterozoic and consolidated 600 million years ago, that is, slightly younger than that of the Yangtze Platform basement (Xiao et al. 2000). The limestone of Dengying Fm, Sinian (Ediacaran) system overlapped the metamorphic rocks. As a whole, the Sinian (Ediacaran)—Cambrian are clastic rock interbedded with minor volcanic rocks, while the Ordovician—Lower Permian are stable carbonate rocks. From the sediment accumulation of Maokou stage in the Middle Permian, the setting became mobile and some breccia limestone of active environment occurred. In the Late Permian, marine mafic lavas erupted in strongly active zones, and finally the Triassic flysch conformably overlaid the Late Permian basalts. Later, some Jurassic-Cretaceous terrestrial basins of diverse sizes were superimposed on folded Triassic aligned on the structural orientation.

Thus, it is deduced that the region had experienced a long period of stable continental crust setting before the Triassic, and flysch was developed in the depression by extension on the continental crust. Contemporaneously with the flysch sedimentation, strong downwarping resulted in some fault oceanic basins of limited sizes, such as the Triassic-dominated ophiolite and blueschist belts on the Ganzi-Litang line in the southeast. Here, the oceanic basin subducted westwards and formed the "trench-arc-basin" pattern. At the end of Late Triassic, folding finished and new suture came into being. Lastly, some intra-continental sinistral ductile shear zones occurred. In the northern part of the belt (southern Kunlun margin fault), especially along the A'nyemaqen line, some small oceanic basins of the Mid-Late Triassic occurred. The oceanic basin of the Late Permian to Middle Triassic subducted northwards, folded, and formed a suture. The intra-continental deformation after the oceanic crust consumption was responsible for strikeslip ductile shear zones and some obvious thrusts.

4.6 Structures Related with Consolidated Crust up to the Beginning of Cretaceous

4.6.1 Karakorum Tectonic Belt

Belt takes "Z" shape and aligns in the NWW direction, with the eastern and western segments located south of the Quanshuigou and southern Western Kunlun marginal faults, respectively (Fig. 14). Its westwards extension can enter into the Central Pamir and the southeastern end is cut by the Altyn Tagh fault. The Precambrian metamorphic sequence, such as schist, marble and minor gneiss occurs in the belt. The Cambrian and Ordovician are clastic rocks and limestones. The Silurian is dominated by thick lowgrade metamorphosed clastic rocks in the lower part and more carbonate rocks in the upper. The Devonian is varicolored terrestrial rocks. The Carboniferous and Permian are sandstone-slate with intercalations of limestone. As to the Mesozoic, minor Triassic and large area of the Jurassic-Cretaceous strata are distributed in the region. Here, the Middle Jurassic—Lower Cretaceous have no conformable contacts with both the underlying Triassic—Lower Jurassic and the overlying Upper Cretaceous. Except the red terrestrial formations of the Upper Cretaceous, all the sequence of the Mesozoic below the unconformity are marine fine clastic rocks with interlayers of bioclastic limestones.

Magmatism activity is dominated by granite intrusions of the Early and Late Mesozoic, some Late Paleozoic, and minor granites of the Early Paleozoic and Proterozoic. The Karakorum tectonic belt extends eastwards and is connected to the Dangla (Tanggula) belt. It should be pointed out that in south-western Karakorum tectonic belt there is important Karakorum fault (Fig. 14). South of the fault, the Carboniferous-Permian sediments have the fossils of cold water fauna, thus the fault can be taken as the northern boundary of the Gondwanaland facies.

4.6.2 Sikhote-Alin Tectonic Belt

In the geological literature, the belt is generally subdivided into three belts; the central belt composed by the Silurian-Devonian and the Carboniferous-Permian and there is a straight ophiolite fault. The eastern belt consists of deep water sediments of the Carboniferous-Permian and Mesozoic while the western belt has diverse settings of deposits. In the Chinese territory, only part of the western Sikhote-Alin tectonic belt is present. This part is located east of the Jiamusi and Xingkai blocks (Fig. 16). In Wanda Mount area, a column of the oceanic ophiolite has been observed (the column is most integrated in Raohe area). The lower part of ophiolite has mafic and ultramafic rocks with intrusions of diabase dykes, the middle part–pillow basalt and jasper rocks and the upper part—radiolarian siliceous rocks overlain by terrestrial fine clastic rocks, siliceous rocks and flysch layers of the Carboniferous-Permian (slide deposits). In light of the assemblages of the radiolarians and conodonts, the ophiolite column formed in the Mid-Late Triassic and the overlying sediments of Early Jurassic. To the west of Wanda Mount and near the Jiamasi arid Xingkai blocks or overlapping the eastern margins of the blocks, shallow sea or littoral facies clastic rocks and alternating marinecontinental pyroclastic deposits have been observed. Folds in the rocks are well developed and linear tight folds are formed, with the axis generally in the NNE or near NS direction. Faults are dominated by the thrust structures. The above ophiolite columns usually occur as tectonic slices and present as serial stacks of slices, which show the typical ophiolite mélange (Zhao et al. 1996).

4.7 Young Plates and Sedimentary Basins

Intraplate sedimentary basins in Northern China and adjacent regions are distributed in the eastern and western domains separated approximately by the Helanshan Mountains—Liupanshan Mountains line. The eastern domain has the NE or NNE tectonic strain and the basins are extensional or extensionshearing. While the western domain has the NW or NWW tectonic strain and the basins are compression or compression-shearing. On the basis of the tectonic positions of the basins and features of the basement, the following types of tectonic basins can be classified.

(1) *Superimposed basins on the platforms (stable cratons).* This type of basins, like the Ordos and Tarim Basins, has old crystalline basements and stable cover sequences which have transitional relationships with the overlying basin deposits since the Mesozoic. Irregular rectangle in shape and long axis in NS direction, the Ordos basin is situated east of the Helanshan Mountains—Liupanshan Mountains line and represents a sedimentary basin developed in the west of the North China block. Besides some margin depressions (like the Fenwei graben) complicated by the extension and extension-shearing of Cenozoic, there is hardly any major deformation inside the basin. The Tarim Basin, located west of the Helanshan Mountains—Liupanshan Mountains line, is almost covered

by the basin sediments over the whole platform. As the NS compression in the Cenozoic, the foreland depressions are developed in the foreland of the Kunlun Mountains in the south and the Tien Shan Mountains foreland in the north, and some giant upwelling and depressions inside the basin (Fig. 11).

(2) *Aulacogens or faulted basins superposed on the cracked craton*, like the North China Basin, Hehuai Basin and Subei Basin.
(3) *Depression basin superposed on the reactivated block*, like the Qaidam Basin.
(4) *Depression basin superposed on the fold belt*, like the Junggar Basin and Turpan–Hami Basin.
(5) *Aulacogen-type faulted basin superposed on the fold belt*, like the Songliao Basin.
(6) *Foreland depression induced by the reworked orogeny*, like the Jiuquan Basin before the Qilian Mountains, the Kuqa depression in southern Tien Shan (northern margin of the Tarim Basin).

Further, we only give a brief introduction to some typical sedimentary basins.

4.7.1 Junggar Basin

Junggar Basin is situated in the southeast part of the Junggar–Balkhash (Barkaxgar–Junggar) fold series. The basin faces the east and west Junggar fold belts in the north and the Northern Tien Shan tectonic belt in the south (Figs. 15 and 16). As to the age and feature of the basement of the basin, granite of the earlier Late Carboniferous and the Pb isotope data show that the post-collisional plutonic granites may be derived from the anatexis of the paleooceanic sediments folded in the Early Paleozoic to earlier Late Paleozoic, not from anatexis of the Precambrian sialic crustal rocks (Xiao and Tang 1991). Xiao Xuchang (Xiao and Tang 1991) also mentioned that the east and west Junggar basins were connected before the Carboniferous and the assumed Precambrian giant horst separating them was actually absent. Sedimentation of the terrestrial coarse clastic rocks and coal-bearing strata indicates that the oceanic crust formations had folded and built up the mountains. The great amount of A-type granites in the folding suggests the post-orogenic extension and collapse. In the Permian, a series of faulted depressions occurred; they were filled by a huge mass of intermediate-mafic-dominated subaerial volcanic rocks, coal-bearing clastic rocks and fluvio-lacustrine facies of sandstone-conglomerate layer. Since the Triassic to Paleogene, the area entered the period of unitary stage of depression development and the fluvio-lacustrine facies conglomerate, sandstone and pelite formed. The sediments thickened southwards and the thickness abruptly increased at the southern margin of the basin, which were coalesced with the coal-bearing molasse of Jurassic and the foreland molasse of Pliocene epoch—Early Pleistocene epoch in the foreland depressions in Northern Tien Shan Mountains. The structure pattern mainly settled since the Miocene and the basin margins resulted from nappes and thrust structures towards the basin. In the west, striking-slip faults accompany and are approximately parallel to the Dalabute (Darbut) shear zone. In the east, alternating upwellings and depressions are controlled by the faults. While the central part of the basin was rather simple, present as monoclines with southern vergence and some gentle nose-like stepped structures (Tian 1990).

4.7.2 Turpan–Hami Basin

Basin is positioned southeast of the Junggar Basin and superposed on the north of Tien Shan tectonic belt, occurring like a long ellipse with its long axis running nearly in the EW direction (Figs. 10 and 15). The properties and ages of both the basement and the overlying basin sediments are similar to that of the Junggar Basin. As the time of marine retrogression in the northern Tien Shan tectonic belt was slightly younger than that of the Junggar fold belt, the Carboniferous here was still of mobile marine setting and thick intermediate-mafic volcanics, carbonate and clastic rocks were developed. By the Early Permian, type of deposit accumulation was changed into alternating marine and aerial facies and formed the intermediate-acid pyroclastic rocks, marlite and mudstone, while in the Late Permian all deposits became of subaerial and aerial deposition setting, i.e., alluvial and lacustrine facies, and the sediments were dominated by sandstone, conglomerate interleaving with mudstones of semi-deep to shallow lacustrine facies. In the Triassic, the basin patterns basically inherited that of the Late Permian, but lakes became shallower and xerothermic and red clastic formation occurred. The Jurassic coal and oil-bearing terrestrial clastic formation accumulated conformably over the Triassic. In the Cretaceous, the lake basins were shrinking, the climate became dry again, and the suite of brownish-red and saffron sandstone-mudstone formed. The lake basins disappeared in the Late Cretaceous and the region was uplifted and being eroded. But in the Cenozoic, basins quickly enlarged and resulted in the ubiquitous overlapping of the Paleogene and Neogene sediments over the underlying sequence. The Cenozoic deposits were red sandstone-conglomerate at the basin margin and siltstone-mudstone in the center. Since the Pliocene epoch, the surrounding mountains strongly uplifted and formed the foreland molasse layer of huge thickness. Compression from both the northern and the southern mountains was responsible for four second-grade structure units, from north to south: the foreland thrust faults in the Bogda, northern depression zone, the central fold-thrust zone, and the slope zone in the south. Thickness of the sediments filling the basins increased from south to north and the maximum thickness was in the northern depression zone.

4.7.3 Tarim Basin

Tarim Basin is a broad region sandwiched between the Altun fault in the east, Chinese South Tien Shan Mountains in the north, and West Kunlun Mountains in the south. The region is situated in western China and is characterized by the basin-range coupling, like the West Kunlun orogen, Tarim Basin, Tien Shan orogen, Junggar Basin, and Altay orogen. In the Tarim Basin, the Tarim continental block has been nearly completely covered with the terrestrial sediments. The cover sequence of the block was deposited in a rather stable setting in the continental block development. Besides the local hiatus between the Devonian and Carboniferous, Permian and Triassic, little deformation occurred. Terrestrial sediments initiated in the Permian. The lower Permian is dominated by the terrestrial clastics with intercalations of shallow sea carbonate rocks and multi-layers of basalts, while the upper Permian is red terrestrial deposits. But the uplift at the end of Permian caused the local absence of the Triassic. In the Bachu region and northeastern Tarim Basin, the upwelling of the mantle was responsible for the wide distribution of basalt and vertical dolerite swarms in the Permian, which is a typical feature of extension. In the Jurassic, the Tarim Basin developed to a great scale, and mainly deposited varicolored sandstones, shales with intercalations of coal measures of fluviolacustrine facies and limnetic facies. Red sanstones and conglomerates of lacustrine facies developed in the Cretaceous. The total thickness of the Jurassic and Cretaceous reaches some 3000 m. The basin was mainly developed in the Tertiary when gypsum and red formations of clastics and sediments up to 9000 m occurred, suggesting a substantial depression of the basin during the period. In the late Cenozoic, especially from Pliocene to Pleistocene, nearly leveled West Kunlun was again uplifted due to intense intracontinent compression from the northwards Indian Plate. Uplifting resulted in foreland depressions occupied by the molasse sequence of giant thickness and foreland thrusts in the northern Kunlun Mts. Meanwhile, the northwards moving rigid Tarim basement made a strong compression along the southern Tien Shan and formed the foreland depressions with huge thickness of the molasse sequence and thrust zone, and several NS strike regional faults in the Tarim Basin.

4.7.4 Qaidam Basin

Basin occupies the area between the Eastern Kunlun tectonic belt in the south and the Qilian tectonic belt in the north. Fillings in the basin are predominantly terrestrial strata of the Mesozoic to Cenozoic (Triassic is absent), among them, the Jurassic consists of coal-bearing clastic formation and the Cretaceous to Cenozoic are represented by red gypsum-bearing clastic formation with many intercalations of conglomerate and their thickness is up to 9000 m. In the Quaternary, the deposition center was in the lake area in the eastern basin and the maximum thickness is 315 m. In the Cenozoic, as the compression from the northern and southern mountains and the influence of the Altyn Tagh striking-slip in the west, the Mesozoic-Cenozoic sequences of the basin demonstrate a series of NW running short Z-shape axis folds. While at the margin, especially the northern margin, isoclinal overturned folds and thrust structures occur.

On the basis of the sporadic outcrops and some drilling records, the Paleo-Mesoproterozoic metamorphic basements exist under the basin's Mesozoic-Cenozoic sediments. Stable cover strata of the Sinian (Ediacaran) to Paleozoic were accumulated over the basement. The zone of the Proterozoic-Paleozoic rocks is sandwiched between two mobile belts, the Southern Qilian belt in the north and the Northern Qaidam margin belt in the south, and Paleoproterozoic basement of the zone is represented by the Dakendaban Group which is overlain with unconformity by clastic and carbonate rocks of the Sinian (Ediacaran), Cambrian, and Lower-Middle Ordovician, while the Upper Ordovician series are absent. The zone is weak in deformation, magmatism, and metamorphism thus may be a part of the Qaidam block. In the northern margin of the Qaidam Basin, there are a series NW running residual mount chains including the Serteng Mount, Liiliang Mount, Xitieshan Mount and Almunik Mount, along which the Paleoproterozoic basement and overlying Meso-Neoproterozoic sequence of various degrees of metamorphism are outcropping. The Lower Paleozoic mainly includes the upper Ordovician series (called Tanjianshan Group), which are the intermediate-mafic volcano-sedimentary associations of mobile feature and zones of the mafic–ultramafic complexes intruding the Ordovician system and older sequence. As volcanic rocks tend to be acid and due to low thickness of siliceous rocks, they do not have the features of a typical ophiolite. So we can consider that the residual mount chain may represent an aulacogen zone penetrating the northern original Qaidam block. If this is true, the aulacogen was folded at the end of the Ordovician, and the Silurian system was absent, while the volcano-sedimentary molasse of the Upper Devonian overlays with unconformity the folded aulacogen and both the northern and southern blocks. Similar situation happened to the Qimantag Mountain area at southern margin of the Qaidam Basin; the difference is that here the aulacogen extension was in the Middle Ordovician and was folded in the Late Ordovician. It should be pointed that before activity of the Altyn Tagh fault, the Qaidam block and the Tarim Platform were of one geological unit and had the basin covers which were similar. So they all have basements and covers of similar time and features, developed Paleozoic aulacogens at the margins, experienced the Early Mesozoic (after Triassic) uplifting and formed similar basin sediments of the Mesozoic-Cenozoic. They all underwent strong compression caused by the uplifting of the QinghaiTibetan Plateau. As this area is smaller than the Tarim Platform, the Qaidam block was more active and mobilized than the Tarim Platform where the stable inherited basins of the Paleozoic developed.

4.7.5 Songliao Basin

Running in NEE direction and semilunar in shape, the sedimentary basin is situated in northeastern China. Its main part is situated to the north and over-laying the Xilinhot and Zhangguangcai ranges fold belts. The southern end superposes the Ondor Sum fold belt. On the basis of the outcrops and drilling records, fillings of the basin consist of a sequence of the Mesozoic-Cenozoic which with unconformity overlies the sequence folded in the Late Paleozoic and granites of the same time. Sedimentation began in the Middle Jurassic, the Triassic—Lower Jurassic series are absent, suggesting the basin was in a state of uplifting in the Triassic to Early Jurassic. Meanwhile, the extension produced a series of NNE and NW running faults which controlled the comparted fault groups of small basins. In these small basins, quick sedimentation of the Middle Jurassic series includes sandstone-conglomerate, sandstone interbedded with mudstone, tuffaceous siltstone and thin coal measures (of alluvial facies and lacustrine facies). From the Late Jurassic to Early Cretaceous, the basins were further subsided, widened, and connected, finally forming a faulted basin of large area. The Upper Jurassic series are dominated by tuffs, tuffaceous conglomerate, and basalt with intercalations of thin layers of mudstone and coal measures of lacustrine facies. The Lower Cretaceous series were grayish-black, grayish-green, interleaving fossil-rich sandstone–siltstone, siltstone and marlite interbedded with sandstone-conglomerate, middle-coarse grained sandstone of various thickness, which were sediments of the subaqueous fan, lake floor fan, deep to semi-deep lake facies and pelagic facies setting. The above strata are thick and volcanic rocks partially have high alkalinity, 15–30%, suggesting the feature of mafic magmatism and deep magmatic activity (Tian 1990). In the early period of Late Cretaceous, the basin was at its flourishing stage and represented a faulted basin of huge area. The lower part of the Upper Cretaceous contains strata of sandstone-conglomerate, sandstone and mudstone, which constitute several sedimentary cycles of mainly lacustrine facies and alluvial facies and have a sharp unconformable contact with the underlying Upper Jurassic—Lower Cretaceous. The overlying Upper Cretaceous strata have an obvious overlapping towards the basin margin. Sediments of the latest Late Cretaceous to Neogene are red clastic formation composed of red sandstone-conglomerate interleaved with varicolored mudstone, implying the deposits of shrinking stage of crust uplifting and basin folding.

Tectonic phases between the Early and Late Cretaceous and between Late Cretaceous and Paleogene demonstrate the structure periods of alternating intrabasin anticlines and synclines formation. The basin folding was evident in the east and weak in the west and the depression center continuously migrated from east to west. The overlying Upper Cretaceous and younger sequence manifest gentle and open folds.

4.7.6 North China Basin

Basin is situated in the eastern part of the North China Platform, that is, the nowadays North China Plain region. It is characterized by the following features:

(1) Low thickness of the crust averaging 30–50 km and a negative gravity anomaly;
(2) The Cenozoic volcanoes were typified by fissure eruption. The Paleogene and Neogene basalts were of transitional tholeiite basalts, while the Quaternary were alkaline basalts. High heat flow has been measured in the basin. The temperature was up to 21–25 °C at the depth of 300 m, exceptional to 30–40 °C and the surrounding mountainous region 17–20 °C (Wang et al. 1982);
(3) Horst and graben structures are widely distributed. Graben structures can usually be traced to the old faults. The structures have obvious asymmetry and occur as dustpan like unilateral grabens and thickness of the sediments can be easily controlled by "the unilateral descending faults of grabens";

(4) In the surrounding mountainous region, some old sequences, especially the Paleo-Mesoproterozoic, commonly have sliding layers along the strongly ductile sheared bedding or foliation. The sliding layers formed in the Late Mesozoic (Yanshanian movement), suggesting the extension and crust thinning in the eastern North China Platform since the Late Mesozoic.

North China Plain is the principal part of the North China Basin and is overlain by the Cenozoic sediments, and the southern part of the basin is dominated by the Late Mesozoic sediments, therefore this is a deposition basin or Hehuai Basin since the Late Yanshanian. The Cenozoic basins over the North China Plain are faulted-depression basins developed on the metamorphic basements of the Archean, Proterozoic, and Paleozoic and their stable cover sequences. Before the faulted-depression basins, a long-term terrestrial open basin could occur. It was filled by the red sandstone-mudstone formations and coal-bearing clastic rock formations of the Permian to Triassic. Due to intensifying of the Tanlu strike-slip fault, the crust upwelled and the Upper Permian and Triassic eroded. The upwelling amplitude controlled the depth of the faulted depression and the sediments thickness. The area once uplifted higher could subside deeper later on and had thicker sediments. Starting in the Middle Jurassic to Cretaceous and Early Paleogene, small basin groups were commonly distributed along the faulted depressions. Fillings in small basins are red coarse clastic formation interbedding with coal-bearing layers and gypsum layers. Volcanic rocks are well developed and of fissure-type of eruption. Alkalinity of basalts is relatively high. This is actually the early development stage of the superposed basins. In the Oligocene to Miocene, the basins were broadened and small basins were connected to one big faulted depression basin and melanocratic mudstone of abyssal to hypoabyssal lacustrine facies deposited. In Pliocene epoch to Quaternary, shallow lake-shoreline facies, bed facies, and alluvial-pluvial facies sediments were accumulated, like the interleaving red sandstone-conglomerate, varicolored sandstone and mudstone, accompanying the eruption of some mafic and intermediate-acid volcanic rocks.

Disjunctive structures of the basins are mostly normal faults, which generally have steep dips at the surface and shallow level and gentle dips or listric faults in the deep. Spatially, the listric faults form a series of dustpan-like unilateral grabens which are composed of steep slope to deep depression induced by descending of the downcast side of a normal fault at one side and a gentle slope at the other side. Along the downcast side of the fault, growing uplift and dragging anticline formed simultaneous with the sedimentation. In a deep depression, anticlines can be formed by dip compression of the boundary faults. At the gentle slope side, eroding unconformity of the overlapping strata or the fault steps was complicated by faulting. Structure styles vary according to the levels, that is, massive structures in the deep, common normal faults, growing faults, dragging anticlines related to the faults, strike-slip faults and salt domes in the middle level and gentle draping structures in the shallow level.

4.7.7 Tectonic Evolution of North China

The oldest crustal age of rocks in north China is 3800 Ma. The basement of the North China (Sino-Korean) Platform underwent at least the following tectono-thermal events and orogenic cycles: the Qianxi or Tiejiashan cycle (3000 Ma), Fuping cycle (2600 Ma), Wutai cycle (2400 Ma) and Zhongtiao cycle (1800 Ma). At 1800 Ma, cratonization was completed. The basements of the Yangtze and Tarim blocks were initially consolidated before 1800 Ma, but ultimately consolidated in the period of 700–800 Ma and connected with the North China (Sino-Korean) Platform, forming the so-called paleo-China Platform. E. V. Khain (Khain et al. 2002) in due time reported the occurrence of the 1020 Ma ophiolite in the southeastern Zaysan Range area and pointed out that the ocean basin opened in the Late Mesoproterozoic. In addition, the 500–600 Ma ophiolites have been noticed in the Zaysan Range area and Western Mongolia (Buchan et al. 2002), suggesting that opening of the oceanic basin had lasted at least to the beginning of the Paleozoic. Ren Jishun (Ren et al. 1999) took the oceanic basin between the paleo-China Platform and Siberian Platform as the Proterozoic Paleoasian Ocean. In the earlier of Cambrian, the Salairian folding made the Siberian Platform accreting southwards and formed the outstanding Altay–Zaysan orogenic belt through Central Mongolia to Ergun. Meanwhile, the paleo-China Platform broke again and formed the Zaysan—Southern Mongolia—Khingan—Tien Shan—Junggar Paleoasian oceans, which were distributed between the northern margin of the paleo-China Platform and the Sayan—Central Mongolia—Ergun orogenic belt. Inside the platform, the Kunlun—Qilian—Qinling aulacogen and small ocean basins occurred and in a broader region, microcontinent blocks of diverse sizes were comparted by the ocean basins or aulacogens. The Caledonian orogenic movement from the end of Early Paleozoic to Middle Devonian epoch made a large part of the Paleoasian Ocean folding and the Kokchetav, Issyk Kul, and Yili microcontinent blocks aggregated to form the Kazakhstan mosaic block, the Bureya–Jiamusi–Xingkai microcontinent blocks to Jihei block. Northwards accreting of the Sino-Korean Platform was responsible for the narrow and long Ondor Sum fold belt. The Kunlun—Qilian—Qinling small oceanic basin and aulacogen closed and formed the Caledonian fold belts. The North China (Sino-Korean), Yangtze, and Tarim blocks aggregated for another time.

Evolution of the Paleoasian Ocean continued till the Late Paleozoic and closed successively from the Early Carboniferous to the Early Permian along the Southern Tien Shan—Beishan—Xilinhot line. While the ocean along the Solonker—Xar Moron line closed in the Late Permian, and the evolution of the Paleoasian Ocean was finally finished.

Closing of the Paleoasian Ocean made the paleoChina Continent a part of Laurasia and entered the evolution stage of joint operations of two dynamic systems, the Tethyan and Pacific systems. Southern Asia is mainly constrained by the Tethyan dynamic system. The Paleotethys was closed in the Late Permian and formed the Permian–Triassic Yushu–Jinshajiang and A'nyemaqen fold belts in southern China. Eastern Asia was controlled by the Paleopacific dynamic system, where the blocks were strongly compressed, overlapped, and welded together. In the Jurassic—Cretaceous period, opening–closing of the Neotethys in southern China continent finally resulted in the Dangla (Tanggula) and Karakorum collision belts. Collision of eastern Asia and the "Pacific continent" shaped the outstanding tectonic–magmatic mobile belt in Eastern China. In Cenozoic, the Indian continental plate and the Eurasian Continent compressed further (continental subduction) and after the Yarlung Zangbo suture, old fold belts like the Kunlun, Tien Shan and Qilian were reworked and rejuvenated as new uplifts. Strong extension effect of the marginal seas in the Western Pacific or Eastern Asia can account for a series of faulted depressions and depression basins on the continent in the Cenozoic (Fig. 18).

Tectonic evolution of North China resulted in unique and numerous lithospheric textures in North China. Great difference of the lithosphere and asthenosphere exist between eastern and western China (Li 2006). The lithosphere and asthenosphere of western China have an obvious "layer" structure, thick lithosphere of 130–200 km and thin asthenosphere of 40–100 km, implying the collisional effect between the Indian and Eurasian plates. The lithosphere and asthenosphere of eastern China show the "block and mosaic" structure, thin lithosphere of 50–100 km and thick asthenosphere of diverse thickness, changing in the range of 200–300 km, suggesting the upwelling of the asthenosphere and extensional thinning of the lithosphere (Fig. 4.19). Taking the NS running Helanshan Mountains—Chuandian tectonic belt as a boundary, the

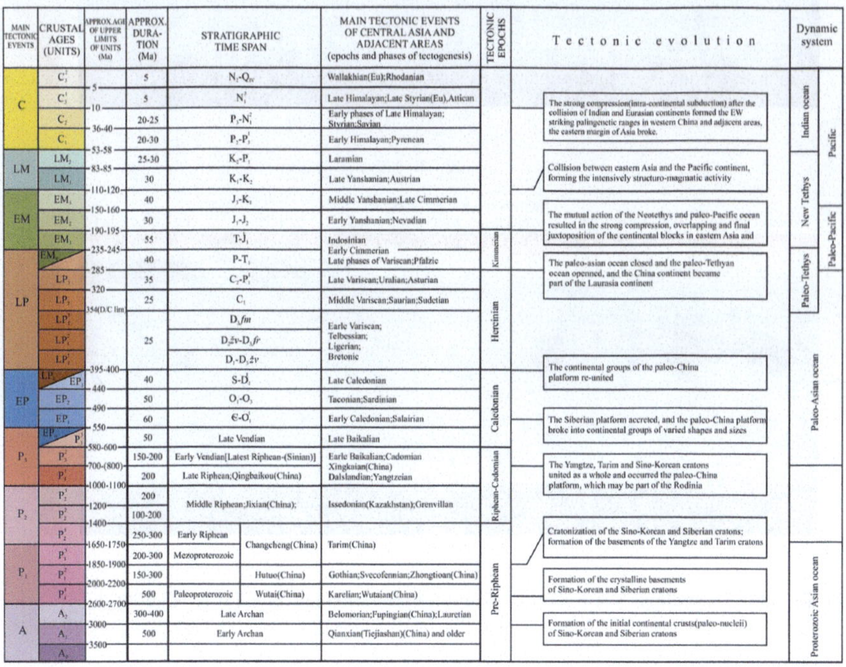

Fig. 18 Sketch table of tectonic evolution of Northern China and adjacent areas

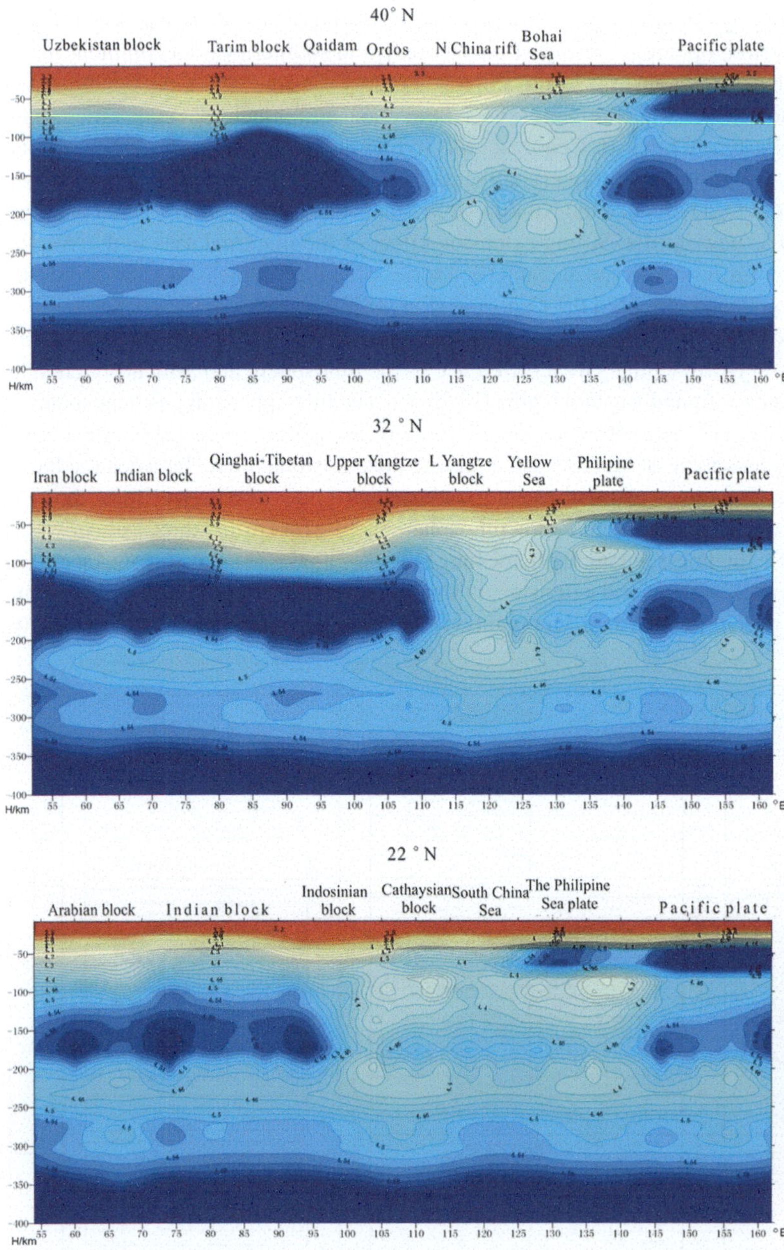

Fig. 19 Three latitudinal S-wave velocity profiles transversing China [after (Li 2006)]

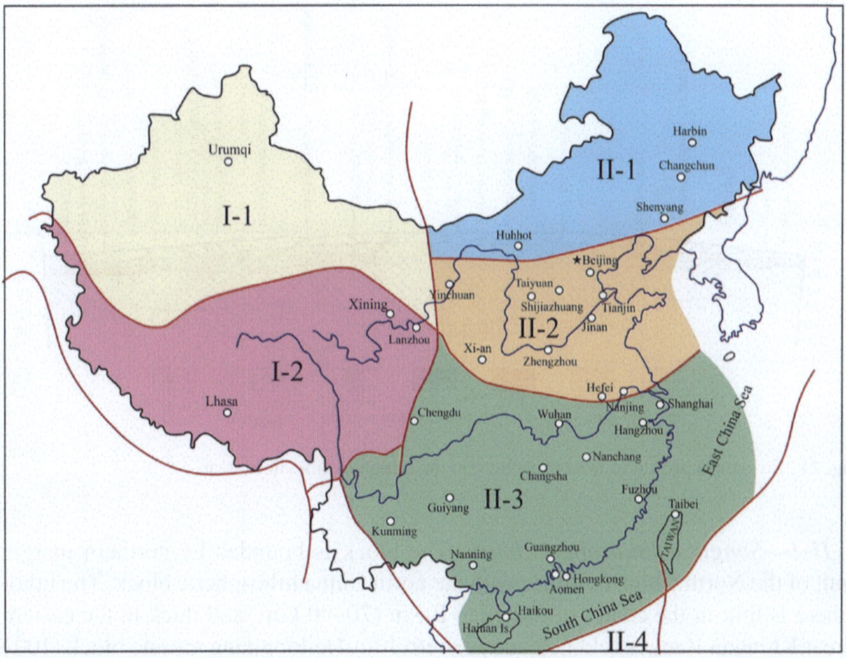

Fig. 20 Lithosphere tectonic units of the China mainland and adjacent sea areas (explanations see the text) (Li 2006)

China Continent its adjacent region can be subdivided into two first-grade lithospheric tectonic units, the eastern and western, as mentioned above, which are further subdivided into 6 second-grade tectonic units: I-1—Xiyu lithospheric block; I-2—Qinghai-Tibet lithospheric block; II-1—Songliao lithospheric block; II-2—North China lithospheric block; II-3—South China lithospheric block and II-4—South China Sea lithospheric block (Fig. 20).

Below we shall give the short characteristics of the lithospheric tectonic units.

I-1—Xiyu lithospheric block. Bounded by the Kengxiwar fault, Altyn Tagh fault, and Qilian northern margin fault in the south and the NS tectonic line in the east, the block is 140–160 km wide in the NS and 100–120 km thick in the lithosphere. With the average thickness of 45–55 km, the crust can be sectored into three layers of thick-crust and thick lithosphere structure. The lithospheric mantle has high shear wave velocity, suggesting rigid feature of the lithosphere.

I-2—Qinghai-Tibet lithospheric block. Situated south of the I-1, both the lithosphere and crust of the block are thick in the central part and thin in the margin, with the average thickness of 140 km (100–200 km). The crust is 70 km in mean thickness (Fig. 21), its density is low, demonstrating the feature of thick crust and lithosphere, hot crust and cold lithosphere, light crust and heavy lithosphere and soft crust and hard lithosphere.

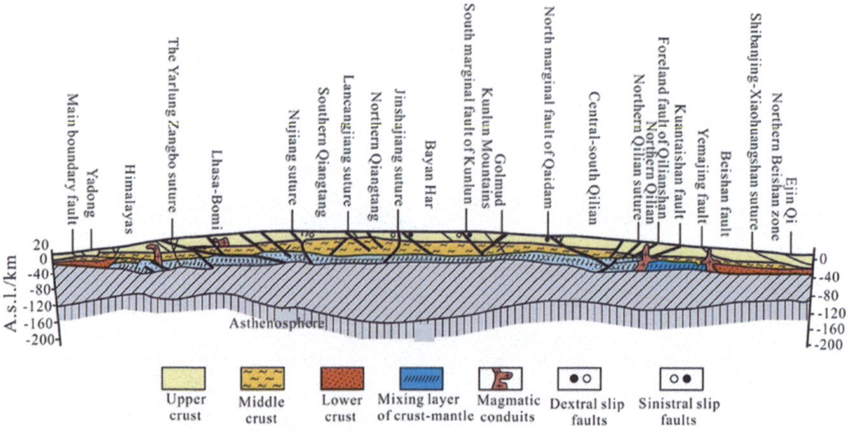

Fig. 21 Structural profile of the lithosphere in the Qinghai-Tibetan Plateau

II-1—Songliao lithospheric block. The block is bounded by northern margin fault of the North China Platform with the north China lithospheric block. The lithosphere is thin in the center of Songliao Basin (70–90 km), and thick in the eastern Great Khingan Ranges fold belt and western Jilin-Heilongjiang mosaic block (100–120 km), crust thickness averaging at 30–40 km, lithosphere some 40 km (Fig. 22), implying the crust and lithosphere are both a thin and light feature.

II-2—North China lithospheric block. Situated south of the Songliao lithospheric block, the North China lithospheric block is bounded by the central Qinling fault with the South China lithospheric block. Thickness of the block is 100–120 km at the margin and in the Ordos Basin, 60–80 km in the North China Basin and only 35–40 km in the Shanxi Plateau, that is thick in the west and thin in the east. The same situation is for the crust: the Ordos Basin 40–45 km in the west, the Shanxi Plateau 35–40 km in the middle and the North China Basin 30–35 km in the east (Fig. 23), showing the crust and lithosphere are both a thin and light feature.

II-3—South China lithospheric block. The block is located south of the North China lithospheric block and the crust shows three layers in structure; it is thick in

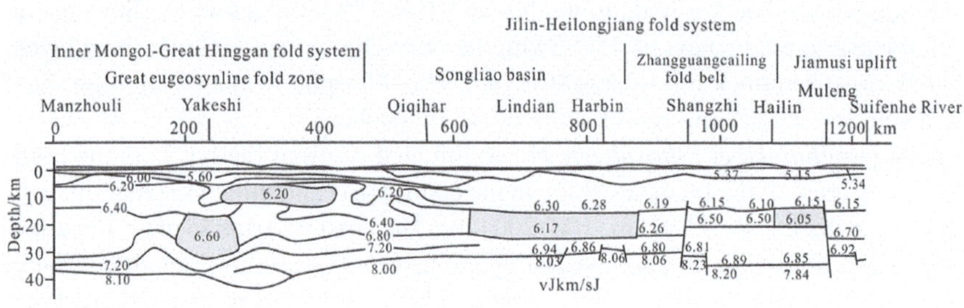

Fig. 22 The Manzhouli-Suifenhe geoprofile (Li 2006)

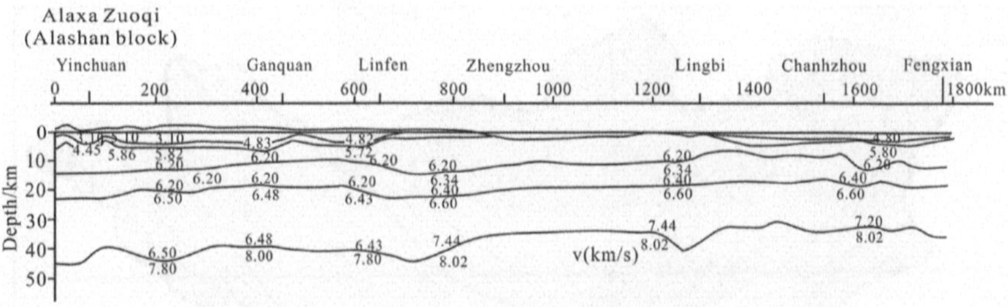

Fig. 23 Seismic profile of Alaxa Zuoqi, Inner Mongolia—Fengxian, Shanghai (Li 2006)

the west, thin in the east, averaging 30–33 km. Thickness of the lithosphere varies greatly: 100–120 km in the southern Qinling-Dabie belt at the northern margin, up to 160–200 km in central Hunan–northern Jiangxi Provinces, 80–100 km in central Sichuan, Jianghao Basin and the southeast coast area, 50–70 km at Taiwan Island and thinner on the East China Sea slope. Lithosphere has a feature of thin crust—thick mantle and light crust—heavy mantle.

5 Tectonics of Mongolia

5.1 Mongolian Part of the Sayan-Baikal Fold System

Mongolian part of the Sayan-Baikal fold system includes the region of Lake Khubsugul, southwestern and northern slopes of the Khangay upland region and the Selenga River basin. Structurally, it consists of the Central Mongolian massif and Eastern Pre-Khubsugul and Bayankhongor nappe-fold megazones (Fig. 24).

5.1.1 Central Mongolian Massif

Central Mongolian massif is composed of the Precambrian accretionary basement and the Vendian (Ediacaran)—Lower Cambrian sedimentary cover.

Structurally, the basement consists of combined Neoarchean and Proterozoic metamorphic and intrusive complexes of different age. The age of the oldest Baydrag "grey" tonalite gneiss is 2.8–2.645 Ga (Kozakov et al. 2005), and the Dariv (Daribi) ultramafic complex is covered with Paleoproterozoic metamorphic complexes (Tomurtogoo 1989). The latter includes amphibolite, leptynite, different crystalline schist, quartzite (often ferruginous), marble, calciphyre (Mitrofanov et al. 1981). Besides, in the western and northern Khangay the massif includes large bodies of gabbro-anorthosite with U–Pb age of 1784 ± 10 Ma (Anisimova et al. 2008). In addition, it in the southwest Khangay there is a Late Mesoproterozoic shelf complex with layers of stromatolite limestone and gold-bearing black schist, which

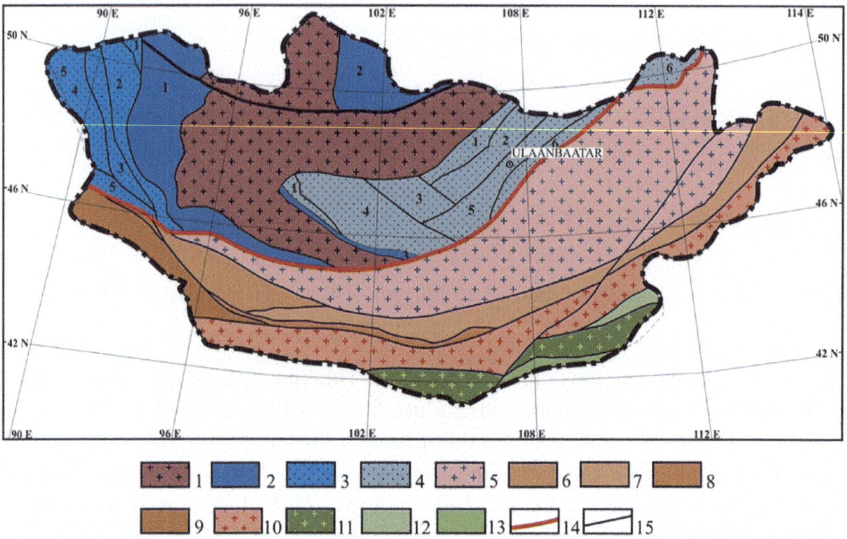

Fig. 24 Map of tectonic zoning for fold areas in Mongolia. 1–4—**North Mongolian fold area**: 1—Central Mongolian massif, 2—Lake (1), Eastern Pre-Khubsugul (2) and Bayankhongor (3) nappe-fold megazones, 3—Mongolian Altay fold system (1–5—terranes), 4—Khangay-Khentey fold system (1–6—terranes); 5–13—**South Mongolian fold area**: 5—Kerulen massif, 6–9—Gobi nappe-fold megazone (6–9—terranes), 10—South Gobi massif, 11—Khatanbulag massif, 12–13—Sulinkheer (Solonker) nappe-fold megazone (12—Zamyn-Uud terrane, 13—other terranes); 14–15—faults: 14—***Mid-Mongolian Tectonic Line***, 15—other faults

stratigraphically covers Paleoproterozoic rocks (Boishenko 1978; Mitrofanov et al. 1981). Similar to these types, there is the Dzhargalant high-temperature metamorphic complex (Tarbagatay Range, Northern Khangay) represented by intercalation of amphibolite, amphibole crystalline schist, marble and biotiteamphibole gneiss with Nd-model age of 1.5–1.3 Ga (Kozakov et al. 2008, 2013).

Neoproterozoic metamorphic and intrusive rocks play an important role in the composition of the massif's basement.

The most widespread Lower-Middle Neoproterozoic stratified rocks are characterized by different types of successions. Two megablocks are clearly identified in the massif basement in accordance with their distribution: External Khangay and West–Pre-Khubsugul, between which the boundary coincides with the sub-latitude Khangay sinistral strike-slip fault.

In the southern External Khangay megablock, the Lower Neoproterozoic is represented by: (1) gneissmigmatite complex including island-arc-type gneissic granitiods with U–Pb age of 856 ± 2 Ma and (2) multiphase amphibolite-quarzite-carbonate-slaty complex with stromatolite limestone; and the Lower-Middle Neoproterozoic consists of (1) metaophiolite complex related to Tas-Uul, Khutul and Onts-Uul allochthons in the basins of the Dzabkhan and Khunguy-Gol rivers (Tomurtogoo

1989) and (2) metavolcanogenicsedimentary complex of the "back-arc" and "island arc" types with an age of 874–790 Ma (Tomurtogoo 1989; Kozakov et al. 2013).

Lower-Middle Neoproterozoic complexes in the Western Khangay are intruded by numerous batholiths of calc-alkaline granites of the Dzabkhan-Mandal complex (gneiss-granite, quartz diorite, tonalite and plagiogranite) with U–Pb age of gneiss-granite of 856 ± 2 Ma (Tomurtogoo 1989; Kozakov et al. 2013). Evidently, metamorphic analogue of this complex in the Northern Khangay belongs to earlier synmetamorphic granitoides with Zr age of 810 ± 2 and 809 ± 4 Ma that intruded the Dzhargalant metamorphic complex in the Tarbagatay Range (Kozakov et al. 2013).

In the northern, West–Pre-Khunsugul megablock of the Central Mongolian massif, the Lower-Middle Neoproterozoic basement includes: (1) terrigenous-carbonate complex (Kharaberin series and its analogues) that covers with transgressive contact the Ulaantaiga block in the Ulaan-Taiga Range, (2) metaturbidites with serpentinite protrusions (Sharga-Gol River valley in the northeastern frame of Darkhat depression), which knowingly are the analogues of Dibin series of the Eastern Sayan, (3) metaophiolite with rhyolite dated as 800 + 5 Ma, forming large Shishged Allochthon, and (4) greenschist complex with HP metamorphism (Khugen-Gol series) located in the Shishged Allochthon (Il'in 1982; Sklyarov et al. 1996; Kuzmichev 2004; Kuzmichev et al. 2005). In the Western Pre-Khubsugul and Western Khangay, Middle Neoproterozoic sedimentary-volcanic formations dominated by successions of subaeral acid volcanic rocks are known. In contrast to Lower-Middle Neoproterozoic complexes, Upper Neoproterozoic acid rocks fill individual depressions and grabens superimposed on them with sharp unconformity. Age of acid lavas was determined as 805 Ma (Dzabkhan Formation) and 790 Ma (Darkhat = Sarkhoy series) (Levasheva et al. 2009; Kuzmichev and Larionov 2011). Typically, subvolcanic quartz porphyries are associated with volcanic rocks of both series and small subalkaline granite intrusions of 774 ± 3 Ma age of crystallization (Kirnozova et al. 2009), alkaline granites of 755 Ma (Yarmolyuk et al. 2008) and riebeckite granite of 752 Ma (Il'in 1982).

The Upper Precambrian—Lower Cambrian sedimentary cover of the Central Mongolian massif is represented by shallow-water carbonate-terrigenous deposits that fill in depressions at the margin of the massif: Khubsugul, Tsagaan-Olom and Buuraltay ones.

The famous sections of Tsaagan-Olom and Khubsugul series are rather similar. Both successions start with tilloides, which then give way to thick (3000–5000 m) carbonate-terrigenous rocks with phosphorite deposits (Il'in1982; Tomurtogoo 1989; Lindsay et al. 1996; Dergunov 2001). Carbonate rocks are represented by dolomite and limestone, often with different fossils including archaeocyatha and trilobites (Marinov et al. 1973). Terrigenous rocks are dominated by siltstone, which composes layers of first hundreds meters of the section. Deposits of both series described above belong to typical sparagmite complex. Its tilloids can be correlated in their age with those of the Nantuo Formation of the Yangtze Platform sedimentary cover, because both of them are conformably overlain by limestone horizon with an age of about 635 Ma (Ovchinnikova et al. 2012).

One of the most important features of the Central Mongolian massif is the presence of Paleozoic magmatic complexes of different age (Gavrilova et al. 1975; Tomurtogoo and Gerel 1999).

The earliest of them are the Khoshim-Gol clinopyroxenite-gabbro complex [588 ± 15 Ma (Badarch et al. 2002)] and batholith calc-alkaline granitoid Telmin complex of diorite, granodiorite and amphibole-biotite granite [K–Ar age from 488 to 426 Ma (Gavrilova et al. 1975)].

Late Devonian alaskite-granite association occupies an essential area (Tes complex and its analogues). Its formation took place after the intrusion of the alkaline-gabbro complex (396–400 Ma) (ijolite-urtite, foyaite and nepheline syenite) in SW Pre-Khubsugul region (Gavrilova et al. 1975).

Clearly superimposed character is typical of the Permian (northward with Lower Triassic) North Mongolian volcano-plutonic belt which includes two magmatic associations: the calc-alkaline volcano-plutonic association and the association of comendite, pantellerite, alkaline basalts and alkaline granites (Gavrilova et al. 1991). The volcanic rocks of the first association usually include continental molasse and fill large volcano-tectonic structures (Orkhon-Selenga, Khan-Tayshir and Buu-Tsagaan troughs) but the second association is strictly confined to narrow fault zones—paleorifts (Dergunov 2001).

Tectonically the Central Mongolian massif is represented by a large superterrane (microcontinent) composed of different-aged and different-type terranes amalgamated by the beginning of sedimentation at its margins of the Late Precambrian—Lower Cambrian sparagmite complex. As a whole, it consists of southern and northern groups of Early Baikalian terranes situated respectively to the south and north from the Khangay strike-slip.

5.1.2 Eastern Pre-Khubsugul Nappe-Fold Megazone

The Eastern Pre-Khubsugul nappe-fold megazone includes the western continuations of the both Khamar-Daban and Dzhida zones in the west of Trans-Baikal region in Russia.

Composition of Mongolian part of the Khamar-Daban zone is studied quite insufficiently. It is known that a gneiss-schist zonal-metamorphic complex probably of the Ediacaran to Lower Cambrian age is distributed in this zone, which from the south is overthrust by a carbonate-slate complex (Bitu-Dzhida series). Small outcrops of Silurian carbonate-terrigenous strata are similar to those from the adjacent area of the Eastern Sayan (Badarch et al. 2002). It is supposed that the Khamar-Daban zone corresponds to the same-name back-arc basin terrane (Al'muhamedov et al. 1996).

In Mongolia, the Dzhida zone surrounds the Khamar-Daban zone from the south and west tracing the rift valley of Lake Khubsugul to the junction with the Ilchir ophiolite zone of the Eastern Sayan. Its boundary with the Khamar-Daban zone is mainly traced marks along the Arigiyn-Gol thrust. In the west and south, the Dzhida zone is cut by the Egiin-Gol and Khangay faults along which it borders the Central Mongolian massif.

Mongolian part of the Dzhida zone (Al'muhamedov et al. 1996; Il'in 1982; Tomurtogoo 1989, 2012; Tomurhuu 1999) is characterized by pronounced thrust nappe-folded structure and includes a system of structural-formational Late Precambrian—Early Cambrian complexes formed in geodynamic environments of ensimatic island arc and seamount.

Island arc structural-formational complexes differ in composition and make up most part of their tectonic nappes. These nappes consist of rock associations belonging to three main types: (1) dismembered and mélanged ophiolite association with boninite pillow-lavas (542 Ma) (D. Tomurkhuu, personal comment) and/or tholeiitic series, (2) volcanic complex of fractionally differentiated calc-alkaline series with limestone lenses containing Yudoma oncolites, and, (3) terrigenous-tuffite complex with minor horizons of olistostrome and/or carbonate rocks with the Lower Cambrian archaeocyatha.

Seamount structural-formational complex is known in region of Egiyn-Gol and Uri-Gol rivers confluence and consists of high-Ti basaltic pillow-lavas and oncolithic limestone (Al'muhamedov et al. 1996; Il'in 1982).

Plutonic rocks widespread in the Eastern Pre-Khubsugul region form numerous batholith-type massifs of Paleozoic granites of diorite-tonalite-granodiorite and alaskite-granite complexes. Concordant U–Pb zircon dating of 480 Ma (Kotov et al. 1997) is known for the first complex (melanocratic tonalite), but for the second complex usually Devonian age is supposed.

5.1.3 Bayankhongor Nappe-Fold Megazone

Bayankhongor nappe-fold megazone includes the extensive but narrow tectonic block of the Southern Khangay between subparallel Galuut normal fault (north) and Ulaan-Tolgoy strike slip fault (south) and is composed of the Middle Neoproterozoic ophiolite and different-types of the Ediacaran—Early Cambrian complexes (Kovalenko et al. 2005; Tomurtogoo 1989). Megazone extends for more than 300 km to the north-west having a width of 20–50 km. It has complicated nappe-folded and imbricated structure with monomictic serpentinite melange.

Ophiolite complex includes serpentinite, websteritepyroxenite-gabbro cumulates, "upper" gabbroides with anorthosite and plagiogranite veins, dyke series of typical "Sheeted Dyke Complex" and pillow-lavas (Tomurtogoo 1989). Composition of dykes and lavas is dominated by plagioporphyrite and dolerite belonging to different petrochemical series (from tholeiitic to alkaline) and mostly correspond by geochemical parameters to MORB. Rocks are metamorphosed to epidoteamphibolite degree and along Ulaan-Tolgoy and other faults are tranformed into mylonites. Anorthosite and cumulative gabbro have the age of 655 Ma by U–Pb dating (Kovach et al. 2005).

Three stratified complexes belong to Ediacaran—Early Cambrian age: carbonate-basaltic and carbonateslate in the south and greenschist metamorphosed volcano-quartzite-slate—in the north. The first is represented by massive and agglomerated lavas of basalts and their tuffs containing sponge specula of probably Early Cambrian age and oncolitic limestones (Kovach et al. 2005; Ryazantsev 1994; Tomurtogoo 1989).

The second, carbonate-slate complex is known as Uldzit-Gol Formation. It is composed of limestone with "black" carbonaceous phyllite-type slates, terrigenous olistostrome horizon with olistoliths of Bayankhongor ophiolite complex and total absence of phosphor-enriched rocks. This carbonate-slate complex includes fossils of Yudoma oncolites and Amga ecardinal brachiopods (Ryazantsev 1994).

The last, volcano-quartzite-slate complex is completed by metabasalts with interlayers of quartzites, followed by metaturbidites (total 3000 m thick), containing sponge spicules of probably Early Cambrian age (Ryazantsev 1994).

Regional metamorphism of ophiolite and other complexes in the Bayankhongor nappe-fold megazone took place about 484.5 ± 5.9 Ma and rocks were exhumed before the beginning of sedimentation of wittingly Upper Ordovician marine molasse (Tomurtogoo and Gerel 1999; Kovalenko et al. 2003).

The Early Paleozoic plutonic rocks occupy minor position in the composition of the Bayankhongor megazone. They form granite plutons with zircon age from 530 to 455 Ma (Kovach et al. 2005; Jahn et al. 2004).

Thus, analyzing the aforesaid facts, in the Bayankhongor nappe-fold megazone, fragments of paleooceanic crust and marine sediments formed during the Late Precambrian—Early Cambrian in low-spreading center, seamount and marginal part of oceanic basin, were structurally combined. As neoautochthon in the zone is represented by Middle Ordovician red granites transgressively covered with Upper Ordovician marine molasse (Tomurtogoo and Gerel 1999), closing of the Bayankhongor oceanic basin and its transformation into suture zone undoubtedly took place between the Middle Cambrian—Early Ordovician.

5.2 *Mongolian Part of the Altay-Sayan Fold System*

Mongolian part of the Altay-Sayan fold system ranges in scale of the mountain system of the Mongolian Altay and adjacent areas of the Depression of Great Lakes and the Valley of Lakes. Tectonically, it includes the Salairian (Early Caledonian) Lake (Ozernaya—widespread Russian name) nappe-fold megazone in the east and the Caledonian fold system of the Mongolian Altay in the west (see Fig. 24).

5.2.1 Lake Nappe-Fold Megazone

The Lake nappe-fold megazone is dominated by the ophiolites and volcano-sedimentary formations of Vendian (Ediacaran)—Lower Cambrian and also by Paleozoic intrusive complexes.

At present, two types of ophiolite sections are known. The first one is the Khan-Taishir ophiolite complex, for the first time identified in the Khan-Taishir

Range (Zonenshain and Kuzmin 1978) and later detailed along the Tsuvraa-Gol canyon (Gibsher et al. 2001). The section includes (upwards): mélanged dunite-harzburgite ultramafic rocks, pyroxenite-gabbro cumulative complex, "upper" gabbroides, boninitediabase dyke series like "Sheeted Dyke Complex", pillow-lavas with basal horizon of eruptive breccias (1000–1500 m) and strata of deep-water siliceous sediments (jasper, siliceous slates—300 m). In total petrological, geochemical and geological data the Khan-Taishir ophiolite corresponds with paleooceanic crust of inter-arc basin (Zonenshain et al. 1985).

The second type is represented by the Bayan-Nuruu ophiolite complex outcropped in the northern slope of the Ikh-Daribi Range (Kheraskova et al. 1985). It consists of dunite-harzburgite ultramafic rocks, dunite-vehrlite-clinopyroxenite-gabbro layered series, diabase dyke series and jasper-spilite formation up to 1500 m. The latter is dominated by pillow-spilite; variolite, diabase, lava breccias and gialoclastics are also recorded. Chemical data indicate that basalts are ranked between oceanic and island arc basalts. In contrast to the Khan-Taishir section, the Bayan-Nuruu ophiolite complex is covered with conformity with rocks of subdivided-differentiated basalt-andesite-rhiolite volcanic series which mark the paleooceanic crust of the island arc basement.

Khan-Taishir and Bayan-Nuruu ophiolite complexes are characterized by the presence of pre-dyke plagiogranite veins dated 570–573 Ma (Sm–Nd and U–Pb methods) in their composition (Gibsher et al. 2001; Dijkstra et al. 2006). Siliceous deposits of the Khan-Taishir complex with eroded but structurally conformable surface are covered with carbonate-terrigenous strata with archaeocyatha limestone (Tomurtogoo 1989).

Internal structure of the Lake megazone is characterized by a system of nappe sheets underlain by serpentinite mélange. The most representative sheets are outcropped in the Bayan-Nuruu Mts. in the northern slope of the Daribi Range, along northern foot of the Naran ultramafic massif and at riverhead of Ichetuyn-Gol at the Khan-Khukhey Range southern slope.

Volcano-sedimentary formations of the Lake megazone include a number of submarine volcanics and as well as terrigenous-siliceous-tuffite with olistostrome horizons and siliceous-carbonate, siliceous-terrigenous and siliceous-carbonate formations (Dergunov 2001; Dergunov et al. 1980).

The largest area in the axial part of the megazone is occupied by volcanic basalt-andesite rocks with reef limestone (Khan-Khukhey, Ulaan-Shand Ranges) and basalt-andesite-rhyolite formation with average thickness of 1500 m. They are of island arc nature and Late Vendian (Ediacaran)—Early Cambrian age (Yudoma complex microphitholiths and archaeocyatha). Volcanic rocks of differentiated basalt-andesite-rhyolite series have everywhere the cramped lateral connection with siliceous-sedimentary rocks with high amount of tephroids or tuffites. Thus in the axial part of the megazone, fragments of island arc volcanoes and adjacent inter-volcano sedimentary depressions are present.

Principally other types of the Vendian (Ediacaran)—Early Cambrian volcano-sedimentary formations are present in NE and SW borders of the megazone. So, if in the former (NE border) tholeiitic formation is conformably covered with siliceous-carbonate formation with archeocyath and algae limestone, in the latter (SW border), the lower position belongs to carbonatesiliceous-basalt formation with archaeocyatha limestone and the upper one, to terrigenous-tuffite formation. The last complexes (sub-alkaline basalts with association of siliceous rocks and reef limestone) are typical of oceanic guyot (Tomurtogoo 2002). The Vendian (Ediacaran)—Early Cambrian volcano-sedimentary formations of the Lake megazone are intensively deformed everywhere and take part in nappe sheets and in numerous isoclinal folded structures.

Younger Paleozoic and Mesozoic-Cenozoic stratified sediments of the Lake megazone everywhere form superimposed structures of different types (Dergunov et al. 1980).

The Middle Cambrian is represented episodically by low-thick carbonate-terrigenous series with trilobite fossils of the "Amga" complex.

The Ordovician is composed of volcanogenic molasse with primary distribution of subaeral andesitic basalts and rare lenses of coral limestone.

For the stratified Silurian, Devonian and Lower Carboniferous, the most representative succession was described in the Chargat superimposed trough situated northwestward of Lake Khar-Us-Nuur (Fig. 25). It is filled with Silurian parti-colored terrigenous, Lower Devonian porphyry, Middle Devonian and Lower Carboniferous carbonate-terrigenous molasse sections.

Plutonic rocks widespread in the Lake megazone are united into Khyargas-Huur, Tokhtogenshil and Tes complexes of Paleozoic age (Gavrilova et al. 1975; Dergunov et al. 1980; Izokh et al. 1990).

Khyargas-Huur complex is represented by concentric zonal plutons of layered gabbroids composed of frequent intermittency of olivine gabbro, gabbro-norite, troktolite, anorthosite, plagio-vehrlite and plagio-lerzolite (Izokh et al. 1990). U–Pb-age of 511 $\pm$ 12 Ma was determined for the rocks of the most representative Khairkhan pluton (Izokh et al. 2009).

Tokhtogenshil complex includes diorite-tonaliteplagiogranite, mesocratic biotite-amphibole diorite, tonalite and high-Na granitoides of different composition (Gavrilova et al. 1975). Radiological age of tonalite and plagiogranite is determined as 494 Ma and that of intruding postcollisional granite as 465 Ma (Kovalenko et al. 2003).

The Devonian Khalzan and Tes complexes include the intrusions of alkali and A-type granitoids, represented by rare-metallic (390–380 Ma), leucocratic and alaskite granites with facies of subalkaline granites, granosyenite and quartz syenite (Gavrilova et al. 1975; Kovalenko 2003).

Tectonically, the Lake megazone in most of the area is represented by collage of island arc, ophiolite and turbidite terranes to which the lengthy Altan-Khukhei seamount terrane is bordered in the west (Tomurtogoo 2002). By all appearances, in the latter, individual fragments of accretion prisms are present. It is evidenced by small outcrops of high-pressure rocks in the Zamtyn-Nuruu Mountains (Hanzl and Krejcu 2008). Now it is very difficult to determine exact boundaries of all terranes. In the Lake megazone, the widespread mass overthrust structures are predominant which are destroyed by vertical faults of different orientation. And the largest part

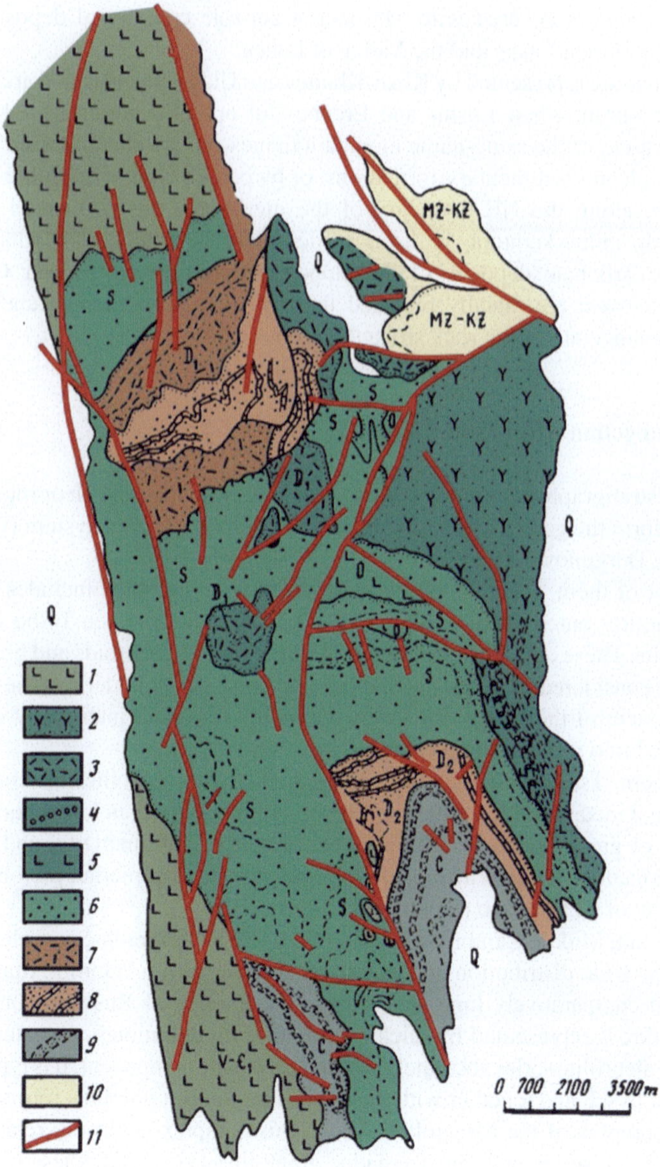

Fig. 25 Schematic geological map of the area around Mt. Sharagat and Mt. Urgat, on the northern slope of Lake Khara-Us-Nur: Chigirtai troughs (see Fig. 24). 1—volcanic and carbonate rocks of the Vendian (Ediacaran)—Lower Cambrian complex; 2–5—Lower-Middle Ordovician rocks: 2—mixed lava and tuff, 3—acid lava, 4—agglomerate, 5 mafic lava and tuff; 6—Silurian red-colored conglomerate and sandstone; 7—Lower Devonian lava and tuff; 8—Middle Devonian limestone-terrigenous deposits; 9—Lower Carboniferous red-colored siltstone and sandstone; 10—unconsolidated Quaternary deposits; 11—faults

of the megazone is covered with Mesozoic-Cenozoic continental deposits in the Depression of Great Lakes and the Valley of Lakes.

Exceptions are represented by Khan-Khuhey and Ulaan-Shand island arc terranes, and Bayan-Nuruu, Khan-Taishir and Erdene-Uul ophiolite terranes. All of them occupy the areas of the same-name mountain ranges and are characterized either by island-arc volcano-sedimentary formations, or by certain type of ophiolite sections. In addition, along the NE boundary of the megazone, the Daagandel back-arc terrane with spilite-keratophyre lavas is definitely determined, in which upper terrigenous-carbonate deposits of the Upper Neoproterozoic—Lower Cambrian from east to west are facially replaced first by siliceous-carbonate and then by terrigenous-siliceous-tuffite rock strata (Dergunov 2001).

5.2.2 Mongolian Altay Fold System

Four large stratigraphic complexes with an interval from the Late Neoproterozoic to Devonian form the geological basis of the Mongolian Altay fold system (Dergunov et al. 1980; Dergunov 2001).

The first of them, Late Neoproterozoic—Cambrian complex includes dismembered ophiolites outcropped mainly in axial part of zone between Tolbo-Huur and Khovd faults. These ophiolites composed of serpentinites, gabbroids and green-stone basalts are considered as relics of paleooceanic crust which underlays the whole of the fold system of the Mongolian Altay. Ophiolite rocks are intensively deformed and changed into serpentinite mélange.

In northern, Tsagaan-Shuvuut subzone, together with ophiolites the Bayrim and Turgen zonal metamorphic complexes are structurally united; in lower part they are composed of green schists with thin layers of marble and quartzite, and in upper part, of terrigenous flysch with sills, floods and intrusions of picrite, picrobasalt and picrodolerite of 513 Ma age (Izokh et al. 2006).

The second, Middle Cambrian—Early Ordovician stratigraphic complex is characterized by wide distribution and usually is compared with "Gorno-Altay" series, determined comparatively long ago in the adjacent area of Russian Gorny Altay. This complex is represented by thick (up to 5000 m) and monotonous terrigenous sandstone-aleurolite series absolutely without carbonate component. It is typical pericontinental flysch and together with the above-mentioned ophiolites forms the main orogenic complex of the Mongolian Altay. This complex is closely related to the sandy aleurolite series with the Tremadoc graptolites (visual thickness is 1220 m), outcropped near Agach-Uul Mt. in the Lake Khulman-Nuur depression (Dergunov et al. 1980).

The third, Middle Ordovician—Early Silurian stratigraphic complex fills a number of depressions and troughs; the largest of them are situated in the limits of the eastern half of the Mongolian Altay. This complex consists of three parts separated by erosion surfaces and structural unconformities (Dergunov et al. 1980; Minjin and Ariunchimeg 2008). Lower part is composed of carbonate-terrigenous strata of 250–1000 m thick either with Arenig and Middle Ordovician graptolites or with a complex of different Middle—Upper Ordovician benthos fossils. Middle part is formed by volcano-terrigenous series (thickness up to 2000 m) distributed in large area which hosts porphyrite, tuffs, reef limestone and "graptolite slates". It includes various fossils of Late Ordovician age. Upper volcano-sedimentary part (thickness 1500 m) has similar composition but a key role in it belongs to polymictic sandstone and siltstone with Lower Silurian fossils.

The forth, Late Silurian—Devonian stratigraphic complex consists of terrigenous and volcano-terrigenous formations located in numerous small troughs and graben-synclines, and also in large Delyun-Sagsay and Achit-Nuur depressions (Dergunov et al. 1980). Small graben-synclines are filled with grey-colored terrigenous Upper Silurian molasses and parti-colored volcano-detrital Lower—Middle Devonian series. Volcanic rocks are characterized by rhyolite, dacite and basaltic andesite associations of subalkaline and rarely alkaline trends. In contrast to them, the most widespread in large Delyun-Sagsay and Achit-Nuur superimposed troughs is thick (at least 2500 m) volcano-terrigenous series represented in lower part by Lower Devonian andesite-rhyolite formation and in larger upper part by sandstone–siltstone flyschoid formation with the Givetan—Fransian fossils of fauna and flora. Fold deformations in the most extended Delyun-Sagsay trough are of liner type.

Upper Paleozoic, Mesozoic and Cenozoic stratified formations have limited distribution and are everywhere located in small superimposed troughs and grabens.

Intrusive rocks of the Mongolian Altay fold system are grouped into two large areals of Paleozoic granite magmatism situated along both sides of the Tolbo-Nuur fault. In northern, Kharkhiraa areal, granites of Khovd polyfacial granodiorite complex and of Kharkhiraa granite—alkaline granite complexe are mainly distributed (Gavrilova et al. 1975). The former includes rocks of diorite group, granites of elevated basicity (tonalite, granodiorite, plagiogranite, microcline-plagioclase granites) and normal granites, and the latter, normal alaskite and leucogranite with facies of granosyenite, quartz syenite and rarely alkaline granite. According to geological data and some radiological (K-Ar method) datings, the Khovd complex formed during the Silurian (426–413 Ma) and the Kharkhiraa complex, in the Middle—Late Devonian (374–342 Ma) (Gavrilova et al. 1975). Except these two complexes, the Kharkhiraa complex includes separate massifs of Turgen diorite-tonalite-granodiorite complex (456–440 Ma) and Khalzan rare-metal alkaline granite complex (378 + 18 Ma) (Kovalenko et al. 2003) as well as small interstice bodies of the Permian gabbro-syenite-granite association.

The Southern Altay intrusive magmatic areal includes massifs of different granitoids of four age groups (Gavrilova et al. 1975). The first group (440 Ma) is represented by tonalite, granodiorite, adamellite, normal and binary-mica granites (Tsagaan-Gol complex). The second group (385–349 Ma) is represented by numerous plutons of large-, and giantporphyry biotite and tourmaline-biotite granites of the Altay complex. The third group includes biotite, binary-mica and muscovite granites of 317–263 Ma. The last group unites small bodies of leucogranites conditionally of Permian age.

Modern structure of the Mongolian Altay fold system is composed of several terranes converged most probably at the end of the Arenig stage in the Early Ordovician.

The northernmost part of the system, situated in Tsagaan-Shuvuut and Turgen-Uul mountainous massifs between Tsagaan-Shuvuut and Bayrim faults, forms the Tsagaan-Shuvuut metamorphic terrane. It corresponds to the same-name subzone (Dergunov et al. 1980) and is composed of Bayrim and Turgen metamorphic complexes. Almost the whole eastern part of the Mongolian Altay is occupied by the Khovd terrane limited from north and east by Bayrim and TsagaanShuvuut faults and from west by Khovd fault. This terrane is composed of the Middle Cambrian—Lower Ordovician turbidite complex from-under which in NW the fragmental ophiolites are disclosed.

Rocks of all tectono-stratigraphic complexes of the Khovd terrane are intensively deformed with formation of isoclinal folds and thrusts, and in the zone of the Khovd fault are metamorphosed to epidoteamphibole stage (Dergunov et al. 1980; Yanshin 1974). Sedimentologically, these turbidites are comparable with deep-water sediments of marginal seas. To the west, the Ulegey terrane is contacting with the Khovd terrane. It is situated between the approaching Khovd and Tolbo-Nuur faults and traced by outcrop chain of dismembered ophiolite complex (probably the Vendian (Ediacaran)—Late Cambrian age) and structurally is joined with the Middle–Upper Cambrian turbidites. In the zone of the Khovd fault, the ophiolites are transformed into monomictic serpentinite mélange.

The Southern Altay terrane makes up the most extreme southwest high-altitude part of the Mongolian Altay. This terrane is composed of the strongly folded Middle Cambrian—Lower Ordovician turbidite complex which was formed under conditions of passive continental margin (Dergunov 2001).

Southward, the Southern Altay terrane is changed by Bodonch metamorphic terrane with high-graded metamorphic rocks intruded by granites of 370 Ma age (Kozakov et al. 2005).

Post-amalgamation complexes of the Mongolian Altay fold system include the Middle Ordovician Turgen and Tsagaan-Gol granitoids, the Middle Ordovician—Lower Silurian volcano-sedimentary stratigraphic complex, the Silurian Khovd granitoid complex, and also widespread Devonian volcanoterrigenous deposits (which fill superimposed orogenic structures such as Delyun-Sagsay, Achit-Nuur troughs and etc.), and the Devonian Altay granite complex. It is evident that all post-amalgamation overlapping and stitch complexes have been formed in environment of Andean-type active continental margin.

5.3 Mongolian Part of the Khangay-Daur Fold System

Mongolian part of the Khangay-Daur fold system envelops the Khangay-Khentey fold system forming large hemi-oval with northeast curvature and cut in the south by Mid-Mongolian Tectonic Line as the prolongation southwestward of the famous Mongol-Okhotsk fault (see Fig. 24).

Tectonically, the Khangay-Khentey fold system is a collage of different-aged terranes including the Dzag-Kharaa turbidite terrane, and changing eastward its five accretionary wedge terranes are distinguished (Tomurtogoo 2012).

Dzag-Kharaa terrane that occupies the northern and western border parts of the fold system (see Fig. 12.1) is confined to joined Galuut and Taryat-Bayangol faults; it is composed of the thick (up to 5000 m) greenschist metamorphosed Cambrian—Lower Ordovician sandy-slate turbidites with lenses of ferrous quartzites and chlorite schists. The latter formed in the environment of continental slope and its foot as a result of intensive display of linear folding, schistosity, considerable succession thickness, presence of conodont fossils (Marinov et al. 1973).

In the next, ***Asralt-Khairkhan*** terrane adjoining the Kharaa segment of the Dzag-Kharaa terrane, there are structurally combined Ordovician—Silurian volcanogenic-slate and volcanic complexes belonging to the Mandal Group and the Khuren-Undur formation respectively. First one took place in autochthon of terrane composed of thick (up to 5000 m) siliceous-sandy-schist complex including separate layers of metavolcanic rocks and quartz-sericite schist (metatuffs) but, in singular occurrences,—small tectonic slices of layered ultramafic–mafic complex. Siliceous rocks are often represented by parti-colored jasper-quartzite and radiolarian jaspers. In contrast, the Khuren-Undur formation in the northern allochthon of the zone is formed by differentiated calc-alkaline volcanic series gradually changed over the area and upward in the succession by tuff-greywacke strata. Predominant basalts are characterized by petrochemical parameters of E-MORB (Gordienko et al. 2012; Tomurtogoo 2002).

The whole central part of the Khangay-Khentey fold system is occupied by Khar-Khorin, Tsetserleg and Ulaanbaatar terranes characterized by similar type of composition. They are formed by thick (not less 4000 m) layered sandy-slate series of Devonian-Carboniferous age (Khar-Khorin, Khangay and Khentey Groups). They host numerous tectonic slices and sheets of red-colored jaspers with the Silurian and Devonian fossils (conodont, radiolarians), basaltic lavas and rarely—ultramafic–mafic cumulate complex, green-violet tuffs and coral limestone (Kurihara et al. 2008; Tomurtogoo et al. 2006). It is recognized that part of basaltic lavas have petrological characteristics of plume "Polynesia-type" volcanic rocks (Kurihara et al. 2008; Safonova et al. 2009).

The next, ***Onon***, terrane immediately adjoining the Mid-Mongolian Tectonic Line, is distinguished from others in structural combination of the Silurian—Devonian island arc volcanic rocks with mélanged Adaatsag ophiolite and the Pennsylvanian—Permian accretionary complex and is covered with the Permian—Lower Triassic Duch-Gol flysch formed probably in fore-arc basin environment.

The last of the above mentioned, Adaatsag ophiolite is composed of serpentinized ultramafic rocks, pyroxenite-gabbro cumulative series, "upper" gabbroids, diabase sill-dyke series and basaltic variolite lavas. The age of this complex (325 Ma) was determined by SHRIMP Zr-dating from pegmatoid gabbro (Tomurtogoo et al. 2005).

Different-aged post-amalgamation overlaped complexes are distinguished in terrane collage of the Khangay-Khentey fold system. They include the Mississippian in marine molasses of Dzag-Kharaa and Asralt-Kharkhan terranes, the Pennsylvanian—Early Permian volcano-terrigenous molasse of the Khar-Khorin terrane, the Permian—Lower Triassic marine molasses of Tsetserleg and Ulaanbaatar terranes and the Triassic (Norian) marine molasse of the Onon terrane. Intrusive complexes are of importance in composition of the Khangay-Khentey fold system where they are represented by granitoid complexes with the age from the Ordovician through the Early Jurassic. The oldest granitoides (Boroo-Gol complex) are located in the Dzag-Kharaa terrane and represented by diorite, biotite granodiorite and amphibole-biotite granites with age of 452 Ma. This complex is intruded by Devonian granite-granosyenite rocks (Sharyn-Gol complex) also widespread in the Asralt-Khairkhan zone where granite erosion surface is covered with Early Carboniferous marine molasse.

Large areas of Khar-Khorin, Tsetserleg and Ulaanbaatar terranes are occupied by Khangay and Khentey polychronous batholiths.

The western, Khangay batholith is composed of subalkaline rocks of two geochemical types: monzonite-granodiorite and granite-granosyenite. These granitoids are supersaturated with alumina and corresponded to potassium subalkaline and alkaline series. Available absolute age data for batholith vary from 269 to 240 Ma (Yarmolyuk et al. 2013). Granitoids from the southern part of the batholith formed from melts which substances were the rocks considerably supersaturated with mantle components (Jahn et al. 2004).

The eastern, Khentey batholith includes a lot of three-phase mesoabyssal plutons composed of porphyry-type granodiorite and granite (I phase), biotite granite (II phase), leucocratic, biotite and alaskite granites (III phase) (Marinov et al. 1973). Plutons are characterized by eruptive relationships with Khentey Group sediments and are intruded by Early Mesozoic granite.

All tectonic terranes of the Khangay-Khentey fold system, including the Onon terrane are intruded by Mesozoic granite. The latter are distributed to the east from 100° meridian and represented by rocks belonging to granodiorite-granite and granite-leucogranite associations and also to association of alkaline granites and syenites (Dergunov 2001).

Rocks of the granodiorite-granite association form a lot of mesoabyssal plutons in the Central Khangay and Northern Khentey. For this association porphyritic, and, less commonly, equigranular biotite-amphibole granodiorite with xenolithes of diorite and quartz-diorite, as well as the later melanocratic biotite granite are typical. The chemical composition of the rocks varies within the calc-alkaline series, from diorite to granite. Small plagiogranite pluton, located on left bank of the Tuul-Gol River in 25 km to southeast from the Nalaikh settlement also adjoins this association. In the adjacent territory of Russia, rocks of the granodiorite-granite association intrude Permian–Triassic volcanic suites, and their K–Ar ages fall in the range of 230–180 Ma (Dergunov 2001; Kovalenko et al. 2003). Besides, U–Pb zircons age of 207 ± 6.8 Ma is also known for the above-mentioned plagiogranite pluton (Kovalenko et al. 2003).

The granitoids of the granite-leucogranite association are characterized by the greatest distribution in the marginal zone of the early Mesozoic igneous region in Mongolia, composing numerous hypoabissal plutons of medium and small sizes.

The leucogranites are the most common rocks in the southwestern part of the studied zone. Their most typical representatives are well known as ore-bearing rare-metal intrusions such as the Janchivlan, Gorikh and Avdar plutons. As a rule, they show two-phase structure. The first phase includes medium-grained and generally equigranular, or, less commonly, porphyritic biotite granites. Muscovite, garnet, and tourmaline are common components of the rocks. The second phase involved dykes of fine-grained, commonly, pegmatoid leucogranites. In places, they form small stocks. Pegmatites are both of schlieren and vein types, and the presence of muscovite, garnet, and tourmaline in the rock composition is particularly typical of the granite.

The granite and leucogranite of the association are assigned to the subalkaline Na–K series. They have high Rb and Sr contents, and, in some plutons elevated Li, Be, and W as well.

Besides, a subgroup within the granite-leucogranite association forms lithium-fluorite granite characterized by domed shape and dykes. Their common feature is the presence of amazonite granites and pegmatites in their composition. Petrochemically they are acid rocks that are very poor in calcium and rich in alumina. They are also characterized by typical unusually high fluorine content (usually 0.2–0.4%).

Some intrusions of the granite-leucogranite association also include subalkaline rocks (monzonite, granosyenite, and syenite) and peralkaline granitoids. The few available Rb–Sr ages for the leucogranites from the studied association fall in the range of 190–230 Ma, and for lithium-fluorite granites, 187 Ma (Dergunov 2001; Kovalenko et al. 2003).

All geological and tectonic data confirm the idea that Khangay-Khentey fold system has been formed at the place of the Mongol-Okhotsk Paleoocean (Zonenshain et al. 1990; Parfenov et al. 1999).

By the Cambrian—Ordovician age of the most ancient formations of the Khangay-Khentey fold system and their metaturbidite composition at the beginning of Mongol-Okhotsk Paleoocean opening, its Khangay-Khentey sector had transform boundary from the Paleo-Siberia continent.

But from the Middle Ordovician to the Early Triassic different accretionary structures formed in this sector—accretionary prisms with various amount of ocean crust fragments and some island arcs and seamounts. Naturally, the established rejuvenation of accretionary structures in the Khangay-Khentey fold system from north to south shows that mechanism of their formation was connected with gradual transition of subduction zone to internal parts of the paleoocean (roll-back of subduction zone). It is characteristic that a similar tendency appears for the overlaid accreted terrane complexes of superimposed troughs and intrusive granitoid complexes with their rejuvenation southward.

Final formation of the Khangay-Khentey fold system as Indosinian orogen was caused by oblique approaching of the Amur subcontinent and the Siberian (North Asian) continent and their continental collision during the Early Mesozoic (Parfenov et al. 1999; Tomurtogoo et al. 2005).

The following tectonic evolution of the Khangay-Khentey fold system was connected with formation of huge mass of intra-plate granites and several continental rifts (Dergunov 2001).

5.4 Tectonics of South Mongolian Variscides

Variscides of South Mongolia include Kerulen and South-Gobi massifs and Gobi nappe-fold megazone located between them (see Fig. 24).

Kerulen massif extends from the southern slope of the Mongolian Altay eastward through the Gobian Altay, North Gobi peneplain to southeast ridge branches of the Khentey Highland and Kerulen River basin. Structurally, it is composed of Ereen-Davaa, Undur-Khaan, Idermeg and Gobian Altay—Baruun Urt terranes converged at the end of the Cambrian—beginning of the Ordovician (Badarch et al. 2002; Tomurtogoo 2002).

The ***Ereen-Davaa terrane*** situated in extreme northeast contains:

(1) metamorphic basement that included the Mesoproterozoic gneiss–crystalline-schist complex (Khaychin-Gol Formation) and metamorphosed in amphibolite or greenschist degree marine volcano-sedimentary complexes with Lower through Middle Neoproterozoic age (Ereen-Davaa Formation and others) and
(2) volcano-sedimentary cover composed of transgressive Upper Neoproterozoic—Lower Cambrian terrigenous-carbonate and volcano-terrigenous formations, which age was determined from stromatolite and microphytolite fossils.

Undur-Khaan terrane is superseded southward of the Ereen-Davaa terrane and situated along narrow zone of the Kerulen fault. It is characterized by complicated structure and composed of dismembered ophiolite (serpentinite, gabbroids and metabasalts) and volcano-greywacke complexes associated with them dated as Late Neoproterozoic—Early Cambrian on the basis of rare microphytolites in limestone from the upper part of the complex.

Next southward ***Idermeg terrane*** occupies large area on the right bank of the Kerulen River, Northern Gobi and Dzuun-Bogd Range. It includes the same tectonostratigraphic complexes as the Ereen-Davaa terrane and differs from it only in appreciable predominance of shelf carbonate-terrigenous deposits with archaeocyatha fossils in sedimentary cover (Idermeg Fm.).

Gobian Altay—Baruun Urt terrane is traced along the Gobian Altay Mts. watershed and further in low-mountainous massifs of Mandal-Ovoo, Baruun-Urt and Matad and is characterized by complicated internal structure. It is totally represented by the Vendian (Ediacaran)—Cambrian greenschist metamorphosed volcano-sedimentary complex with spilite-keratophyre lavas, tuffaceous turbidites and oncolitic limestone. It is characterized by protrusions of serpentinites and gabbro of 518 Ma (Tomurkhuu and Otgonbaatar 2013). This terrane also includes the Tseel block with polymetamorphic crystalline rocks containing granulite lenses (738 ± 20 Ma), which is distinguished by the presence of detrital zircons of Middle Ordovician age and experienced the doubleply high-temperature metamorphism in the time interval of 390–350 Ma (Kozakov et al. 2011; Kroner et al. 2010).

The oldest post-amalgamation complexes of the Kerulen massif (microcontinent) are represented by batholiths of the Kerulen complex of the Early Paleozoic (from 536 to 410 Ma) palingenous calc-alkaline granites numerous massifs of which equally intrude all terranes and unite them into unified consolidated crust mass. Other post-amalgamation complexes are represented by the following:

- Late Ordovician—Lower Silurian carbonateterrigenous deposits with chain of reef limestone in the south,
- Silurian marine molasse with Tuvella fossils (Bayan-Uldzit graben-syncline),
- polyfacial volcano-sedimentary series with rich fossils of the whole Devonian,
- Lower Devonian "black-shale" complex,
- Middle–Upper Devonian volcanogenic molasse that also filled separate superimposed troughs (Pre-Kerulen, Salkhit etc.).

The Permian volcano-plutonic association of plume origin plays the independent role in overlapped complexes of this massif; it builds up the large Eastern Mongolian volcano-plutonic belt (Yarmolyuk et al. 2013).

Gobi nappe-fold megazone is a lengthy linear structure of sublatitudinal orientation traced through axial part of Gobi Desert from Baruun-Khuuray Depression in the west to Lake Buyr-Nuur in the east. Tectonics of this megazone is described in numerous publications (Ruzhentsev 1985; Ruzhentsev et al. 1987, 1989, 1992; Ruzhentsev and Pospelov 1992).

In general, the Gobi megazone includes four terranes: Edren and Mandakh in the north, and GurvanSaykhan and Baruun-Khuuray—in the south (see Fig. 24).

Edren terrane occupies Edren and Aj Bogd Ranges of Trans-Altay Gobi. Structurally it is divided into two subterranes, of which the northern one (Khuviyn Khar) is composed of probably Silurian—Devonian basalts, cherts and thin-grained sedimentary rocks. Second subterrane (Aj Bogd) is characterized by widespread distribution of the Devonian basalt-andesitic rocks of calc-alkaline series and tuff-sedimentary complexes.

Mandakh terrane is traces to the east from the Edren terrane and differs from it in nappe-fold structure and presence of serpentinite melange and the Late Silurian—Devonian volcano-sedimentary complex with high-Na basalts, andesites, their tuffs, jaspers, greywacke sandstone and rarely limestone containing different fossils (Marinov et al. 1973).

Gurvan-Saykhan terrane is a linear sublatitudinal structure which western end is situated along the southern foothill of the Mongolian Altay and eastern, far in region of Lake Buyr-Nuur. Recent data on tectonic structure are mainly based upon research materials in Baruun-Saykhan, Dzolen and Ongon-Ulaan Mts. in Trans-Altay Gobi (Zonenshain et al. 1975; Ruzhentsev and Pospelov 1992; Ruzhentsev et al. 1989; Tomurtogoo 1989).

Four structural-formation complexes that overthust one another are determined there. The base of each of them is usually composed of serpentinite mélange with blocks of dunite-harzburgite serpentinites, pyroxenite, troctolite, gabbro and diabases. Lately, an anorthosite block with Zr-age of 480 Ma in was found in the mélange (D. Tomurhuu, pers. com.). Allochthon complexes (Berkh-Uul, Khairkhan, Gurvan-Saykhan and Dzolen) are represented by facially very monotonous successions which differ in details (Dergunov 2001).

So, the first of them includes the basal horizon of ophicalcite breccias (1–25 m) which transgressively covers ultramafic rocks. This horizon soon gives way to strata of layered siliceous rocks with red jasper with the Late Silurian—Early Devonian radiolarians (60 m). Higher in the succession, basaltic pillow-lavas (200–800 m) occur followed by tuffitegreywacke series (1300–1400 m).

In the second (Khairkhan) succession the basal layer of volcanomictic sandstones with pebbles of ultramafic rocks, spilite, acid lavas and tuffs (70 m) is overlaid by thick (up to 2000 m) series of aphiric basalt, epiclasts and polymictic sandstone encompassed in lower part of the clayey-siliceous and chert horizon with the Early Devonian conodonts.

The third (Gurvan-Saykhan) complex includes jasper-volcanogenic rocks (jaspers with the Late Silurian—Early Devonian radiolarians) differentiated from basalts to andesite pillow-lavas, and also tuffs ant tuffites and overlaid with conformity rhythmically layered tuff-terrigenous series with total thickness up to 3000 m.

The last (Dzolen) complex is composed of volcano-greywacke strata in which jasper and subalkaline basaltic pillow-lavas (300 m) above is first increased by a thick layer (400–500 m) of rhyolite tuffs and tuffites and thereupon by rhythmically layered greywacke (up to 1500 m).

Geological and petrological studies show that basaltoids of all the above-mentioned allochthon complexes of the Gurvan-Saykhan terrane formed in the geodynamic environment of zone spreading center crossed by transform fault (Tomurtogoo 2012). In different parts of the above-mentioned terrane collage of the Gobi megazone, the lower marine terrigenous and volcanogenic Tournaisian–Visean molasses are overlaid by neoautochthonous complex filled in system of graben-synclines. Numerous conformed Carboniferous granite intrusions penetrate the overthrust structure of its tectonic collage (Tomurtogoo 2002). Extremely west ***Baruun-Khuuray terrane*** occupies the same-named depression that geographically belongs to the East Junggar area and is subdivided into three subterranes: Baaran in the north, Ulaan Us in the center and Baytag in the south. Brief analysis given below is based on the numerous publications of S. V. Ruzhentsev and co-authors (Ruzhentsev et al. 1992; Dergunov 2001).

Baaran subterrane is composed of the Devonian—Early Carboniferous volcanic rocks, varying in facies from basalts to rhyodacites, all of them showing high alkalinity. Lavas, various tuffs, and epiclastics are common, and hypabyssal bodies are prominent. Whole trachyandesitic (south) and shoshonite-latite (north) volcanic associations are distinguished in this complex. There, limestone bands, enclosed in lava and tuff, yielded remains of brachiopods, which give the age from the Middle Devonian to the Tournaisian–Visean.

Next, *Ulaan Us subterrane* includes the lower volcanic and upper flysch formations. The first one (up to 1500 m) is a combination of acid and mixed volcanic, rhyolitic tuffs, tephroids and tuffites, containing rare and thin layers of limestone.

Volcanics make up a differentiated calc-alkaline series, from basalt to rhyolite. The age of formation is Eifelian to Frasnian based on remains of brachiopods, tabulate corals etc. The upper flysch formation is a monotonous sequence of rhythmically bedded rocks. Its lower section (400–500 m) is dominated by polymictic sandy flysch, containing lenses (up to 100 m) of the Lower Carboniferous crinoids-brachiopod calcarenite. The upper section of the formation is composed of fairly homogeneous sandy-clayey flysch.

The last, *Baytag subterrane* in southern frame of Baruun Khuuray Depression also consists of volcanic and terrigenous formations. The lower (Devonian) one is chiefly represented by volcanic rocks which make up two tectonic sheets. The lower sheet includes aphiric basalt, andesites, their tuffs, and epiclastics (up to 1000 m). They form a tholeiitic andesite-basalt association, which probably formed within an ensimatic island arc. The volcanics of the upper sheet are represented by pillow lavas of basalts as the products of differentiation of oceanic tholeiites. They probably mark the initiation of emplacement of the ensimatic island arc. Volcanics of both above-mentioned sheets are covered with the Lower Carboniferous flysch which is similar to that in the Ulaan Us subterrane. At the same time, in autochthon of the subterrane there is a grey-coloured carbonaceous formation of the same age with lenses of paralic coal (Olonbulag Fm.).

So, the whole of the Baruun-Khuuray terrane is a relic of complicated tectonic combination of two island arcs, and recent structure of which includes fragments of ensialic (autochthon) and ensimatic (allochthon) island arcs.

South-Gobi massif is separated from Gobi nappefold megazone by the Gobian Tien Shan—Nukht-Davaa fault and situated in the Gobian Tien Shan, Khan-Bogd, Ulaan-Uul and Nukht-Davaa mountains (see Fig. 24).

Compositional features of the South Gobi massif resemble nearly the Idermeg terrane of the Kerulen massif in the presence in its composition of the Paleoproterozoic gneiss complex (Maan't Group), the Upper Mesoproterozoic—Middle Neoproterozoic carbonate-quartzite-schist complex (Bargyn-Obo Group) and the Upper Neoproterozoic—Middle Cambrian carbonate-terrigenous complex (Suuzh-Khudag and Tugrig-Khudag Series) (Marinov et al. 1973). But in contrast to the Idermeg terrane, the Late Silurian through Devonian marine volcano-sedimentary and carbonate-terrigenous complexes and Late Silurian and Devonian granite intrusions are widely distributed there. Another difference consists in distribution of the Carboniferous—Early Permian volcano-plutonic complex to which in some linear structures typical of continental rifts comendite-pantellerite association with alkaline granites is corresponded (South Mongolian volcano-plutonic belt) (Dergunov 2001). Correlation of terrane and overlaid complexes of the South Gobi massif with similar complexes of the Kerulen massif and the Gobi nappe-fold megazone shows that in the Pre-Ordovician time, the South Gobi massif formed the united continental block with the Kerulen massif. But later, in the time interval between the Ordovician and Devonian, the former was separated from the latter by Trans-Altay oceanic basin afterwards in which place the Gobi nappe-fold megazone formed which represented the Andean-type active continental margin that continued up to the Early Permian exclusively.

5.5 Tectonics of South Mongolian Indosinides

Indosinides of South Mongolia occupy two lowmountain massifs along the state boundary with China. The western mountain massif (Ikh Khongozh, Tsaagan Uul, Baruun-Tsokhiot, Onch Khairkhan and others) topographically belongs to Gobian Beishan. The second, eastern massif, separated from the first one by large Galbyn-Gobi Depression, is usually named Toto-Shan.

Tectonically South Mongolian Indosinides are subdivided into Khatanbulag massif (microcontinent) and Sulinkheer (Solonker) nappe-fold megazone (see Fig. 24).

Khatanbulag massif is situated in low-mountains of Tsagaan Uul (Gobian Beishan) and Khutag Uul (Toto-Shan) separated by Mesozoic Galbiyn Gobi Depression. The Basement of the massif is represented by gneisses and crystalline schists which contains granite-gneiss (1781 Ma) and by greenschist metamorphosed volcano-carbonate-slate complex with metabasalt, metarhyolite, jaspers, reef limestone and intrusions of gneiss-like granitoides (952 Ma) (Tomurhuu et al. 2008; Yarmolyuk et al. 2005). In addition, the massif contains post-collisional granite (455 Ma) (Yarmolyuk et al. 2005), as well as a series of superimposed Paleozoic troughs filled with riftogenic volcanics with pillow lavas and rhyolite dated as 422 $\pm$ 18 Ma (D. Tomurhuu, pers.com.), carbonate-terrigenous strata containing the Early Devonian benthos fossils, the Mississippian flyschoid series (Zhirem-Uul Formation, 2100 m), Late Devonian—Mississippianian volcano-sedimentary strata with benthos fossils (Khangi-Uul Formation), Pennsylvanian—Lower Permian carbonate-volcanic and carbonate complexes with bryozoans-foraminiferan limestone bioherms (Bayrim-Ovoo and Aguy-Uul Formations, thickness up to 2500 m) and the thick (up to 3000 m) Permian Lugiyn-Gol flysch complex (Dergunov 2001; Ruzhentsev et al. 1989; Yarmolyuk et al. 2005). Four last of the above named complexes are overlaid by volcanogenic molasse as a member of the Upper Permian volcano-plutonic belt. The entire of the Khatanbulag massif belongs to the Grenville fragment on which complexes of Paleotethys active continental margin were superimposed during the Late Silurian to Permian (Tomurtogoo 2002; Yarmoluk et al. 2005). It is worthy to note that in adjacent region of Inner Mongolia of China, the Xilinhot massif is a structural analogue of the Khatanbulag massif of South Mongolian Indosinides (Ping et al. 2010).

Sulinkheer (Solonker) megazone occupies the eastern and southern foothills of Toto Shan Mts. and eastward extends to Inner Mongolia of China. At present, it consists of seven terranes: newly determined Zamyn-Uud terrane in east and six others (Southern terrane collage) which correspond to the southern part of the former Solonker thrust-fold zone (Ruzhentsev et al. 1989) (see Fig. 24).

Zamyn-Uud terrane extends along NE foothills of Toto Shan Mts. like a narrow tectonic wedge of sublatitudinal orientation and is situated between the Zamyn-Uud and Ulaan-Badrakh faults (Tomurtogoo 1997). Structurally, this terrane is the prolongation of the Hegenshan ophiolite zone (Inner Mongolia, China) (Ping et al. 2010), but being covered in the largest area by Meso-Cenozoic sediments it is studied very poorly. It is known that greenschist metamorphosed complex with metabasalt, siliceous and different greenschist protruded by listvenitized serpentinite and rarely gabbro prevails in it. The age of this accretionary complex is unknown and probably it is Devonian—Mississippian. Furthermore, a flyschoid formation of fore-arc origin containing faunal fossils of Pennsylvanian age is located there (Tomurtogoo 2012). In some publications, the Zamyn-Uud terrane has another name—Enshoo terrane (Badarch et al. 2002).

Southern terrane collage of the Sulinkheer megazone includes Talyn Khuren, Duulgant, Nomt-Uul, Tavan-Khar, Ereelzh and Suuzh terranes (Ruzhentsev et al. 1989; Tomurtogoo 2002, 2012).

Extremely northwestern ***Talyn Khuren terrane*** that overthrusts the southern edge of the Toto Shan block of the Khatanbulag massif, is composed of monomictic serpentinite mélange with large blocks of ultramafic rocks and listvenites (Dzuun Togoo, Baruun Togoo and Talyn Khuren Mts.), as well as numerous small blocks or boudins of trondjhemite (323 Ma), gabbro (296 Ma), tonalite (294 Ma), anorthosite (252 Ma), diabase, basalt, black and multi-coloured cherts (Ping et al. 2010; Tomurtogoo 2012).

Duulgant terrane traced southeastward from Talyn Khuren terrane, is composed of intensively deformed and metamorphosed greenschist turbidite complex of presumably Late Paleozoic age (Tomurtogoo 2002). Previous researchers attributed this complex to the independent metamorphic block or terrane (Ruzhentsev et al. 1989; Badarch et al. 2002). But in its lithological and textural features (dense saturation of the synsedimental quartz veins) it is comparable with deep-water trench deposits (Tomurtogoo 2002).

Nomt-Uul terrane occurs in the southeast foothill of Toto-Shan Mts. south of sublatitudinal Sulinkheer fault and consists of basaltic pillow-lavas, jaspers and terrigenous mixtites (up to 1100 m) with boulders and blocks of fusulinida limestone, ranging in age from the Pennsylvanian to Early Permian (Ruzhentsev et al. 1989).

Tavan-Khar terrane follows the Nomt-Uul terrane to south in same-name mountain and is made up of Pennsylvanian—Lower Permian island arc carbonate-andesite-graywacke complex (1500–2000 m thick) with massive andesites, lithic tuffs, tephroids, epilastics and brachiopod limestone (Ruzhentsev et al. 1989).

Ereelzh terrane located further to the south, has imbricated structure and includes intensively deformed siliceous turbidite containing slices of basalt and limestone (with conodonts, fusulinids and brachiopods of Pennsylvanian—Lower Permian age), as well as serpentinite bodies. In addition, this accretionary complex is covered with flyschoid strata with lenses of limestone containing remains of the Upper Permian brachiopods and bryozoans (Ruzhentsev et al. 1989; Tomurtogoo 2012).

The last, ***Suuzh terrane*** is situated along state boundary with China and presented by a chain of narrow ultramafic allochthons in which there are determined small blocks of vehrlite and massive gabbro with Rb–Sr age of 367–344 Ma (Ruzhentsev et al. 1989; Tomurtogoo 2012).

Inside the terranes, separate small plutons of subduction granitoides with age of 294–292 Ma (Tomurhuu et al. 2008) are located. Neoautochthon is represented by Triassic—Jurassic coarse-detrital continental molasse.

In general structure of South Mongolia, the whole of the above-mentioned terrane collage of the Sulinkheer megazone is a well-preserved fragment of the Indosinian Paleotethys suture.

The presence of the Zamyn-Uud accretionary prism in the tectonic collage of the Toto-Shan fold system shows that origination of the Sulinkheer (Solonker) oceanic basin (Indosinian Paleotethys) has started in the middle of the Devonian, when as a result of the Variscan orogenesis the closing of the Trans-Altay oceanic basin (Variscan Paleotethys) occurred to form the South Mongolian Variscides. In addition, the opening of the Sulinkheer (Solonker) oceanic basin was accompanied by the destruction of the northern borderland (in present-day coordinates) of the Tarim—North China craton (paleocontinent) with separating of the Khatanbulag massif as microcontinent.

In general, the tectonic evolution of the Toto Shan fold system took place in the time interval from the Late Devonian to the Permian, when in the Sulinkheer (Solonker) basin the Zamyn-Uud accretionary prism (and the same name terrane), Nomt-Uul back-arc basin (Talyn-Khuren and Nomt-Uul terranes), Tavan-Khar island arc (the same name of terrane), Duulgant-Ereelzh accretionary prism with Khar-Erdene fore-arc trough (Duulgant and Ereelzh terranes) and submarine rise (oceanic range or plateau) of piling-up ocean crust (Suuzh terrane) sequentially formed. Characteristically, events of this interval are reflected for the Khatanbulag massif in formation and build-up of its volcano-sedimentary cover in rift troughs and sedimentation. Final formation of the fold system as Indosinides occurred at the very end of the Permian when the entire area was involved into collisional processes connected with the northward drift of the North China Craton that caused the closing of the Sulinkheer (Solonker) oceanic basin and formation of its terrane collage.

References

Al'muhamedov AI, Gordienko IV, Kuzmin MI, Tomurtogoo O, Tomurhuu D (1996) The Dzhida zone: a fragment of the Paleoasian ocean. Geotectonics 4:25–42

Alexeiev DV, Aristov VA, Degtyarev KE (2007) Age and tectonic position of volcanogenic-siliceous series in ophiolite complex of the Atbashi Range (South Tien Shan). Doklady Rus Acad Sci 413(6):1–4 (in Russian)

Alexeiev DV, Ryazantsev AV, Kroner A, Tretyakov AA, Xia X, Liu DY (2011) Early Ordovician high-pressure belt in the Chu-Yili Mountains: implications for the earliest stages of accretion in Kazakhstan and the Tien Shan. J Asian Earth Sci 42:805–820

Anisimova IV, Kozakov IK, Yarmoluk VV et al (2008) New results of geochronological (U-Pb method by zircons) and Nd isotope study of anorthosites in the Khunjilin Gol massif (Tarbagatay Block). In: Sklyarov Eu V (ed) Geodynamic evolution of lithosphery of the Central Asian Mobile Belt (from ocean to continent). IZK SO RAN, Irkutsk, vol 6, pp 18–19 (in Russian)

Atlas of the Lithology-Paleogeographical, Structural, Palinspastic and Geoenvironmental Maps of Central Eurasia (2002) Alma-Ata: YUGGEO, vol 26, p + 37 map sheets (in Russian)

Badarch G, Cunningham WD, Windley BF (2002) A new terrane subdivision for Mongolia: implications for the Phanerozoic crustal growth of Central Asia. J Asian Earth Sci 21:87–110

Bakirov AA, Bykov RI, Gavrilov VP (1970) Basement, main faults of Turan Plate in connection with its oil-bearing. Nedra , Moscow, 240 p (in Russian)

Baratov RB (1976) (Ed.-in-chief) Subdivisions of stratified and intrusive rocks of Tajikistan. Donish, Dushanbe, 276 pp (in Russian)
Biske Yu S (1987) Lower Paleozoic of South Fergana: structural position, stratigraphy, geological history. Vestnik of Leningrad State University, series 7, no 7, pp 3–11 (in Russian)
Biske Yu S (1996) Paleozoic structure and history of the Southern Tien Shan. St. Petersburg University, SPb, 192 p. (in Russian)
Biske Yu S, Tabuns EV (1996) Pre-collision basalts in Atbashi-Kokshaal Hercynides (Central Tien Shan) and their geodynamic nature. Dokl Acad Sci 348, pp 81–84 (in Russian)
Boishenko AF (1978) On age of rocks from metamorphic complex of the Bayankhongor zone (MPR). Izv Akad Nauk SSSR, Ser Geol (1):48–55 (in Russian)
Buchan C et al (2002) Timing of accretion and collisional deformation in the Central Asian orogenic belt: implications of granite geochronology in the Bayan-Khongor ophiolite zone. Chem Geol 192:23–45
Bukharin AK, Maslennikova IA, Pyatkov AK (1985) Pre-Mesozoic structural-formation zones of the Western Tien Shan. Fan, Tashkent, 152 p (in Russian)
Burtman VS (1976) Structural evolution of the Paleozoic fold systems. Nauka, Moscow, 164 pp (in Russian)
Burtman VS (2006) Tien Shan and high Asia: tectonics and geodynamics in the Paleozoic. GEOS, Moscow, 216 p (in Russian)
Chen B et al (1996) Terrane tectonics and evolution in the northern Qinghai-Tibetan Plateau. Geological Publishing House, Beijing, pp 1–120
Cheng Y et al (eds) (1994) Introduction to geology of China. Geological Publishing House, Beijing, 480 p
Dalimov TN, Shayakubov T Sh, Troitskiy VI et al (1998) Geology and mineral deposits of the Republic of Uzbekistan. University, Tashkent, 724 p (in Russian)
Degtyarev KE, Shatagin KN, Kotov AB et al (2008) Late Precambrian volcano-plutonic Association of Aktau-Junggar Massif (Central Kazakhstan): structural position and age. Dokl Acad Sci 421(4) (in Russian)
Degtyarev KE, Ryazantsev AV, Tretyakov AA, Yakubchuk AS, Tolmacheva TY, Kotov AB, Salnikova EB, Yakovleva SZ, Gorokhovskii BM (2012) Structure, age substantiation and tectonic setting of the lower-middle ordovician volcanic-sedimentary and plutonic complexes of the western part of the Kyrgyz Range (Northern Tien Shan). Stratigr Geol Correl 20(4)317–345
Demina LI, Lomize MG, Avdonin AB (1995) Geodynamic characteristics of peridotites of the Pre-Sonkul. Vestnik Moscow State University. Geology, (1):91–99 (in Russian)
Deng W et al (1995) Geological feature and tectonic implication of the ophiolite in the Karakorum region. Acta Petrol Sin 11(SI):98–111
Dergunov AB (ed) (2001) Tectonics, magmatism and metallogeny of Mongolia. Routledge, London-New York, 288 p
Dergunov AB, Luvsandanzan B, Pavlenko VS (1980) Geology of West Mongolia. Nauka, Moscow, 195 p (in Russian)
Dijkstra AH, Brouwer FM, Cunningham WD, Buchan C, Badarch G, Mason PRDl (2006) Late Neoproterozoic proto-arc ocean crust in the Dariv Range, Western Mongolia: a super-subduction zone end-member ophiolite. J Geol Soc, London 163:363–373
Dong Y et al (2003) Tectonic affinity and Proterozoic evolution of the northern Qinling. Acta Geosci Sin 24(1):3–10
Fang T et al (2002) Mafic magma intrusive event in the Proterozoic in the Altyn Tagh, Xinjiang. Chin Geol 29(1):48–54
Gan X et al (1996) Information Archean-Paleoproterozoic U-Pb ages of the inherited zircons from the plutonic rocks, South China. Geochimica 31(3):112–120
Gao Y et al (1990) Progress on study of the Kunlun-Qinling suture, the boundary between the northern and southern plates of China. In: Proceedings of the third structure conference of China: international symposium on the structure evolution and dynamics of the continental lithosphere, vol 1. Geological Publishing House, Beijing, pp 7–15

Gao Z et al (1993) Precambrian of the northern Xinjiang autonomous region. Geological Publishing House, Beijing, p 42

Garetsky RG, Shlesinger AE, Yanshin AL (1971) Problem of basement of young platforms. Bull MOIP Geology (3):146–147 (in Russian)

Gavrilova SP, Yashin RM, Zaitsev NS (1975) Granitoid and alkaline formations in the structures of Western and Northern Mongolia. Nauka, Moscow, 288 p (in Russian)

Gavrilova SP, Luchitskaya AI, Frihhar DI, Orolmaa D, Badamgarav J (1991) Volcano-plutonic associations of Central Mongolia. Nauka, Moscow, 232 p (in Russian)

Geological Map of the Tajik SSR and Adjacent Areas, scale 1:500,000 (1991) Vlasov NG, Diakov Yu A (eds-in-chief). Leningrad, VSEGEI

Geology and Metallogeny of Karatau (1986) (ed) (Nikitin NF) vol 1. Nauka, Alma-Ata, 240 p (in Russian)

Geology and Mineral Resources of Afghanistan (2008) Book 1: Geology (eds-in-chief) Abdullach SH, Chmyriov VM. BGS, London, 488 p (English version)

Ges' MD (1999) Magmatism and geodynamic evolution of the Tien Shan Caledonian orogeny. In: Bakirov AB, Dikikh AN (eds) Problems of geology and geography in Kyrgyzstan. Ilim, Bishkek, pp 33–42 (in Russian)

Ges' MD, Makarychev GI (1985) Riphean and Cambrian basaltoides of the Northern Tien Shan and their significance for tectonic analyses. Izv Acad Sci USSR, Geological serial (3):33–42 (in Russian)

Gibsher AS, Khain Eu V, Kotov AB, Sal'nikova EB et al (2001) Late vendian age of the Khantaishir Ophiolite complex in Western Mongolia: new U-Pb data. Geol Geophy 2(8):1179–1185 (in Russian)

Gordienko IV, Medvedev A Ya, Gornova MA, Tomurtogoo O, Goneger TA (2012) The Haraa Gol terrane in the Western Khentey Mountains (Northern Mongolia): geochemistry, geochronology, and geodynamics. Geol Geophys 53(3):365–379 (in Russian)

Hanzl P, Krejcu Z (eds) (2008) Geological map of the Trans-Altay Gobi, scale 1:500,000. Czech Geological Survey, Prague

Hu A et al (1992) The first occurrence of the 3.3 Ga old rocks at the northern margin of the Qaidam Craton, Xinjiang. Chin Sci Bull (7):624–630

Hu A et al (2001) Progress on the study of isotope geochemistry of the Precambrian basement of the continent in northern Xinjiang. In: Theoretical research on the metal mineral resources of Xinjiang. Science Publishing House, Beijing, pp 137–156

Huang G, Zhang ZW, Dong ZH, Zhang WF (2011) Zircon LA-ICP-MS U-Pb age of plagiogranite from Tonghuashan ophiolite in Southern Tien Shan Mountains and its geological implications. Geol China 38(1):94–102

Il'in AV (1982) Geological development of Southern Siberia and Mongolia in the Late Precambrian. Early Paleozoic. Nauka, Moscow, 114 p (in Russian)

Izokh AE, Polyakov GV, Krivenko AP, Bognibov VI, Bayarbileg L (1990) The Gabbroid Formations of Western Mongolia. Nauka, Novosibirsk, 269 p (in Russian)

Izokh AV, Polyakov GV, Travin AV et al (2006) Cambrian age of the Uregnur platinum-bearing picritebasalt volcano-plutonic association from the Mongolian Altay (western Mongolia). In: Sklyarov Eu V (ed) Geodynamic evolution of lithosphere of the Central Asian Mobile Belt (from ocean to continent). IZK SO RAN, Irkutsk, pp 4–95 (in Russian)

Izokh AV, Polyakov GV, Vishnevsky AV, Shelepaev RA, Oyunchimeg T (2009) Age of ultramafic-mafic magmatism of Western Mongolia. In: Proceedings of the International Symposium on Large Igneous Provinces of Asia: Mantle Plumes and Metallogeny, 6–9 Aug 2009. Sibprint, Novosibirsk, pp 137–140

Jahn B-M, Capdevia R, Liu D, Vernon A, Badarch G (2004) Sources of Phanerozoic granitoids in the transect Bayankhongor-Ulaanbaatar, Mongolia: geochemical and Nd isotopic evidence, and implications for Phanerozoic crustal growth. J Asian Earth Sci 23:629–653

Jiang C (1996) Research report of China tectonics. In: Ren J (ed)

Jiang C et al (1992) On the opening-closing tectonics of the Kunlun Mountains. Geological Publishing House, Beijing, 224 p

Jin C et al (2001) The Hongguleleng-Almantai ophiolite and mineralization. In: Theoretical research on the metal mineral resources of Xinjiang. Science Publishing House, Beijing

Khain VE, Leonov Yu G (Eds.-in-chief) (1979) Tectonic map of Europe and adjacent areas, scale 1:10,000,000. GUGK, Moscow

Khain EV et al (2002) The most ancient ophiolite of the Central Asian fold belt: U-Pb and Pb-Pb zircon ages for the Dunzhugur Complex, Eastern Sayan, Siberia and geodynamic implications. Earth Planet Sci Lett 199:311–325

Kheraskova TN, Tomurtogoo O, Khain Eu V (1985) Ophiolites and upper Precambrian to lower Paleozoic rocks of the Ozernaya (Lake) zone of the Daribi range of Western Mongolia. Izvestiya Academii Nauk SSSR, Series Geology 6:25–31 (in Russian)

Kheraskova TN, Dubinina SV, Orlova AP, Cerezhnikova EA (1997) Early Paleozoic accretion complex of Northern Tien Shan. In: Perfiliev AS, Raznitsyn Yu N (eds) Tectonic and geodynamic phenomena. Nauka, Moscow, pp 221–242 (in Russian)

Kheraskova TN, Bush VA, Didenko AN, Samygin SG (2010) Rodinia break-up and earlier stages of Paleoasian ocean evolution. Geotektonika (1):5–22 (in Russian)

Khristov SE, Chernyshuk VP (1987) Stratigraphy and structural elements of the lower Paleozoic deposits in Dgumgal range. In: Korolev VG (ed) Tien Shan Caledonides. Ilim, Frunze, pp 79–92 (in Russian)

Khristov EV, Ivanov KC, Mikolaichuk AV et al (1999) Structural position and age of ophiolites of Karaterek massif. In: Bakirov AB, Dikikh AN (eds) Problems of geology and geography in Kyrgyzstan. Ilim, Bishkek, pp 65–69 (in Russian)

Kiriukhin LG (1974) Tectonics and oil-bearing of epi-geosynclinal Paleozoic-Triassic complex of the young plate of Eurasia. VNIGNI, Moscow, 36 p (in Russian)

Kirnozova TI, Kozakov IK, Yarmoluk VV, Kovach VP, Fugzan MM (2009) Late Riphean age of postcollisional subalkaline granites of the Tarbagatay uplift from Early Caledonian superterrane in Central Asia. In: Proceedings of IV Russian conference on isotopic geochronology "Isotopic systems and time of geological processes", vol I, pp 237–240 (in Russian)

Kiselev VV, Apayarov FH, Komarevtsev VT, Tsyganok EL (1987) Zircon isotopic age of metamorphic complexes of Tien Shan. In: Isotopic dating of metamorphism and metasomatism processes. Nauka, Moscow, pp 180–184 (in Russian)

Kiselev VV, Apayarov FH, Komarevtsev VT et al (1993) Isotopic age of zircons in crystalline complexes of Tien Shan. In: Kozakov IK (ed) Early precambrian of Central Asian fold belt. Nauka, SPb, pp 99–115 (in Russian)

Klishevich VL, Sobolevskaya RF (1993) Ichkebash series of Dgetymtau Range. In: Mambetov AM (ed) New data on the precambrian and Paleozoic biostratigraphy of Kyrgyzstan. Ilim, Bishkek, pp 102–107 (in Russian)

Knyazev VS, Florensky PV, Charygin AM (1967) Geological composition of Pre-Jurassic surface of Turan Plate. Sovetskaya Geol (6):48– 58 (in Russian)

Knyazev VS, Florensky PV, Charygin AM, Ship OA (1970) Folded basement and intermediate complex of Turan Plate. In: Basement, main faults of Turan Plate in connection with its oil-bearing. Nedra, Moscow, 240 p (in Russian)

Konopelko D, Kullerud K, Apayarov F, Sakiev K, Baruleva O, Ravna E, Lepekhina E (2012) SHRIMP zircon chronology of HP-UHP rocks of the Makbal metamorphic complex in the Northern Tien Shan, Kyrgyzstan. Gondwana Res 22:300–309

Korolev BG, Maksumova RA (1984) Pre-Cambrian tillites and tilloides of Tien Shan. Ilim, Frunze, 190 p (in Russian)

Kotov AB, Sal'nikova EB, Kovach VP, Reznitskiy LZ (1997) On the age of metamorphism of the Slyudyanka crystalline complex, Southern Baikal area. Petrology 5(4):380–393

Kovach VP, Jian P, Yarmolyuk VV et al (2005) Magmatism and geodynamics of early stages of the Paleoasian ocean formation: geochronological and geochemical data on ophiolite of the Bayanhongor zone. Dokl Earth Sci 404(7):1072–1077

Kovalenko DV (2003) Geological interpretation of the paleomagnetic data from Cretaceous—Paleogene rocks of the southern Koryak Highland, Doctoral Thesis, Institute of Lithosphere. Ac Sci Rus 25 p (in Russian)
Kovalenko VI, Yarmolyuk VV, Kovach VP, Kotov AB, Sal'nikova EB (2003) Magmatism and geodynamics of Early Caledonian structures of the Central Asian Foldbeld (isotopic and geological data). Geol Geophys 44(12):1280–1293 (in Russian)
Kovalenko VI, Yarmolyuk VV, Tomurtogoo O et al (2005) Geodynamics and crust-formation processes in the Early Caledonides of the Bayankhongor zone (Central Mongolia). Geotectonics 39(4):298–316
Kozakov IK, Sal'nikova E, Kotov AB, Glebovitsky VA, Bibikova EV, Kirnozova TI, Kovach V Ya (2005) Age constraints and geodynamic settings of crystalline complexes formation in eastern segment of the Central Asian foldbelt. In: Leonov MG (ed) Problems of the Central Asian tectonics. GEOS, Moscow, pp 137–170 (in Russian)
Kozakov IK, Salnikova EB, Kovach VP, Anisimova IV, Plotkina YuV, Myskova TA, Fedoseenko AM, Yarmolyuk VV, Kozlovskii AM, Yakovleva SZ, Sugorakova AM (2008) Vendian stage in formation of the early caledonian superterrane in Central Asia. Stratigr Geol Correl 16(4):360–382
Kozakov IK, Didenko AN, Azimov P Ya, Kirnozova TI, Sal'nikova EB, Anisimova IV, Erdenejargal Ch (2011) Geodynamic settings and formation conditions of crystalline complexes in the South Altai and South Gobi metamorphic belts. Geotectonics 3:7–30
Kozakov IK, Sal'nikova EB, Yarmolyuk VV, Kovach VP, Kozlovsky AM, Anisimova IV, Plotkina Yu V, Fedoseenko AM, Yukovleva SZ, Erdenejargal Ch (2013) Formation stages of the Songino Block continental crust in Early Caledonian superterane of Central Asia: I. Geological and geochronological date. Petrology 21(2):1–19 (in Russian)
Kroner A, Lehmann J, Schulmann K, Demoux A, Lexa O, Tomurhuu D, Liu D, Wingate MTD (2010) Lithostratigraphic and geochronological constraints on the evolution of the Central Asian Orogenic Belt in SW Mongolia: early Paleozoic rifting followed by Late Paleozoic accretion. Am J Sci 310:523–574
Kroner A, Alexeiev DV, Hegner E, Rojas-Agramonte Y, Corsini M, Chao Y, Wong J, Windley BF, Liu D, Tretyakov AA (2011) Zircon and muscovite ages, geochemistry, and Nd-Hf isotopes for the Aktyuz metamorphic terrane: evidence for an Early Ordovician collisional belt in the Northern Tien Shan of Kyrgyzstan. Gondwana Res 21:901–927
Kunin N Ya (1974) Intermediate structural stage of Turan Plate. Nedra, Moscow, 264 p (in Russian)
Kurchavov AM, Yarmoliuk VV (1984) Placement of continental Permian volcanics of Central Asia and its tectonic interpretation. Geotektonika (4):75–89 (in Russian)
Kurenkov SA (1983) Tectonics of ophiolite complexes of South Tien Shan (Alay and Atbashi Ranges). Nauka, Moscow, 96 p. (in Russian)
Kurenkov SA, Didenko AN, Simonov VA (2002) Geodynamics of spreading. GEOS, Moscow, 294 p (in Russian)
Kurihara T, Tsukada K, Otoh Sh et al (2008) Upper Silurian and Devonian pelagic deep-water radiolarian chert from the Khangai-Khentei belt of Central Mongolia: evidence for middle Paleozoic subduction-accretion activity in the Central Asian Orogenic Belt. J Asian Earth Sci 34:209–225
Kuzmichev AB (2004) Tectonic history of the Tuva-Mongolian Massif: early Baikalian, late Baikalian and early Caledonian stages. PROBEL-2000, Moscow, 192 p (in Russian)
Kuzmichev AB, Larionov AN (2011) The Sarkhoy series in the East Sayan: neoproterozoic (~770–800 Ma) Andean-type volcanic belt. Geol Geophys 52:875–895 (in Russian)
Kuzmichev A, Kroner A, Hegner E, Dunyi L, Yushng W (2005) The Shishkhid ophiolite, Northern Mongolia: a key to the reconstruction of a Neoproterozoic island-arc system in Central Asia. Precambrian Res 138:125–150
Levasheva NM, Gibsher AS, Meert JG (2009) Origin and tectonic evolution of Precambrian microcontinents of the Urals-Mongolian belt. In: Sklyarov Eu V (Ed) Geodynamic evolution of lithosphere of the Central Asian Mobile Belt (from ocean to continent). IZK SO RAN, Irkutsk
Li Z (1992) Time determination and geological implication of the metamorphic rocks in the western Gansu Province. Research report of the Gansu Institute of Geology

Li J (2004a) "Mosaicking and superimposition" characteristics of continental crust of China and its tectonic evolution. Geol Bull China 23:986–1004
Li J (2004b) Tectonic pattern and evolution in the Late Neoproterozoic to Paleozoic of eastern Xinjiang. Geol Rev 50(3):304–322
Li J (2001) Collision time and style of the Sino-Korean and Yangtze blocks—the evolution of the sedimentary setting in the Sinian to Jurassic, middle and lower reaches of the Yangtze River. Acta Geol Sin 75(1):5–34
Li T (2006) Structure units of China lithosphere. Chin Geol 33(4):700–710
Li S et al (1991) Granitoids geochemistry of the Tongbai-Dabie Mountains. China Geoscience University Press, Beijing
Li Q et al (2000) Explosive seismic detect over the contact between western Kunlun and Tarim. Chin Sci (D edition) 30(51)
Lindsay JF, Brasier MD, Shields G, Khomentovsky VV, Bat-Ireedui YA (1996) Glacial facies associations in Neoproterozoic back-arc setting, Zavkhan basin, western Mongolia. Geol Mag 133(4):391–415
Liu D et al (1991) Occurrence of the 3.8 Ga old crustal rocks. Geol China (5)5–17
Liu C et al (2003) Geochemistry and $^{40}Ar/^{39}Ar$ constraints of the post-collisional granites in the Eastern Kunlun orogenic belt. Bull Eastern China Geol College 26(4)
Lomize MG, Demina LI, Zartschikov AA (1997) Kirgiz-Terskey paleooceanic basin. Geotektonika (6):35–55 (in Russian)
Lu X et al (1999) Time and structural implication of the A-type granite at Tuwushan Mount, eastern Qinling Mountains. Chin Sci Bull 44(9):975–978
Makarychev GI, Ges' MD (1981) Tectonic nature of junction zone of the Northern and Middle Tien Shan. Geotektonika (4):57–72 (in Russian)
Maksumova RA (1999) New type of lower Paleozoic section in Northern Tien Shan. In: Bakirov AB, Dikikh AN (eds) Problems of geology and geography in Kyrgyzstan.Ilim, Bishkek, pp 92–99 (in Russian)
Maksumova RA, Zakharov IL, Zima MB et al (1988) Overthrust-napped structure of the Tien Shan Early Caledonides by the new data on stratigraphy of Lower Paleozoic series. In: Bakirov AB, Kiselev VV (eds) Precambrian and Lower Paleozoic of Tien Shan. Ilim , Frunze, pp 144–152 (in Russian)
Marinov NA et al (eds) (1973) Geology of the Mongolian People's Republic. Nedra, Moscow, vol I. Stratigraphy, 582 p.; vol 2. Magmatism, Metamorphism, Tectonics, 752 p. (in Russian)
Mikolaichuk AB (1998) Riphean—lower Paleozoic deposits of Issyk-Kul massif. Izv Kyrgys Natl Acad Sci (2/3):50–53 (in Russian)
Mikolaichuk AB, Kurenkov SA, Degtyarev KE, Rubtsov BI (1997a) Main stages of geodynamic evolution of Northern Tien Shan in the Late Precambrian—early Paleozoic. Geotektonika (6):16–34 (in Russian)
Mikolaichuk AB, Mambetov AM, Misius PP (1997b) On stratigraphic volume and age of Kapkatac series of Northern Tien Shan. Stratigr Geol Correl 5(1):79–85 (in Russian)
Minjin Ch, Ariunchimeg Y (2008) Ordovician and Silurian sedimentary basins of Western Mongolia. In: Development of Early Paleozoic Biodiversity: role of biotic and abiotic factors, and event correlation. Abstracts of Conference. KMK Scientific Press, Moscow, pp 68–70
Mitrofanov FP, Kozakov IK, Palei IP (1981) Precambrian of Western Mongolia and Southern Tuva. Nauka, Leningrad, 156 p (in Russian)
Mossakovsky AA (1976) Orogenic structures and volcanism of Eurasia Paleozoides. Nauka, Moscow, 127 p (in Russian)
Osmonbetov KO, Knauf VI, Korolev BG (1982) (eds) Stratified and intrusive formations of Kyrgyzstan. Ilim, Frunze, vol 1—357 p.; vol 2—245 p (in Russian)
Ovchinnikova GV, Kuznetsov AB, Vasil'eva IM, Gorokhov IM, Letnikova EF, Gorokhovsky BM (2012) U-Pb age and Sr isotopic characteristics of supratillite limestone from Neoproterozoic TsagaanOlom Formtion, the Dzabhan River Basin, Western Mongolia. Stratigr Geol Correl 20(6):28–40 (in Russian)

Parfenov LM, Popeko LI, Tomurtogoo O (1999) The problems of tectonics of the Mongol-Okhotsk orogenic belt. Pac Oceanic Geol 18(5):24–43 (in Russian)

Pashkov BP, Shvolman VA (1979) Mesozoic continental margins in the Pamirs. Geotectonika (6):42–56 (in Russian)

Peive AV, Zonenshain LP, Knipper AL, Markov MS, Mossakovsky AA, Perfiliev AS, Pushcharovsky Yu M, Fedorovsky VS, Shtreis NA, Yanshin AL (1980) Tectonics of Northern Eurasia (Explanatory note to the Tectonic Map of Northern Eurasia, scale 1:5,000,000). Nauka, Moscow, 141 p. (in Russian)

Ping J, Dunyi L, Kroner A,Windley BF, Shi Y, Zhang W, Zhang F, Miao L, Zhang L, Tomurhuu D (2010) Evolution of a Permian intraoceanic arctrench system in the Solonker suture zone, Central Asian Orogenic Belt, China and Mongolia. Lithos 118:169–190

Pospelov II (1987) Formations and Tectonic evolution of late Variscides in the Southern Tien Shan and Northern Pamirs. In: Early geosynclinal formations and structures. Nauka, Moscow, 149–178 (in Russian)

Puchkov VN, Khristov EV, Ivanov KS, Mikolaichuk AV (1987) New data on the age of green-schist series of Southern Tien Shan. Dokl Acad Sci USSR 294:435–438 (in Russian)

Ren J et al (1999) The tectonics of China from a global view—a guide to the tectonic map of China and adjacent regions. Geological Publishing House, Beijing, 46 p

Ruzhentsev SV (1985) Geology of Gobi Tien Shan and aspects of the south Mongolian ocean. Izvestiya VUZ"ov. Geologiya i Razvedka 6:12–19 (in Russian)

Ruzhentsev SV, Pospelov II (1992) The South Mongolian Variscan fold system. Geotectonics 30(5):383–395

Ruzhentsev SV, Badarch G, Voznesenskaya TA, Sharkova TT (1987) Formation and structures of Variscides in Southern Mongolia. In: Pustcharovsky Yu M, Khvorova IV (eds) Early geosynclinal formations and structures. Nauka, Moscow, pp 101–137 (in Russian)

Ruzhentsev SV, Pospeleov II, Badarch G (1989) Tectonics of the Mongolian Indosinides. Geotectonics 23:476–487

Ruzhentsev SV, Pospeleov II, Badarch G (1992) Tectonics of the Baruunkhurai Basin, Mongolia. Geotectonics 22(1):67–77

Ryazantsev AV (1994) Ophiolites of the Bayanhongor zone in structure of the Mongolian Paleozoides. Dokl Akad Nauk 336(5):651–654

Safonova I Yu, Utsunomiya A, Kojima S, Nakae S, Tomurtogoo O, Filippov AN, Koizumi K (2009) Pacific superplume-related oceanic basalts hosted by accretionary complexes of Central Asia, Russian Far East and Japan: age, geological occurrence and geochemistry. Gondwana Res 16:587–608

Shlesinger AE (1974) Late geosynclinal and early platform structures in Eurasia Hercynides. Nauka, Moscow, 233 p (in Russian)

Sklyarov EV, Postnikov AA, Posokhov VF (1996) Structural setting, metamorphism, and petrology of the Hugeyn group (Northern Mongolia). Geolo Geophys 37(6):69–78

Song B et al (1994) Ages of the Chentaigou supracrustal rocks at Anshan area, Liaoning Province. In: Geology of Liaoning Province. Science Publishing House, Beijing, pp 21–57

State Geological Map of USSR (new series). Scale 1:1,000,000 (1983) Map of pre-quaternary formations. Sheet J-42, 43 (VSEGEI, Leningrad)

Sudorgin AA (1990) Lithology and ore-bearing of Kashkasuy and Zhargala series of the Upper Riphean—Vendian of the eastern part of Middle Tien Shan. In: Sagyndykov KS, Turdukeev ID (eds) Lithology and ore-bearing carbonaceous formations of upper Precambrian—lower Paleozoic in Tien Shan. Ilim, Frunze, pp 33–50 (in Russian)

Tangry M, Yano T, Bakirov A, Nakajima T, Uchimi S (1995) Mineral parageneses and metamorphic P-T paths of ultrahigh-pressure eclogites from Kyrghyzstan, Tien Shan. The Island Arc, vol 4, pp 280–292

Tian Z (1990) Formation mechanism and distribution of the Mesozoic-Cenozoic basins of China. In: Proceedings of the third structure conference of China: international symposium on the structure evolution and dynamics of the continental lithosphere, vol 1. Geological Publishing House, Beijing, pp 22–28

Tomurhuu D (1999) Geodynamics of formation preorogenic magmatic complexes of the Dzhida zone, Mongolia. Ph.D. Thesis. Ulaanbaatar, 100 p. (in Russian)

Tomurhuu D, Otgonbaatar D (2013) Age, composition and geodynamic setting of intrusive rocks in the Ikh Bogd—Ongon Ulaan transect. Mongolian Geosci 38:9–31 (in Mongolian)

Tomurhuu D, Enkhjargal B, Otgonbaatar D, Ariunchimeg Ya, Enkhtuvshin R (2008) Geochronology and geochemistry of the granitoid complexes from Sulinheer region and their geodynamic setting. Institute of Geology and Mineral Resources, MAS, Ulaanbaatar. Transactions, vol 18, pp 29–49 (in Mongolian)

Tomurtogoo O (1989) Ophiolites and formation of folded belts in Mongolia. D.Sc. Thesis. Geological Institute, Academy of Sciences of the USSR, Moscow, 423 p. (in Russian)

Tomurtogoo O (1997) A new tectonic scheme of the Paleozoides in Mongolia. In: Zhaqin X, Yufeng R, Xiaoping Q (eds) Proceedings of the 30th international geological congress, VSP 7, pp 75–82

Tomurtogoo O (2002) Tectonic map of Mongolia, Scale 1:1,000,000 (with Brief Explanatory Notes, 23 pp.). Geological Information Centre MRAM—15 sheets, Ulaanbaatar

Tomurtogoo O (2012) A tectonic subdivision for the orogenic domains of Mongolia. Proceedings of the IGMR, MAS, Ulaanbaatar, vol 21, pp 5–25 (in Mongolian)

Tomurtogoo O, Gerel O (1999) Geotraverse through a terrane collage in southern Khangay. In: Badarch G, Jahn B-M, Tomurhuu D (eds) IGCP-420 continental growth in the Phanerozoic: evidence from Central Asia. Second workshop excursion guidebook, Ulaanbaatar, Mongolia, Geosciences Rennes. Hors serie 2, pp 3–91

Tomurtogoo O, Windley BF, Kroner A, Badarch G, Liu DY (2005) Zircon age and occurrence of the Adaatsag ophiolite and Muron shear zone, Central Mongolia: constraints on the evolution of the Mongol-Okhotsk ocean, suture and orogeny. J Geol Soc London, vol 162, pp 125–134

Tomurtogoo O, Tomurhuu D, Erdenesaikhan G (2006) Guidebook for the excursion in the Central and Southern Mongolia. Abstracts and excursion guidebook for the second international workshop of the IGCP-480. Ulaanbaatar, pp 107–131

Travin AV, Simonov VA, Kurenkov SA et al (2002) First data on absolute age of ophiolites of Northern Tien Shan. In: Semikhatov MA, Kurenkov SA (eds) Geology, geochemistry, geophysics at the turn of the XX and XXI centuries, vol 1. Sviaz'-Print, Moscow, pp 105–106 (in Russian)

Volozh Yu A, Pilifosov VM, Sapozhnikov RB (1981) Tectonics of Turan Plate and Pre-Caspian depression by the data of regional geophysics. In: Problems of Kazakhstan tectonics. Nauka, Alma-Ata, pp 70–178 (in Russian)

Wang J (2002) Geological feature of the Fuping basement complex in the Qinling orogenic belt. China Northwest Geol 35(2):1–8

Wang L et al (1982) Expansion in the deep and its relation with strong earthquakes in the North China Plain and Bohai Sea region. In: Proceedings of the second structure conference of China: international symposium on the structure evolution and dynamics of the continentallithosphere, vol 2. Geological Publishing House, Beijing, pp 38–53

Wang Z et al (1990) Tien Shan tectonic cycles and mineralization. Science Publishing House, 214 p

Wang Q et al (1994) Stratigraphical classification and formation setting of the Archaean supracrustal rocks in eastern Hebei Province, China. J Geol Miner North China 9(l)

Wu J et al (1991) Major geological events of the North China platform in the Early Precambrian. Geological Publishing House, Beijing, 114 p

Wu J et al (1998) Geological features and structural evolution of the Archaean Sino-Korean paleocontinent. Geological Publishing House, Beijing

Xiao X, Tang Y (eds) (1991) Structural evolution of the southern margin of the Central Asian paleo-megasuture [Proceedings of the Sino-USA cooperation geological survey: international communication on geological sciences (13)]. Science Technology Press, Beijing, 150 p

Xiao X et al (2000) Structural evolution and uplift mechanism of the Qinghai-Tibetan Plateau. Guangzhou Science and Technology Press, 311 p

Xiao X et al (2003) Re-examination of the Kudi ophiolite and its geological implication. Geol Bull China 22(10):745–750

Xing F et al (1993) Occurrence of the Paleoproterozoic basement in the middle and lower reaches of the Yangtze River. Chin Sci Bull 38:1883–1886

Xiong X et al (2006) Mobility of trace elements in the subduction plates: constraints of the eclogite-facies transporting veins in the western Tianshan Mountains. Acta Geol Sin 80(1):51–60

Yang J et al (2004) Determination of the Neoproterozoic ophiolite at the northern margin of the Qaidam basin–evidence of break-up of the Rodinia? Geol Bull China 23(9–10)

Yanshin AL (Ed. in-chief) (1966) Tectonic map of Eurasia, scale 1:5,000,000. GUGK, Moscow

Yanshin AL (ed) (1974) Tectonics of the Mongolian People's Republic. Nauka, Moscow, 284 p. (in Russian)

Yarmolyuk VV, Kovalenko VI, Sal'nikova EB, Kozakov IK, Kotov AB, Kovach VP, Vladykin NV, Yakovleva SZ (2005) U-Pb age of sinand postmetamorphic granitoids of South Mongolia: evidence for the presence of Grenvillides in the Central Asian Foldbelt. Dokl Earth Sci 404:986–990

Yarmolyuk VV, Kovalenko VI, Anisimova IV, Sal'nikova EB, Kovach VP, Kozakov IK, Kozlovsky AM et al (2008) Late Riphean alkaline granites of the Dzabkhan microcontinent: evidence for the timing of Rodinia breakup and formation of microcontinents in Central Asian Foldbelt. Procced Earth Sci 420:583–588

Yarmolyuk VV, Kuzmin MI, Kozlovsky AM (2013) Late Paleozoic-Early Mesozoic intraplate magmatism in North Asia: traps, rifts, giant Batholithes, and the geodynamics of their origin. Petrology 21(4):101–126

Yuan X et al (1994) Reflection seismic profile in the crust of eastern Qinling Mountains. J Geophys 37(6):749–758

Zhang Y et al (1994) Introduction to geology of Hoh Xil and adjacent region. Seismology Publishing House, Qinghai Province, pp 6–155

Zhang C et al (1999) Occurrence of the Proterozoic in the western Qinling Mountains. J Stratigr 23(2):211–216

Zhao C et al (1996) Structure framework and crustal evolution in the eastern. Liaoning University Press, Jilin-Heilongjiang Provinces, 186 p

Zhu Y et al (2000) Jinningian magmatic activity and evolution at the eastern segment of the eastern Kunlun Mountains. Earth Sci 25(3):231–266

Zonenshain LP, Kuzmin MI (1978) Khan-Taishir ophiolite complex in Western Mongolia and problems of ophiolites. Geotectonics 1:19–42

Zonenshain LP, Suetenko OD, Jamyandamba L, Eenjin G (1975) Structure of the axial part of South Mongolian eugeosyncline in the Dzolen Range. Geotectonika 4:28–44 (in Russian)

Zonenshain LP, Kuzmin MI, Kononov MV (1985) Absolute reconstructions of the Paleozoic oceans. Earth Planet Sci Lett 74:103–116

Zonenshain LP, Kuzmin MI, Natapov LM (1990) Geology of the USSR: a plate tectonic synthesis. American Geophysical Union, Washington, D.C. Geodynamic Series, vol 21, 242 p

Tectonic Domains of Eastern Asia

O. V. Petrov, I. I. Pospelov, S. D. Sokolov, Chen Bingwei, Ren Liudong, Kim Bok Chul, and Hwang Jae Ha

Abstract The chapter provides description of the tectonic structure and evolution of orogenic structures of Chukotka and Koryak areas, Southern China and Korean Peninsula—main structures within Eastern Asia, which define the present-day morphology of the region. For these structures, the authors present paleo-tectonic reconstructions from the Pre-Cambrian to the Mesozoic and Cenozoic and describe stages of magmatism and tectonic rebuilding.

1 Tectonics of the Chukotka and Koryak Fold Areas

To the northeast and east of the Verkhoyansk-Kolyma fold area there are structures of Chukotka and West-Koryak fold areas (Fig. 1). The boundaries of these areas in most part are overlapped by Okhotsk-Chukotka volcanic belt. Structures of the Koryak fold area can be traced closer to and along the Pacific.

O. V. Petrov (✉) · I. I. Pospelov
Russian Geological Research Institute (VSEGEI), 74 Sredny Prospect, St. Peterburg 199106, Russia
e-mail: vsgdir@vsegei.ru

S. D. Sokolov
Geological Institute, Russian Academy of Sciences (GIN RAS), 7 Pyzhevsky Pereulok, Moscow 119017, Russia

C. Bingwei · R. Liudong
Institute of Geology, Chinese Academy of Geological Sciences (CAGS), No. 26 Baiwanzhuang Road, Beijing 100037, P.R. China

K. B. Chul · H. J. Ha
Korea Institute of Geoscience and Mineral Resources, 124 Yuseong-gu, Daejeon 305-350, Gwahangno, Korea

O. V. Petrov and S. Dong (eds.), *Tectonics of Asia (Northern, Central and Eastern Asia)*, Springer Geology, https://doi.org/10.1007/978-3-030-62001-1_5

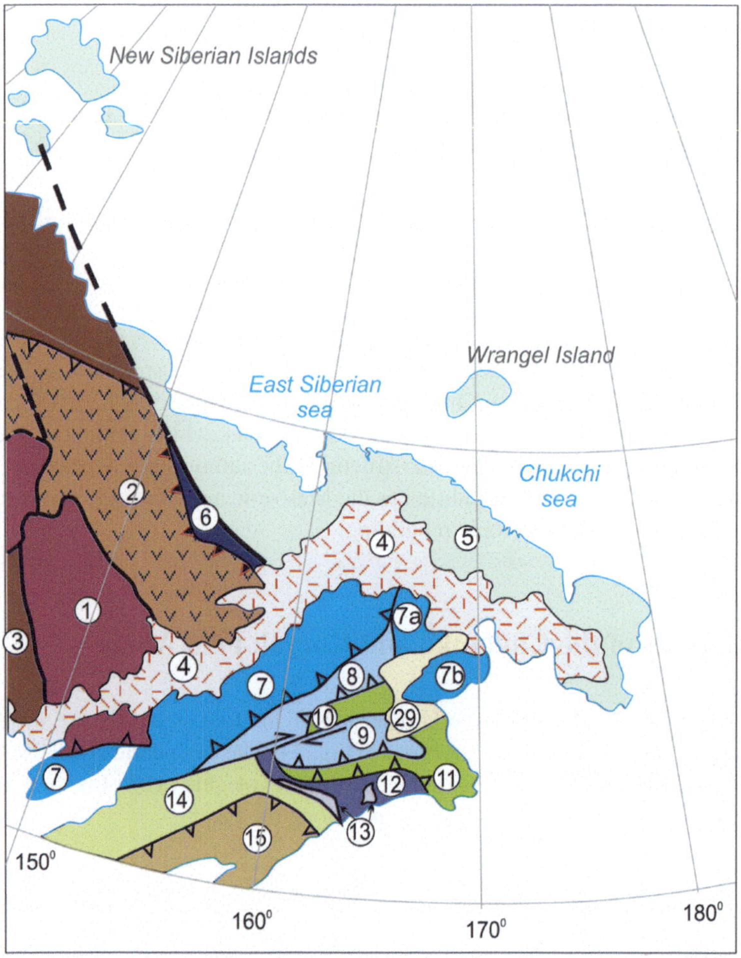

Fig. 1 Tectonic map of the Chukotka and Koryak fold areas. 1–3—structures of the Verkhoyansk-Kolyma fold area. 4—Okhotsk-Chukotka volcanic belt. Chukotka fold area: 5—Anyui-Chukotka fold system, 6—South Anyui suture; Koryak fold area: 7—West-Koryak fold area and its prolongation (7a—Kanchalan zone, 7b—Zolotoy ridge zone); 8–15—zones of the Koryak fold area: 8—Algan; 9—Mainitz; 10—Velikorachenskaya; 11—Alkatvaam; 12—Ekonay; 13—Yanranay; 14—Ukelayat; 15—Olyutor

1.1 Chukotka Fold Area

The Chukotka fold area occupies the Arctic continental margin of North-Eastern Asia and includes New Siberian Islands, Wrangel Island and others. Therefore, in scientific literature, the name *New Siberian-Chukotka* fold area can often be found. On the continent, it consists of the Anyui-Chukotka fold system and South-Anyui suture. Correlation scheme of tectonic events and litho-stratigraphic complexes is shown in Fig. 2.

South Anyui suture (SAS) is located along the southern margin of the Chukotka fold area with structures of the Alazeya–Oloy fold zone that is apart of the Verkhoyansk–Kolyma fold area. The SAS extends from Bolshoy Lyakhovsky Island in the eastern part of the Laptev Sea into western Chukotka where it is exposed in the catchment areas of the Bolshoy Anyui and Maly Anyui rivers. In the latter, the suture was first identified as the South Anyui zone. To the east, the continuation of the suture is recorded from "eugeosynclinal" complexes of the Velmai terrane of East Chukotka

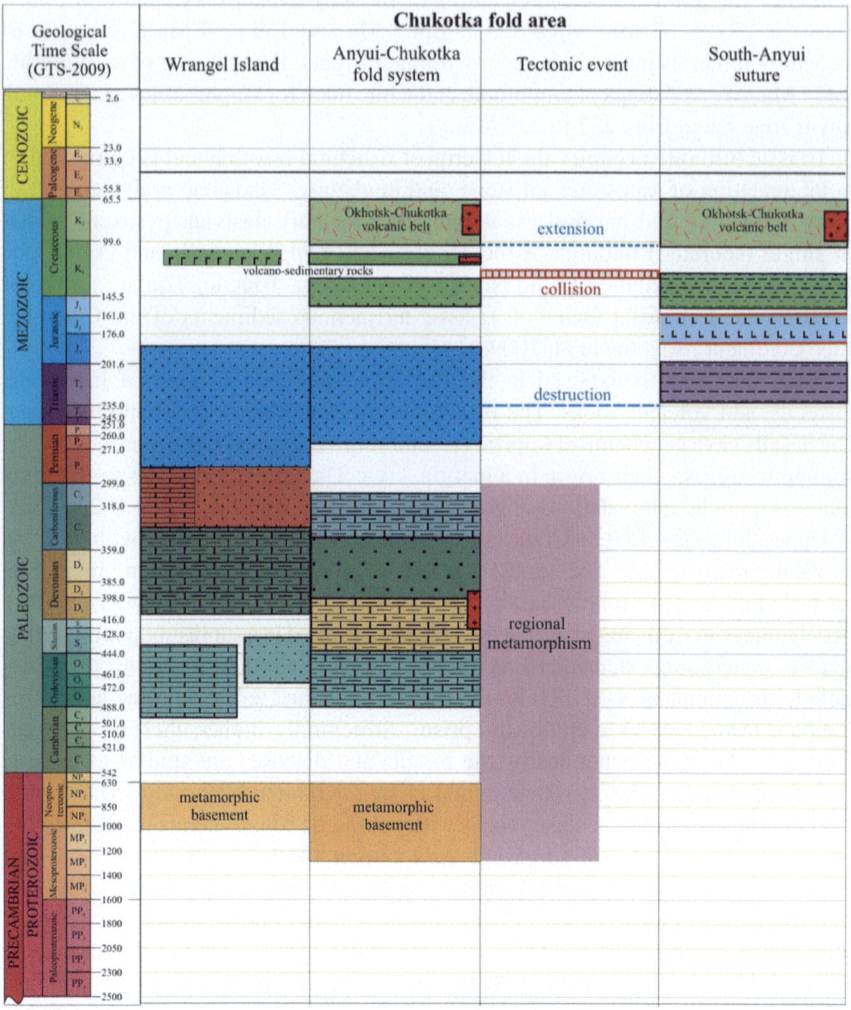

Fig. 2 Correlation chart of tectonic events and litho-stratigraphic complexes of the Chukotka fold area. See legend in Fig. 11

(Parfenov et al. 1993). SAS formed as a result of the collision of the Chukotka microcontinent with the structures of active margin of the Siberian continent.

SAS represents a system of intricately deformed tectonic slices and sheets (Sokolov et al. 2009a) composed of different tectono-stratigraphic units, which originated in diverse geodynamic environments: Upper Triassic distal turbidite, Lower Jurassic—Kimmeridgian oceanic crust, Upper Jurassic—Lower Cretaceous ensimatic island arc, accretionary prism, and flysch basin. Ophiolites are known on Bolshoy Lyakhovsky Island, South Anyui zone, and Velmai terrane. Ophiolites consist of peridotite, gabbro, diabase, basalt, and rare chert. Amphibolites and glaucophane schists are associated with the ophiolites. Geochronological data are not available. The Sm–Nd whole-rock dating of Bolshoy Lyakhovsky Island for pillow basalts is 291 $\pm$ 62 Ma, while 133.5 $\pm$ 4.5 Ma and 139 $\pm$ 8 Ma K–Ar ages of secondary minerals most likely record younger events. K–Ar dating of amphibolite is 473 Ma. Ar–Ar datings of amphibole and white mica for amphibolites of the South Anyui zone range from 257 to 229 Ma.

Triassic turbidite occupies the lowermost structural position and is composed of an intercalation of sandstone, siltstone, and mudstone. Sandstone grains consist of quartz (70–90%), feldspar, and less abundant sedimentary clasts and pyroxene. Based on single (sporadic) findings of the *Monotis* and *Otapiria* shells and conodonts, this sequence is attributed to the Norian of the Upper Triassic. Triassic turbidites are considered as distal facies of Triassic terrigenous sediments of the Chukotka microcontinent (Sokolov et al. 2009a).

The oldest rock is exposed in the Polyarny uplift and consists of limestone, siliceous, and volcanic rocks. The limestones include corals of Carboniferous age. The basalts have geochemical imprints resembling MORBs, and the presence of Nb-minimum suggests their origin in a marginal sea. The upper part of the stratigraphy contains sporadic intercalations of gray, black, and greenish cherts up to 1–2 m thick.

Upper Jurassic—Lower Cretaceous flysch consists of turbidites with Volgian to Valanginian faunas. The sandstones are of polymict composition. In places, the turbidites include pillow basalts, picritic basalts and volcanic layers (in situ). Geochemical analysis suggests that the volcanics formed in a suprasubduction setting that was undergoing extension (Sokolov et al. 2009a). In southern part, the turbidites includes fragments of N-MORB basalt, bedded chert, and terrigenous mélange. This complex formed in an accretionary prism. Structurally higher, there is a basalt-chert assemblage. Geochemically, the basalts and diabases are similar to oceanic or back-arc basalts. The cherts yield the Liassic—Kimmeridgian radiolarians.

Hauterivian-Barremian clastic rocks with polimictic conglomerates overlie highly deformed structure. ***Anyui-Chukotka fold system*** (ACH) is composed of Upper Proterozoic crystalline basement with volcanic and sedimentary rocks with horizons of marble, granite gneiss, and ultramafite, metamorphosed under conditions of amphibolite and greenschist grades. The oldest datings are obtained for crystalline rocks by the Rb–Sr (2565 and 1990 Ma) and K–Ar (1570 and 1680 Ma) methods. The U–Pb zircon age datings of granite intrusions are ranged between 609–677 Ma and 650–540 Ma.

Paleozoic–Mesozoic platform and shelf deposits are comparable in composition but differ in facies aspect. They consist of carbonate, carbonate and terrigenous deposits spanning the interval from the Middle Ordovician to the Upper Carboniferous. In most areas the units are metamorphosed under conditions of amphibolite and greenschist grades. The Paleozoic rocks are discordantly overlain by Triassic strata, predominantly turbidites that accumulated on the shelf, continental slope and rise (Tuchkova 2011). In the lower part of the Triassic sequence there are numerous sills and small igneous bodies of gabbrodiabase, diabase, and dolerite, which are deformed into folds together with host rocks. Zircons separated from the rocks are dated at 252 ± 4 Ma [U–Pb method, (Ledneva et al. 2011)] and 233–218 Ma [K–Ar method]. The highest strata of the sedimentary cover comprise Late Jurassic– Early Cretaceous terrigenous deposits that overlie unconformity and consist of conglomerate, sandstone and shale horizons with fauna and plant detritus.

Structures of granite-metamorphic domes established in the Chukotka terrane formed during the late Early Cretaceous. The U–Pb zircon data of 108 to 104 Ma suggest the Cretaceous age of orthogneisses sampled from the Koolen dome. Zircons from granite of the Alyarmaut dome yield dates between 117 and 115 Ma.

Paleotectonically, the Anyui-Chukotka terrane is considered as Chukotka microcontinent (Parfenov et al. 1993) or part of Chukotka—Arctic Alaska microplate (Grantz et al. 2011).

Main stages of tectonic evolution. The Chukotka fold area formed during Early Cretaceous closure of the Proto-Arctic Ocean and collision of the Chukotka microcontinent with active margin of the Siberian continent (Parfenov et al. 1993; Sokolov et al. 2009a). The Chukotka microcontinent includes structures of ACH, shelf, and the Arctic Islands of the East-Siberian and Laptev Sea. Neoproterosoic basement overlaps sediments of platform and shelf types. Fragments of oceanic crust are known in the SAS, where they crop out as dismembered ophiolites and basalt-chert assemblages. Their age indicates the existence of Proto-Arctic Ocean in the Late Paleozoic, although its opening time remains unknown. The youngest fragments of oceanic crust are of the Oxfordian–Kimmeridgian age.

Late Paleozoic—Early Mesozoic. In the Paleozoic the Proto-Arctic Ocean connected Paleopacific and Paleoasian oceans. After the closure of the Carboniferous period of the Ural branch of the Paleoasian Ocean, the Proto-Arctic Ocean corresponded to a large gulf of the Paleopacific. The southern Siberian edge of the ocean was active. The Aluchin and Gromadnyi-Vurguveem ophiolites of the suprasubduction origin suggest that the ocean–continent transition was of the West Pacific type with ensimatic island arcs and back-arc basins (Ganelin 2011).

Northern North-American oceanic margin was passive. At that time, the Chukotka was part of the North American continent (Grantz et al. 2011; Parfenov et al. 1993). In the Triassic, it was a passive margin with terrigenous sediments and turbidites (Tuchkova 2011). The distal facies of respective turbidites are known in the Ustieva unit of the SAS. During the Late Permian and the Early Triassic, continental crust of East Arctic experienced destruction (see Fig. 2) which caused its thinning (Ledneva et al. 2011). Triassic deposits of the Chukotka microcontinent were impregnated by numerous sills and small hypabyssal diabase, gabbro, and dolerite bodies with local intercalations of tuffs and basalts comparable in geochemical characteristics with traps of the Siberian platform.

Rifting in the North American continental margin commenced in the Early Jurassic and subsequently resulted in opening of the Amerasian Basin and detachment of the Chukotka–Arctic Alaska continental block (Grantz et al. 2011). Part of this continental mass corresponded to the Chukotka microcontinent. The uplifted southern rift shoulder of Chukotka was deformed and subjected to erosion (Tuchkova 2011). The deformation was characterized by normal fault, local fold, and cleavage and they were responsible for the lacking of the Lower–Middle Jurassic deposits over a considerable part of ACH.

Late Jurassic—Early Cretaceous. The spreading in the Proto-Arctic Ocean ceased in the Kimmeridgian time. The Prot-Arctic Ocean was transformed in the South Anyui remnant basin. The basin was filled in with the Tithonian turbidite (Fig. 3). The oceanic crust subducted under the Oloy volcanic belt. Magmatism of the Late Jurassic–Early Cretaceous time was associated with development of subduction zone inclined under the continental margin of Siberia that was consuming first the oceanic crust of the South Anyui Basin and afterward the continental crust of the Chukotka microcontinent. Formation of accretionary wedges with basaltic and cherty blocks was confined to the Oloy convergent boundary (Sokolov et al. 2009a).

Collision between the microcontinent and Siberia culminated in the Hauterivian—Barremian. Tectonic deformations associated with the collision initially formed the northward vergent fold-and-thrust structures and then dextral lateral strike-slip and thrust faults with southern vergence (Sokolov et al. 2009a). Post-collisional granites are dated at 117–115 Ma.

At the post-collision stage granite-metamorphic domes and associated depressions filled in with the Aptian–Albian volcano-sedimentary deposits formed under conditions of crustal extension.

1.2 West-Koryak Fold Area

The Late Mesozoic West-Koryak fold area is located on the junction of the Verkhoyansk-Kolyma, Chukotka and Koryak fold areas (see Fig. 1). Its structures are greatly discordant to structural trend of the Verkhoyansk-Chukotka fold area and also exhibit some discordance with structures of the Koryak fold area. It consists mainly of island arc complexes and accreted ophiolitic and oceanic complexes (Fig. 4) (Sokolov 1992; Parfenov et al. 1993; Sokolov et al. 2009b).

Island arc complexes are traced along the coast of the Sea of Okhotsk to Penzhina region and further to the Pekulney Ridge and the southern slope of the Chukotka Ridge. They are ranging in age from late Paleozoic to Early Cretaceous. They extended from Taigonos Peninsula. The island arc complexes belong to two convergent margins of different age: Koni-Taigonos one in the Late Paleozoic—Early Mesozoic and Uda-Murgal arc in the Late Jurassic—Early Cretaceous (Sokolov 1992; Sokolov et al. 2009b). The structures of accretionary wedges of the Uda–Murgal

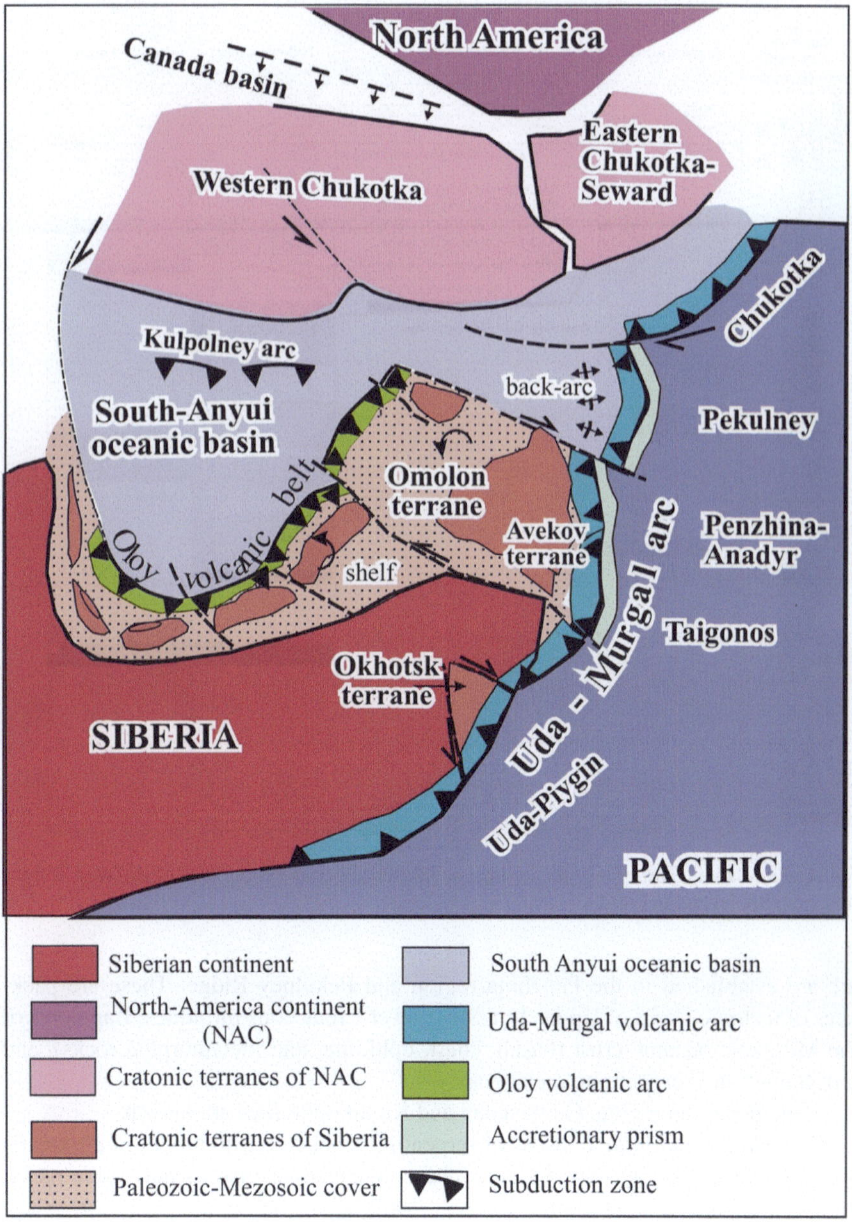

Fig. 3 Paleotectonic reconstruction of convergent margins of the Late Jurassic—Early Cretaceous

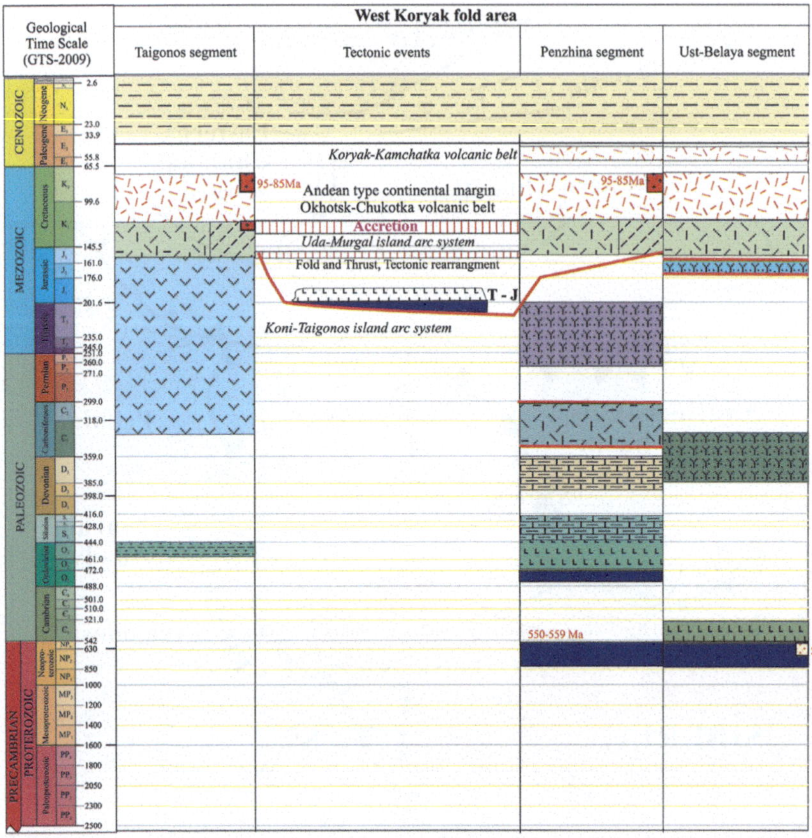

Fig. 4 Correlation chart of tectonic events and litho-stratigraphic complexes of the West-Koryak fold area. See legend in Fig. 11

arc are established in the Penzhina region and Pekulney Ridge. These are packages of tectonic slices of Upper Jurassic–Lower Cretaceous turbidites, fragments of the Mesozoic oceanic crust (basalt, chert, ophiolite, and metamorphic rocks), and terrigenous and serpentinite melanges.

In the Penzhina region, Ganychalan and Kuyul ophiolites are known.

Ganychalan ophiolite consists of serpentinite melange with blocks of ultramafic rocks, gabbro, plagiogranite, basaltic, cherty, terrigenous, and metamorphic rocks; gabbro and gabbro-amphibolite (Essay… 1982; Khanchuk et al. 1992). Ar/Ar age of hornblende from slightly altered gabbro is 559 ± 3 Ma. The volcanic, cherty and sedimentary rocks are Ordovician–Lower Silurian. The geochemical characteristics indicate the oceanic nature of the ophiolites. The metamorphic rocks, including glaucophane schist, are spatially associated with the ophiolite. The isotopic data give concordant results: 325 ± 5 Ma (K–Ar age of volcanic rocks) and 327 ± 5 Ma (Rb–Sr age of dynamoschists). The attributes of metamorphism correspond to its subduction related origin consistent with the age of island arc volcanism of the Koni-Taigonos island arc. ***Kuyul ophiolite complex*** is composed of serpentinite mélange with blocks of ultramafic rocks, gabbro, basalt, chert, limestone, and terrigenous and metamorphic rocks (Sokolov et al. 1996). Fragments of oceanic, suprasubduction, and within-plate ophiolites make up a system of tectonic sheets deformed in intricately deformed folds.

Ust-Belaya ophiolite complex is located in northern part of the Ust-Belaya Mountains in the middle course of the Anadyr River. Ophiolite composes several tectonic sheets among Middle Jurassic and Upper Jurassic–Valanginian terrigenous and tuffaceous rocks. The ophiolite consists of ultramafic rocks, gabbro, dolerite, and basalt, which are overlain by the Middle–Upper Devonian and Lower Carboniferous sandstone, siltstone, tuff, and limestone. The U–Pb age of the zircon from the gabbro is 799 $\pm$ 15 Ma and from the diorite 575 $\pm$ 10 Ma, indicating several stages of magmatic activity.

Main stages of tectonic evolution Structures of the West Koryak fold area formed during several stages of accretion and deformation (see Fig. 4). The earliest stages are poorly understood and are associated with geodynamic environment of Ganychalan and Ust-Belaya ophiolites.

The ultramafic-gabbro complexes of the Ust-Belaya ophiolite formed in the continental crust rifting which caused further development of ocean basin. Since the late Paleozoic, two stages of tectonic evolution were recognized.

Late Paleozoic—Early Mesozoic. At that time between the North-Western Pacific and the Eurasian Plate there was a convergent boundary, along which the structure of the Koni-Taigonos island-arc system formed. The polarity of the subduction zone remains unknown. The island-arc volcanism commenced during the Early Carboniferous (Sokolov 1992; Parfenov et al. 1993) or even the Middle Devonian. Volcanism lasted until the beginning of the Late Jurassic. Oceanic complexes are known in the Kuyul and Ust-Belaya ophiolites and on the Taigonos Peninsula. Their age is determined in the interval from the Early Paleozoic to Early Cretaceous. ***Late Jurassic—Early Cretaceous***. At the boundary between the Middle and Late Jurassic the tectonic transformations took place; they were accompanied by restructuring on continental margins, continental-oceanic transition zones and readjustments of oceanic plates in the Pacific (Sokolov 1992). The Volgian sediments overlie older complexes with sharp angular unconformity. New convergent margin of Eurasia and the Pacific commenced during the Volgian. Structures of the Uda-Murgal island-arc system appeared along this margin (see Fig. 3). In Taigonos, Penzhina, and Pekulney segments, the accretionary prism formed in frontal parts of the island arcs. During the Aptian-Albian, volcanic activity came to an end. Completion of volcanism coincided with the accretion of oceanic complexes, deformation, and formation of a new continental margin. A new stage of tectonic development of the continental margin began during the late Albian. The Okhotsk–Chukotka volcanic belt (OCVB) arose along the new convergent boundary. Late Albian to Late Cretaceous OCVB overlies structures of the Western Koryak fold area with significant angular unconformity. Therefore, OCVB may be considered as a post-accretion complex that integrated Verkhoyansk-Kolyma, Chukotka, and West-Koryak fold areas.

1.3 Koryak Fold Area

The Koryak fold area includes North-Koryak (Anadyr-Koryak) and South-Koryak (Olyutor-Kamchatka) fold systems (see Fig. 1). The North Koryak fold system consists of Algan, Mainitz, Velikorechenskaya, Alkatvaam, and Ekonay tectonic zones. The South Koryak fold system consists of Ukelayat and Olyutor tectonic zones.

North-Koryak fold system

The North-Koryak fold system consists of several nappes and tectonic sheets of south-eastern vergence (Fig. 5). Correlation of tectonic events and lithostratigraphic complexes is shown at Fig. 6.

Algan and Mainits zones have a lot in common in their structure and composition of the constituent complexes. They are of imbricated structure and consist of ophiolite, extended zones of serpentinite melange, the Middle Jurassic–Lower Cretaceous tuffaceous and terrigenous rocks of the Pekul'nei Formation and Chirynai Group, as well as postamalgamation Albian–Upper Cretaceous flyschoid sequence (Sokolov 1992). As a rule, serpentinite mélange occupies lower structural position and is generally built in a similar way. The blocks (up to 4 × 15 km) are of diverse composition: ultramafic and gabbroic rocks, plagiogranite, dike suites, volcanic rocks varying from basalt to rhyolite, cherts, Paleozoic and Mesozoic limestone, amphibolite, green and glaucophane schists. Permian limestone in the Mainitz zone contains Tethyan fauna. The ophiolite (Sokolov 1992) formed in oceanic and suprasubduction geodynamic settings during the Late Paleozoic–Early Mesozoic and Late Mesozoic. Basalt, chert, turbidite, graywacke were deposited in the geodynamic regimes of marginal sea, ensimatic island arcs, and accretionary prisms.

The wedge-shaped ***Velikorechenskaya zone*** is located between the Algan and Main zones. This zone consists of terrigenous rocks up to 5 km thick (Sokolov 1992). The Albian–Coniacian flysch is overlain by shallow-water and continental deposits

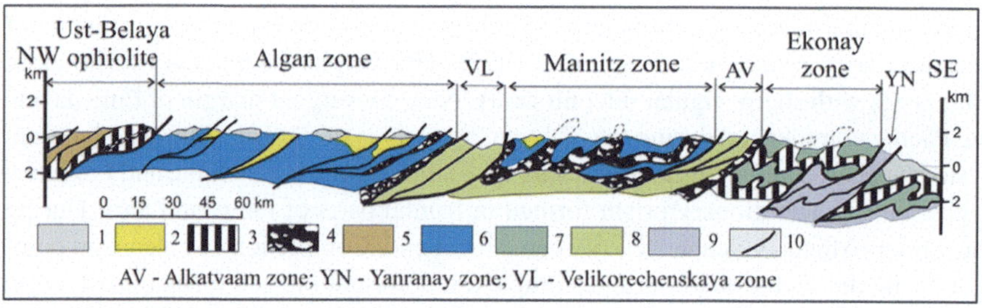

Fig. 5 Geological cross-section of the North-Koryak fold system. 1—overlap sequence; 2—Albian–Campanian clastic complex; 3—ophiolite; 4—serpentinite mélange; 5—Paleozoic carbonate and terrigenous sediments; 6—Mesozoic island arc and back-arc complexes; 7—Upper Jurassic–Cretaceous turbidite; 8—Jurassic–Cretaceous tuff-terrigenous complex; 9—accretionary prism complex; 10—thrusts

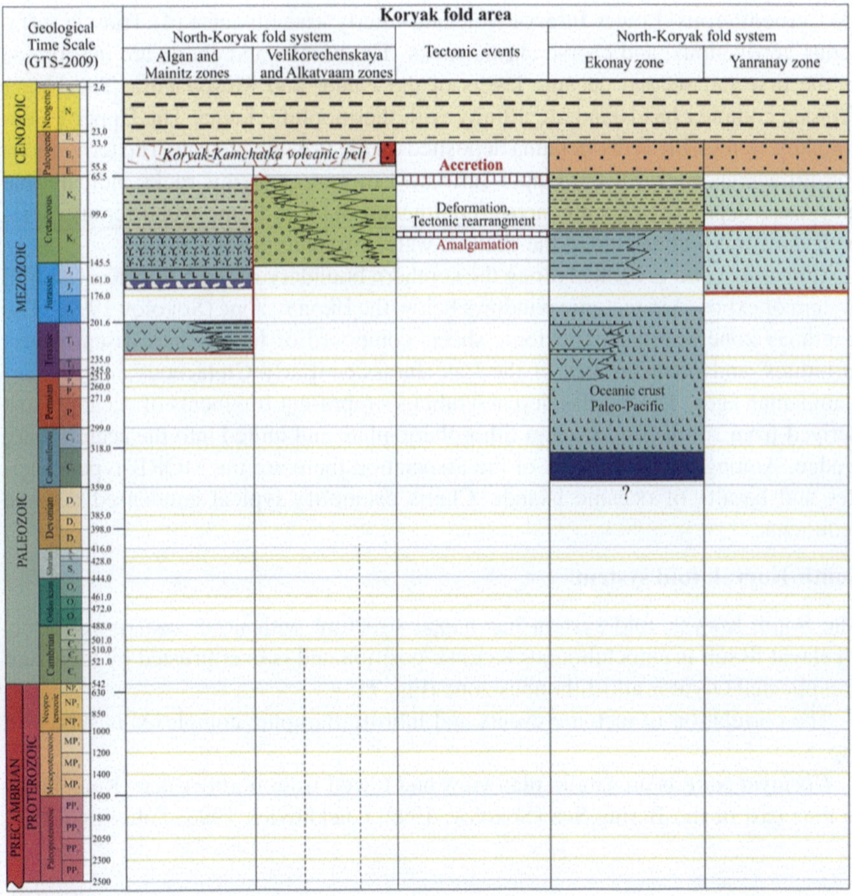

Fig. 6 Correlation scheme of tectonic events and litho-stratigraphic complexes of the North-Koryak fold system. See legend in Fig. 11

of Senonian–Danian age. Volcanogenic-sedimentary horizons are in Maastrichtian–Paleocene stratigraphic intervals.

Alkatvaam zone is composed of thick (up to 5–6 km) Upper Jurassic–Lower Cretaceous and Upper Cretaceous–Paleocene terrigenous, often flyschoid sequences. Their internal structure is rather complex. Several tectonic sheets differ in stratigraphy, lithology, and facies features. Island-arc volcanic and sedimentary rocks occur in some sheets at the Hauterivian and Maastrichtian–Paleocene levels.

Ekonay zone has a complex fold–nappe structure with recumbent and overturned folds, which are unconformably overlain by the post-accretionary Upper Maastrichtian—Paleocene rocks (Sokolov 1992). Several nappes can be regarded as separate lithotectonic complexes: (1) Upper Jurassic– Cretaceous tuffaceous terrigenous rocks; (2) ophiolites: ultramafic and gabbroic rocks; gabbro, plagiogranite, and dikes;

(3) Carboniferous–Lower Jurassic volcanic–cherty assemblages; (4) Triassic tuffaceous, terrigenous, and coarse clastic rocks. The basalts and cherts are interpreted as the first and second oceanic layers of the Paleo- and Mesopacific. The Upper Triassic complex has island arc affinity with thermophilic fauna. The Upper Paleozoic limestone with Tethyan fauna deposited on the oceanic plateau. The terrigenous and tuffaceous rocks of the Upper Jurassic–Lower Cretaceous rocks indicate that they were deposited near the active continental margin. Some fossiliferous facies with shellstones characterize the shallow water, probably shelf setting.

Yanranay zone is located along the southern boundary of the North-Koryak fold system or exposed in tectonic windows below the Ekonay zone (Sokolov 1992). The Yanranay zone consists of tectonic sheets composed of basalt-chert assemblages, turbidites, and olistostromes of the Late Jurassic—Early Cretaceous and Albian-Campanian ages. The basalt-chert assemblages represent fragments of oceanic crust derived from subducting Izanagi lithospheric plate and united into the accretionary wedge. Among volcanic rocks of the association there are the MORB-type tholeiites and basalts of oceanic islands. Cherts exemplify typical condensed oceanic sediments.

South-Koryak fold system

The South-Koryak fold system has nappe structure with northwestern vergence. Ukelayat flysch is autochthonous tectonic position and is overthrusted by large and complicated Olyutor allochthonous zone (Fig. 7).

The correlation of tectonic events and lithostratigraphic complexes is shown in Fig. 8.

Ukelayat zone is arcuate in plan view and traced from northwestern Kamchatka to the coast of the Bering Sea (Sokolov. 1992; Chekhovich 1993; Solov'ev 2008).

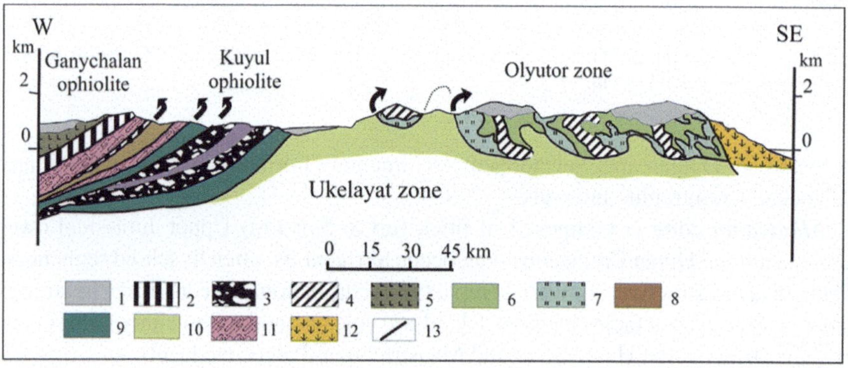

Fig. 7 Geological cross-section of the South-Koryak fold system. 1—Neogene–Quaternary overlap sequence; 2—ophiolite; 3—serpentinite mélange; 4—zonal ultramafic–mafic complex; 5—Ordovician basalt, chert, limestone, turbidite; 6—Achaivayam island arc complex; 7—Vatyna basalt-chert complex; 8—Upper Paleozoic—Lower Mesozoic limestone and tuff-clastic complex; 9—Upper Jurassic—Lower Cretaceous turbidite; 10—Upper Cretaceous—Eocene flysch; 11—metamorphic complex; 12—Paleogene island arc complex; 13—tectonic contacts

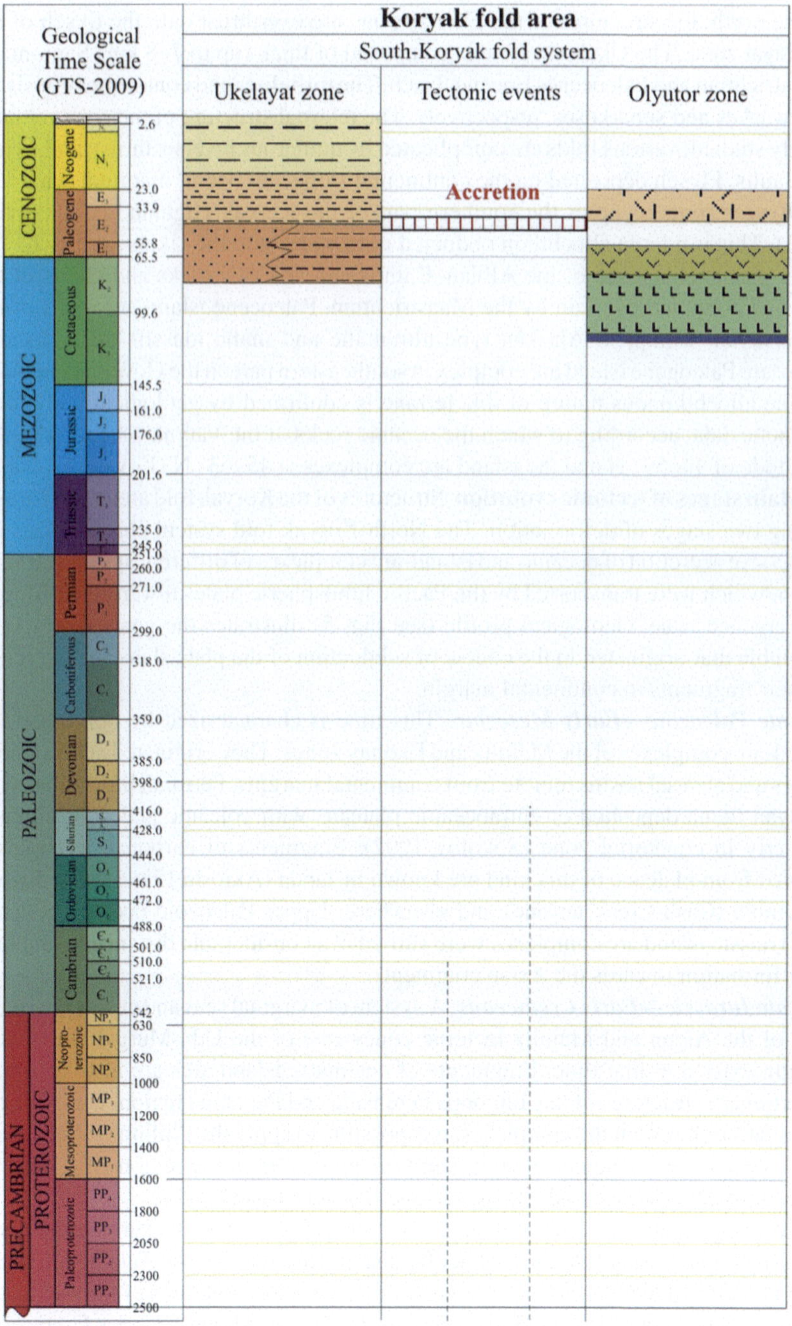

Fig. 8 Correlation scheme of tectonic events and litho-stratigraphic complexes of the South-Koryak fold system. See legend in Fig. 11

To the north, the structures of the Ekonay zone are overthrust onto the flysch of the Ukelayat zone. The Ukelayat zone is composed of thick (up to 7–8 km) Santonian–Maastrichtian and Paleocene–Eocene flysch. The turbidites and contourites consist of graywackes and subarkoses, respectively. The internal structure of the zone remains poorly studied; various folds are complicated by numerous reverse, thrust, and strike-slip faults. Flysch deposited on the continental slope and rise of marginal sea.

Olyutor zone occupies the southern part of the Koryak Highland (Chekhovich 1993). This is a large allochthon obducted northward onto the Ukelayat flysch. The lower part is composed of the Albian–Campanian volcanic rocks and cherts of the Vatyna Group and overlain by the Maastrichtian–Paleocene island arc rocks of the Achaivayam Complex. Alaskan type ultramafic and mafic massifs occur locally. There are Paleogene island arc complex in southeastern part on the Govena Peninsula.

The allochthonous nature of this terrane is confirmed by geological and paleomagnetic data, according to which the oceanic rocks of the Vatyna Group formed at a latitude of 44–59° N and the island arc complexes at 45–63° N (Kovalenko 2003).

Main stages of tectonic evolution. Structures of the Koryak fold area were formed during two stages of deformation. The North-Koryak fold system characterizes the successive accretion of oceanic and island-arc complexes of different age and composition, which were transported by the Pacific lithospheric plates toward Uda-Murgal convergence zone. Geological profile (see Fig. 5) illustrates the general structural ensemble that originated in the course of subduction of the plate slabs and accretion of their fragments to continental margin.

Late Paleozoic—Early Mesozoic. This time is characterized by ophiolitic and island-arc complexes of the Mainitz and Ekonay zones. They originated at mid-ocean ridges and abyssal basins remote from continental margins. Permian limestones with Tethyan fauna deposited on intraoceanic plateaus with volcanic pedestals situated formerly in equatorial zone (Sokolov 1992). Fragments of carbonate complexes derived from plateaus of this kind are known in Japan (Akioshi Limestone), British Columbia (Cash-Creek terrane) and elsewhere. Upper Paleozoic oceanic plateaus and Triassic island arc complexes were shifted at a considerable distance from place their formation towards the Asian continent.

Late Jurassic—Early Cretaceous. A system of marginal seas and ensimatic island arcs of the Algan and Mainitz tectonic zones east of the Uda-Murgal convergent margin existed at that time. Fragments of ensimatic island arcs are also known in accretionary structures of the Taigonos Peninsula and Penzhina region. Based on the age relations between the complexes, it is possible to apply the Philippine Sea model for development of the ocean-continent transition zone in the region (Fig. 9). It is characterized by short-lived island arcs and intra-arc basins.

Rearrangements in structural patterns took place in the Aptian–Albian time. Tectonic movements of that time that affected all structures of the Asian continental margin included the North-Koryak fold system. The paleostuctures of the Algan and Mainitz zones were integrated and overlain by Upper Albian—Upper Cretaceous sediments which included turbidites.

Late Cretaceous—Paleogene. The Okhotsk–Chukotka volcanic belt was located along the continental margin of Andean type. Terrigenous sediments with tuff layers

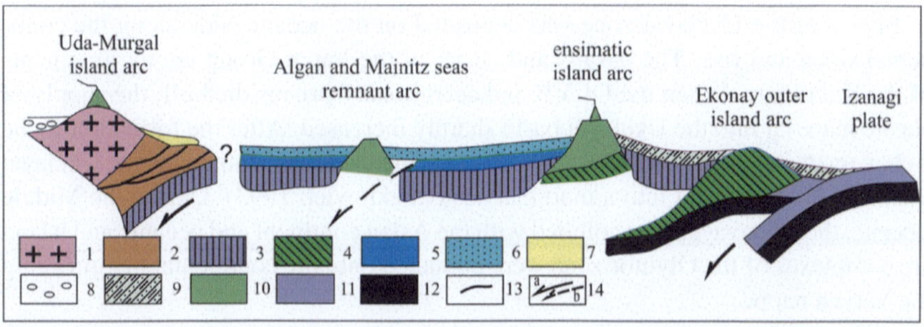

Fig. 9 Paleogeodynamic reconstructions of the North-Koryak fold system for the Late Jurassic—Early Cretaceous. 1—Asian continent; 2—accreted complexes (Ganychalan, Kuyul, Ust-Belaya ophiolites, and others); 3—oceanic crust; 4—blocks of Paleozoic and Early Mesozoic crust; 5—basalt-chert assemblage; 6—greywacke; 7—shelf sediment; 8—molasse; 9—turbidite; 10—island arc; 11—oceanic crust of the Izanagi plate; 12—upper mantle; 13—faults; 14—subduction: a—active, b—inactive, frozen

accumulated in the Penzhina forearc basin. Post amalgamated sedimentary cover of the Algan–Mainitz Basin and Velikorechenskaya zone deposited on a vast shelf. The flysch of the Alkatvaam zone was deposited in the basin between two uplifts. The continent–ocean transition zone of the Albian—Coniacian time resembles in general the Bering Sea (Fig. 10). Upper Maastrichtian shelf sediments overlay with unconformity the structures of the Ekonay and Yanranay zones. Their age determines the time of their accretion to the Asian continental margin. The external boundary of a vast shelf is traced from Taigonos to Penzhina shelf and then abruptly turns to the east where the Yanranay accretionary wedge was in progress at the foot of the continental rise.

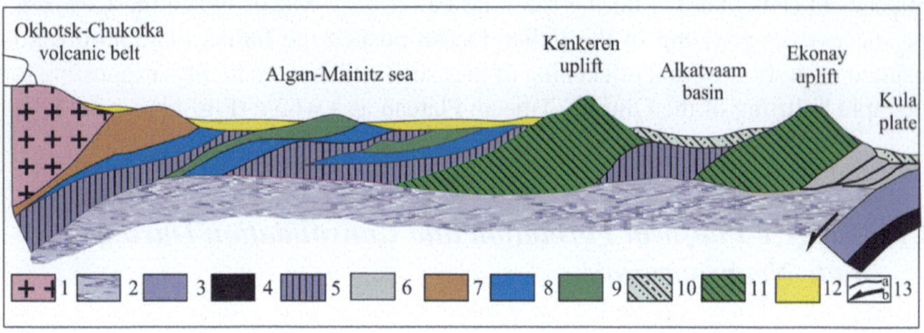

Fig. 10 Paleogeodynamic reconstructions of the North-Koryak fold system for the Late Cretaceous. 1—Asian continent; 2—lower crust and upper mantle of the continent-ocean transition zone; 3—oceanic crust of the Kula plate; 4—upper mantle; 5—deformed oceanic crust; 6—Yanranay accretionary prism; 7—accreted complexes (Ganychalan, Kuyul, Ust-Belaya ophiolites, and others); 8—basalt-chert-greywacke complex; 9—island arc complex; 10—turbidite; 11—Paleozoic and Lower Cretaceous complexes of the Ekonay zone; 12—clastic rock; 13—a—faults; b—subduction zone

Flysch of the Ukelayat zone was deposited on the oceanic side along the continental slope and rise. The basalts and cherts of the Vatyn Group are the fragments of the Kula Plate. When the OCVB and accretionary prisms died off, the supply of clastic material into the Ukelayat basin sharply increased. After the formation of the Achaivayam island arc on the oceanic crust during the Late Campanian, the Ukelayat Basin was transformed into a marginal sea (Chekhovich 1993). During the Middle Eocene, the Achavayam arc collided with the Asian continent and oceanic and island arc complexes of the Olyutor zone were obducted onto the continental margin along the Vatyna nappe.

2 Tectonics of Southern China and Adjacent Areas

Introduction

Tectonically southern China is situated south of the Kunlun-Qinling-Dabie orogenic belt and can be subdivided into three domains. The central part is the Yangtze Platform and two stages, the initially consolidated proto-platform stage (c 800 Ma) and completely consolidated platform period (700 Ma), have been identified. The southeastern part is called the Southern China orogenic belt and represents the southeastern accretionary continental margin of the Yangtze Platform, i.e., the Early Paleozoic Southern China fold belt, Late Paleozoic Southeastern Coastal fold belt. The above mentioned belts of Southeast China were strongly reworked and reactivated during the Mesozoic-Cenozoic due to the activity of the Pacific system. The southwestern part is called Dianzangchuan (Yunnan–Tibet–Sichuan) Tethyan orogenic belt which was the southwestern accreted continental margin of the Yangtze Platform in the Early Paleozoic and characterized by the opening–closing of the Tethyan Paleoocean (Paleotethys) during the Late Paleozoic—Mesozoic. In the Cenozoic, the successive spreading of the Indian Ocean pushed the Indian Plate northwards resulted in shortening and thickening of the crust, which were finally responsible for the rapid uplifting of the Qinghai-Tibetan Plateau as a whole (Fig. 11).

2.1 Yangtze Platform Formation and Consolidation During Late Neoproterozoic

Yangtze Platform

The platform covers a large area nearly of the whole reaches of the Yangtze River from the eastern Yunnan Province in the west to the South China and Yellow seas in the east, namely, the Qinling-Dabie fold belts in the north, the Longmenshan Mts. and Red River faults in the west neighboring the Dianzangchuan Tethyan orogenic belt, and the Southern China fold belts in the southeast (Fig. 12).

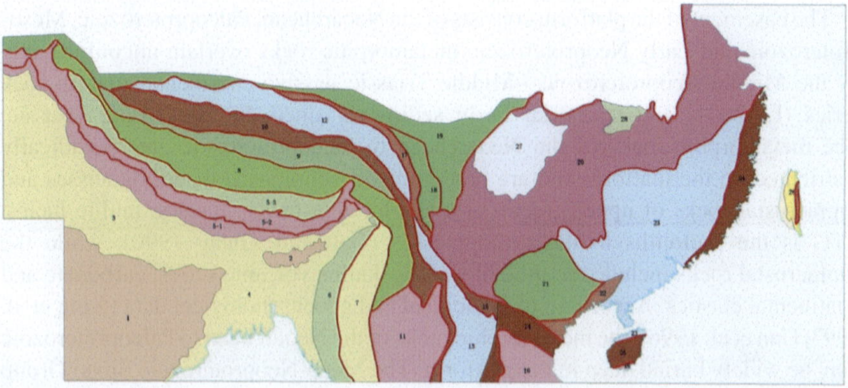

Fig. 11 Major tectonic units of Southern China and adjacent regions. 1. Indian Shield; 2. Shillong promontory; 3, 4. Siwalik foreland basins and Bengal Bay basin; 5. Himalayan orogenic belt (5–1. Lesser Himalayan nappes, 5–2. High Himalayan uplifted belt, 5–3. Tethys Himalayan fold zone); 6. Arakan-Yoma orogenic belt; 7. Irrawaddy Basin; 8. Lhasa block; 9. Nagqu fold zone; 10. Southern Qiangtang block and N. Lancangjang fold zone; 11. Sibumasu block; 12. Northern Qiangtang-Qamdo block; 13. Langping-Puer block; 14. Truongson fold zone; 15. Tentiojiang fold zone; 16. Indochina—South China Sea Platform; 17. Jinshajiang fold zone; 18. Litong fold zone; 19. Songpan-Garze fold zone; 20. Yangtze Platform; 21. Youjiang fold zone; 22. Qinzhou fold zone; South China fold zone; 24, 25. Southeastern Coast fold zone and Hainan fold zone; 26. Taiwan fold belts; 27. Sichuan basin; 28. Jinghan basin

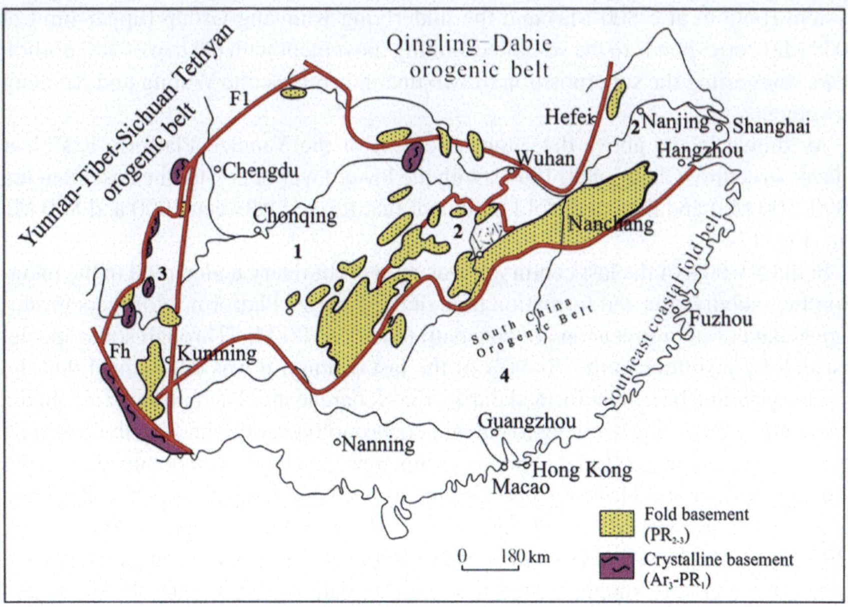

Fig. 12 Basement distribution of the Yangtze Platform and its structural portioning (revised from Fig. 5–1 of (Cheng et al. 1994). 1. Upper Yangtze block, 2. Lower Yangtze block, 3. Kangdian block, 4. Jiangnan block, Fl—Longmenshan Fault, Fh—Red River Fault

The basement of the platform consists of the Neoarchean, Paleoproterozoic, Mesoproterozoic and Early Neoproterozoic metamorphic rocks overlain unconformably by the Middle Neoproterozoic—Middle Triassic passive continental margin rock series (Li 2001) and terrestrial basin sediments since the latest Late Triassic. The metamorphic rocks of the Neoarchean to Paleoproterozoic are sporadically distributed on the platform and are predominantly composed of orthogneisses and supracrustal rocks of upper greenschist–amphibolite facies, local granulite facies. TTG is the protoliths to the orthogneisses (Gao and Zhang 1990), while the supracrustal rocks include the tholeiite, calc-alkaline volcanic series, carbonate and continental clastics. According to geological and geochronological data (Xing et al. 1993; Gan et al. 1996), the metamorphic rocks of the Neoarchean to Paleoproterozoic may be widely buried deep in the platform. The Meso-Neoproterozoic Sibao Group in the northern Guangxi Province includes the mafic and ultramafic rocks. The active continental margin formations of the same period occur in southern Anhui, northern Jiangxi Provinces, Kangdian block and Sichuan Province, ophiolite or residual ocean crust are sliced along the zone from the Dahongshan, Hubei Province, to southern Anhui, northern Jiangxi Provinces, northern Guangxi Province and Longmenshan in Sichuan. Dolerite dykes were reported in Neoproterozoic sequence in southern Anhui (Li 2004).

According to Ren Jishun (Ren et al. 1980, 1999), the basement of the Yangtze Platform has been formed through two major tectonic movements at 1000 and 800 Ma ago. In fact, the unconformity between the upper Chengjiang sandstone of the Nanhua System (bottom at c 800 Ma) and the underlying Kunyang Group (upper limit at 1000 Ma) corresponds to the so-called Jinning movement with the gap of 200 million years, suggesting the superposition of two unconformities, the Wuling and Xuefeng movements.

As shown in the above data, the basement of the Yangtze Platform has clear bilayer structure (Cheng et al. 1994), with the lower layer of crystalline basement (ca 2900–100 Ma) and the upper fold basement that formed between 1000 and 800 Ma ago (Fig. 12).

In the 50–60's of the last century, it was only preliminary understood in the metamorphics subdivision and formation time of the Yangtze Platform. Most geochronological data of poor precision are concentrated at ca 800 Ma. Through some special research by institutes in the 70–90's of the last century, it was determined that the lower crystalline basement formed during the Neoarchean—Paleoproterozoic. In the new century, particularly owing to recent geochronological research on the Jiangnan Uplift by Professor Gao Linzhi and his group, new data have been obtained: e.g., the Shuangqiaoshan and Heshangzhen Groups in northern Jiangxi—western Zhejiang gave the zircon SHRIMP U–Pb age of 831 Ma $\pm$ 5 Ma (Hengyong Formation), 829 Ma (Anlelin Fm.) and 767 Ma $\pm$ 5 Ma (Shangshu Fm.). These data are inspiring as they have greatly supplemented the research data of the Yangtze Platform and demonstrated the significant role of the ca 800 Ma tectono-thermal event in the evolution of the platform.

The sedimentary sequence over the ca 800 Ma unconformity diverges over places, such as the purple sandstone and conglomerate of the Chengjiang Formation in the Kunming, Yimen, Dongchuan, Yunnan Province, sandstone and conglomerate of the Nantuo Formation in the boundary area of Hubei and Hunan Provinces, the molasse-like formation of the conglomerate interbedded with acid volcanic layers in the Kaijianqiao Formation in western Sichuan Province, flysch formation of the Chang'an-Fulu Formations in the southeastern Guizhou—northern Guangxi Provinces in the southeastern Jiangnan Uplift and the successive keratophyre to flysch formations of Shuangxiwu–Luojiamen Formations in western Zhejiang Province. So, the ca 800 Ma Chengjiang movement started the initial consolidation of the Yangtze Platform and complete consolidation after the beginning of the tills to the overlapping of the limestone of the Dengying Formation during the Neoproterozoic. The stable cover sequence of the Yangtze Platform formed from the tillites (700–600 Ma) to Middle Triassic, while the Mesozoic-Cenozoic over the Upper Triassic is a terrestrial basin sequence.

The different features of parts of the Yangtze Platform are shown in Fig. 12 and four first-grade structure subunits can be discerned.

1. *Upper Yangtze block* is distributed over Sichuan Province and covered with terrestrial basin sequence of Mesozoic to Cenozoic and hiatus of Devonian and Carboniferous. The region was wholly uplifted after the crustal movement during the Late Mesozoic to Cenozoic. At the depth, there is an approximately rhombic rigid crystalline block which constitutes the stable core of the platform. The surrounding cover sequence of shallow sea facies from Paleozoic to Middle Triassic was well developed. Folding occurred in the movement during the Middle to Late Mesozoic and formed a ring arc fold zone around the core.
2. *Lower Yangtze block* is well developed in the cover sequence of Paleozoic to Middle Triassic and strong folding (Indosinian movement) in Early Mesozoic, and intensive tectono-magmatism in Middle–Late Mesozoic. A major volcanic-porphyry belt formed along Yangtze River and the Mesozoic Luzong, Ningwu volcanic basins and the Jianghan, Subei fault basins existed during the Cretaceous to the Cenozoic.
3. *Kangdian uplift block* is a NS narrow uplifted zone that was eroded during the Late Neoproterozoic to Middle Triassic, and exposed the basement and magmatic arc rocks after folding. The important Panxi mafic belt and the widespread Emei basalt between the block and the southwestern upper Yangtze Block formed in strong extensive movement during the Late Paleozoic. The Mesozoic to Cenozoic fault basin (Chuxiong basin) is situated to the west of the uplift block.
4. *Jiangnan uplift block* is a long-term mobile uplifting zone at the southeastern margin of the Yangtze Platform and transitional tectonically to the Southern China orogenic belt. The change from the basement to the cover is also transitional, not abrupt and the basement was poorly consolidated. The uplift block was reworked several times during the Mesozoic to the Cenozoic.

Pre-middle Devonian structures associated with consolidation of the Yangtze Platform Southern China fold belt

The belt is the major part of the Southern China orogenic belt and is adjacent to the Yangtze Platform in the northwest, the Lishu-Haifeng Fault as a boundary to the southeastern coastal fold belt in the east (Fig. 11). In the Southern China orogenic belt, the Early Paleozoic with the underlying Meso-Neoproterozoic sequence is a successive flysch formation of great thickness. At the bottom of the Cambrian there are the P-, U-, V-bearing carbonaceous shale formations (stone coal), over which a slightly metamorphosed sandstone flysch with intercalations of carbonate rocks (thickness of more than 3000 m) is situated. The Ordovician–Silurian graptolite-bearing shale and flysch formation is about 5000 m thick. Folding, granite intrusions and low-grade metamorphism occurred during the Late Silurian and were overlapped with unconformity by the Middle Devonian sandstone and conglomerate. All sequences of the whole Upper Paleozoic from the Devonian unconformity to the Middle Triassic are stable shallow facies, alternating sea-continental facies carbonate, and Meso-Cenozoic terrestrial basins initially filled with clastics and coal-bearing clastic deposits.

Therefore, the Southern China fold belt is actually the crustal unit accreted to the southeastern margin of the Yangtze Platform and was consolidated during the pre-Devonian.

It is noteworthy that two special secondary units, the Youjiang and Qinzhou-Fangcheng fold zones, were developed at the southwestern ends of the Southern China fold belt.

Youjiang fold zone is situated at the Youjiang River area between the Red River and Baise faults, and is located at the southwestern end of the Southern China fold belt (Fig. 11). During the Early Paleozoic, the area was a transitional zone between the south accretional belt and the Yangtze Platform. As the Caledonian folding at the end of Early Paleozoic was not so strong, the Silurian to Devonian sequence was successive and continuous, and the crust had not been completely consolidated. During the Late Paleozoic the fold zone tended to be more mobile and eventually the flysch of huge thickness deposited during the Early Mesozoic, especially Middle Triassic. Locally, basic volcanic rocks and hypabyssal intrusives occur, even a sporadical ophiolite in the Napo area. The fold zone was formed as a result of the Indosinian movement during the Late Triassic.

Qinzhou-Fangcheng fold zone is located south of the Hepu-Beiliujiang Fault (Fig. 12) as a long narrow stripe along the Shiwandashan Mts. There, the graptolite-bearing shale section is successive from the Silurian to the Devonian. The siliceous layer of huge thickness is of Upper Devonian to Middle Permian age, and the most important structural movement occurred between the Middle and Late Permian, and in the latter period the thick molasse sequence deposited and with unconformity overlaid the earlier strata including the Middle Permian.

Accretional fold belt on the western margin of the Yangtze Platform

The belt is situated south of the Qinling orogenic belt and west of the Yangtze Platform, and to the south, the Lancangjiang suture is a boundary to the southern Qiangtang block. The belt includes the northern Qiangtang–Changdu block and Bayan Har—Songpan-Garzê block and geographically is situated in central Qinghai-Tibetan Plateau. Tectonically, the belt belongs to the northern Dianzangchuan Tethyan orogenic belt (Fig. 13). The Bayan Har—Songpan-Garzê block is nearly covered with Triassic flysch, while the northern Qiangtang-Changdu block is mostly occupied by the Mesozoic to Cenozoic terrestrial basin deposits since Jurassic. Along the northern and southern sides of the basin, i.e., near the Lancangjiang and Jinshajiang-Yushu sutures, there is a long narrow Late Paleozoic fold belt.

Along the eastern margin of the Songpan-Garzê block and adjacent area, and from back Longmenshan to Motianling Mts., Precambrian metamorphics of the

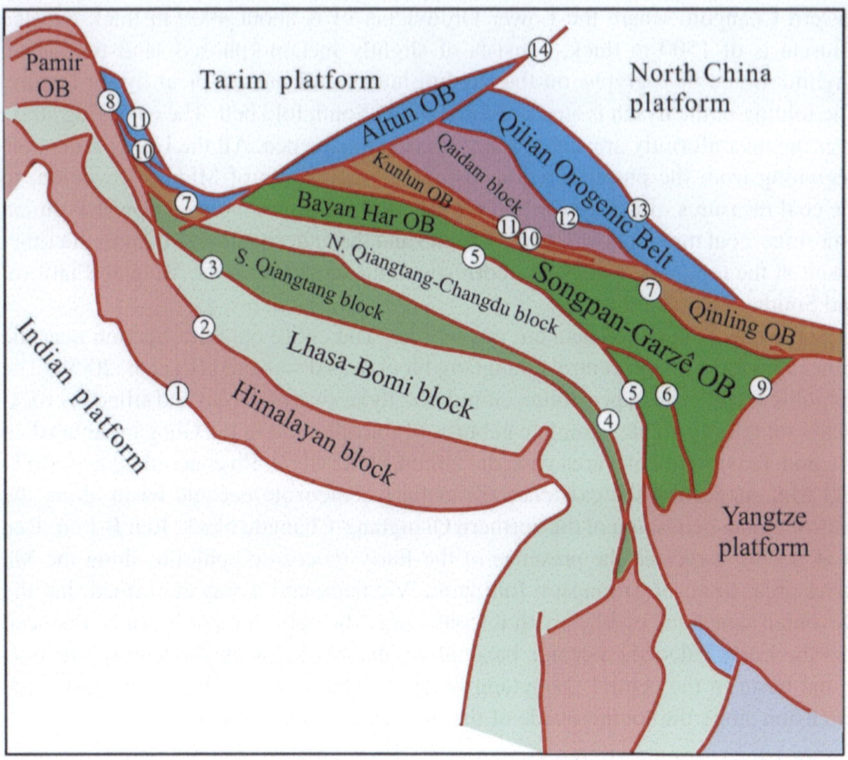

Fig. 13 Tectonic domains of the Qinghai-Tibetan Plateau and adjacent orogenic belts. 1. Main Boundary Thrust (MBT); 2. Yarlung Zangbo suture; 3. Bangong-Nujiang suture; 4. Lancangjiang suture; 5. Jinshajiang suture; 6. Garze-Litang suture; 7. South Kunlun fault; 8. Karakorum fault; 9. Longmenshan-Jinping fault; 10. Central Kunlun fault; 11. North Kunlun fault; 12. North Qaidam fault; 13. Qilianshan foreland fault; 14. Altyn Tagh fault

Bikou Group occur. Its lower part is composed of metabasic volcanic rocks intercalated with slightly metamorphosed clastics, dolomite and siliceous rock which bear Mesoproterozoic stromatolite and microplant fossils; the upper part consists of flysch interbedded with minor intermediate-acid volcanics which with unconformity overlies a fossiliferous dolomite layer of the Sinian Qiaozhuang Formation. Cambrian P-, Mn-bearing carbonaceous slates and Ordovician flysch-dominanted sequence, sandstone, Silurian shale and thin limestone layers deposited during the Early Paleozoic. The Lower Paleozoic sequence underlying the Meso-Cenozoic of the above two blocks is basically similar to that of the southeastern margin of the Yangtze Platform. The overlying Devonian and the whole Upper Paleozoic sequence are stable sea facies and alternating sea and terrestrial facies deposits which are successive from the underlying Silurian and older sequence of the area.

In the northern Qiangtang-Changdu block, some well-developed Paleozoic successions are preserved along the margins of Meso-Cenozoic basins and interior reliefs. A typical section is exposed in the Mangkang-Haitong area, southeastern Changdu, where the Lower Ordovician of is about 4000 m thick and the Silurian is of 1500 m thick, consists of slightly metamorphosed sandstone-slate, phyllite. Some hieroglyphs on the rhythm layers indicate a typical flysch feature. The folding of the flysch is similar to that of the South fold belt. The overlying strata over the unconformity are stable Late Paleozoic sequence. All the Upper Paleozoic beginning from the purplish red sandstone-conglomerate of Middle Devonian, to the coal measures of Lower Carboniferous, Upper Carboniferous—Lower Permian limestone, coal measures of Upper Permian and the andesite flow similar to the Emei basalt at the top, almost finds the corresponding sequence in the Yangtze Platform and Southern China fold belt.

Some new results and data are noteworthy. The entire ophiolite section near the Lancangjiang suture in central Qiangtang block was discovered (Li et al. 2009). The ophiolite includes metaperidotite, cumulates, dyke swarm, basalt and siliceous rock. Many samples from the cumulate gabbros of the ophiolite in the Guoganjianianshan Mt. and Taoxinghu Lake area gave the zircon SHRIMP U–Pb concordant ages 467–423 Ma, suggesting the existence of an Early Paleozoic oceanic basin along the Lancangjiang belt, south of the northern Qiangtang-Changdu block. Ren Jishun (Ren et al. 1999) suggested the presence of the Early Paleozoic ophiolite along the Ma River area, south of Truongson fold zone, Vietnam, and it was confirmed that the Devonian sandstone overlies with unconformity the ophiolite. So it can be deduced that the Early Paleozoic oceanic basin along the Ma River may extend northwards to the basin in the central Qiangtang block, but there is no relics of its eastwards extension along the southern side of the Southern China fold belt.

Pre-late Carboniferous—Permian structures associated with consolidation of the crust

Southeastern Coastal fold belt is situated along the southeastern coastal region of China, east of the Lishui-Haifeng fault and bound by the Southern China fold belt in the west and is actually the Late Paleozoic fold belt formed as a result of accretion to the Southern China fold belt (Fig. 11).

The region is largely occupied by Late Jurassic—Early Cretaceous volcanic rocks and intrusions with possible sporadical occurrence of pre-Sinian (preEdiacaran) metamorphics; this is one of the reasons for existence of the so-called Cathaysian ancient continent. Many years of geological mapping, mineral exploration and research data have strengthen our understanding and re-examination of the structural feature of the orogenic belt.

At the end of the last century, the Zhejiang Geological Survey has first discovered a section of Carboniferous clastics near Nanxi, Fujian Province. We have observed the section. According to the rock association and fossils, the section can be subdivided into 3 parts.

The lower part is mainly composed of microclastic rocks, such as sericite phyllite, carbonaceous phyllite, metasandstone, fine sandstone, with the thickness of 650 m but the bottom is not exposed. Fossils has been discerned in the rocks, and according to Professor Gao Lianda's identification, the microfossils of *Densosporites* sp., *Cyclogranisporites* sp., *Leiofusa* sp. A., *Leiofusa* sp. B., *Punctatisporites* sp., *Leiotriletes* sp., of Late Devonian to Early Carboniferous age are present. The sequence has typical rhythms and hieroglyphs of flysch, indicating the flysch deposition on the passive continental margin.

Middle part is the conglomerate and pebble-bearing sandstone, 400 m in thickness. The lower layers with coarser pebbles and tendency to be fine-grained in the upper layers and siliceous siltstone may be present. Compositions of the pebble are complex, such as phyllite, slate, marl and subordinate porphyblastic biotite-plagioclase gneiss (migmatite), basic volcanic rocks, marble and schists.

The upper part is the phyllite with intercalations of siltstone and fine sandstone, and lenses of siliceous limestone with Fusulinida, Brachiopoda and Foraminifera fossils. The carbonaceous fine sandstone intercalations contain clasts of Late Carboniferous fossil plants.

The conglomerate layer in the middle part is transitional and successive and conformable to the strata of the upper part, but may be unconformable to the Early Carboniferous underlying flysch. In addition, we have observed many drill samples of the Yangshan Iron deposit in Dehua County in the southern part of the belt, and geological excursions to some sporadical outcrops were carried out. We have found that the so-called Yangshan Group of Devonian–Carboniferous is actually a series of phyllite, sandstone slate with interlayers of metalimestone with flysch feature. The difference from section A consists in occasional occurrence of intermediate-basic volcanic interlayer in the slate.

In the Fujian realm east of the Lishui-Haifeng fault, there occur scattered metamorphic rocks, such as at Pantian, Gande in Anxi. The sequence was previously considered as pre-Sinian because of highgrade metamorphism and ubiquitous migmatization, but due to the occurrence of such fossils as *Conochitina* sp., *Cythochitina* sp., and *Rhabdochitina* sp., *Fungochitina* sp. the strata were determined as Ordovician. The Middle Permian Qixia limestone widely overlaps the Yangshan Group and older metamorphics of Early Paleozoic at Pantian. The data suggest the existence of the mobile depositing environment from the Late Devonian (or older) to the Early

Carboniferous along the Southeastern Coastal belt of China. The deposits are represented by flysch and siliceous-volcanic formation. Stable cover deposits began from the Late Carboniferous. So, we can say, the Southeastern Coastal fold belt was actually consolidated up to the Late Carboniferous and was a result of southeastward accretion of the Southern China fold belt.

The Southeastern Coastal fold belt can be trailed to the Carboniferous-Permian of the Taiwan Island across the Taiwan Strait. As Zuyi (1992) described, heavy grey-brownish massive limestone occurring at a depth of 803 m below the surface from Well 1of Jiali was correlated to Carboniferous-Permian by lithological features. To the east of the Central Mountain Range, mainly schist, marble and sporadic gneisses and amphibolite are exposed; generally they are called Late Paleozoic—Mesozoic Great Nan'ao schist. In addition, some mafic and ultramafic rocks of the ophiolite are also present in the rock series. Meanwhile the Yuli HP-LT and Tailuge HP-LP metamorphic zones can be discerned. The formation of the ophiolite and coupled metamorphic zones took place mainly in the latest Late Paleozoic, even going down to Cenozoic. So, the formation and evolution of the Taiwan Arc fold mainly took lace during the Late Cretaceous to the Cenozoic. The Taiwan Arc fold may extend and correlate northwards to the southwest Japan in the north and southwards to the Philippines in the south.

Hence, though there is some difference between the structure feature of the Southeastern Coastal fold belt and Taiwan Island, the two domains evolved in the same tectonic environment and share intimate relations between them, such as the folding of Taiwan Arc fold at certain stage of Carboniferous–Permian may be similar to that of the Southeastern Coastal fold belt. The difference is that the Southeastern Coastal fold belt was a passive continental margin while the Taiwan Island Arc was in the oceanic environment before the consolidation. During the Mesozoic-Cenozoic, both domains experienced strong superimposition and reactivation of the Pacific tectonic realm, and the Southeastern Coastal fold belt in continental setting and the Taiwan Island Arc fold continental margin setting.

Some problems on the structure of the Hainan Island and its adjacent area

According to the geological data available, the Hainan Island can be subdivided into three structural units from north to south (Fig. 11).

(1) ***Northern margin of the Hainan Island*** includes the region north of the Anding fault, the Qiongzhou Strait and Leizhou Peninsula in the north. The region is occupied by large area of Cenozoic sandstone, conglomerate, mudstone and basalt. The Cenozoic cover is mainly underlain by slightly metamorphosed rocks, like Early Paleozoic phyllites and schists, and minor stable Devonian-Carboniferous sequence and terrestrial basin deposits of Cretaceous age. Thus the structure feature of the region actually belongs to Southern China fold belt.
(2) ***Major part of the Hainan Island*** is situated between the Anding and Lingshui faults. The fold basement of the region includes two rock associations; the lower part is a metamorphic series of Meso-Neoproterozoic to Early Paleozoic age, such as the main schist and intercalated sandstone slate, siliceous rock and

metamorphic volcanic rocks with migmatization. The upper part is a low-grade metamorphic rocks of Devonian—earliest Early Carboniferous age, metaclastics, pyroclastics of mainly basic eruption and siliceous rock. The upper and lower parts are divided by unconformity. Both parts are overlain with unconformity by the stable cover sequence of late Early Carboniferous age. According to the fossil data, the unconformity may be a boundary between the Tournaisian and Visean stages.

The viewpoint of Ziye (1989) is noteworthy. He mentioned that the Carboniferous-Permian strata in the Dongfang County contain diamictite of glaciomarine facies feature, such as the profile association, lithology and geochronology all are similar or identical to that of the northern Tibet, western Yunnan and southeastern Asia, and of the same formation time as the Gondwanaland tillite. This point should be emphasized and needs further study. Though the westwards extension of the main Hainan Island is cut and displaced by the Cenozoic Red River—Yinggehai fault depression, the Truongson fold belt in Vietnam may correlate with the extension. Its further westwards extension may be connect with the northern Tibet-Lancanjiang belt. It is still unclear whether it runs eastwards to the South China Sea to the Southeastern Coastal fold belt or not.

(3) ***Southernmost Hainan Island*** is situated to the south of the Lingshui fault and includes large part of the South China Sea to the south. This small block may be a massif consolidated during the Late Neoproterozoic to the beginning of Paleozoic (late Pan-African event) but dismembered during the Cenozoic. At Yaxian County, southernmost Hainan Island, the P-, Mn-bearing siliceous carbonate rocks occur, and the upper part is dominated by limestone, dolomitic limestone and quartzite of platform cover sequence. While at the Xisha archipelago in the South China Sea, the Cenozoic coral reef (about 1400 m thick) is mainly exposed. At Zhankong (Xiyong No. 1 Well), the bottom of the coral reef is of Miocene age and is directly overlying the metasedimentary sequence of weathered rocks. The metamorphic rocks may be trails of the Precambrian basement underlying the platform cover sequence. The basement consists of quartz-mica schist and alternating leucocratic and melanocratic granitic gneisses which gave the ages of 627 Ma (Rb–Sr whole rock) and 77 Ma (Rb–Sr mineral isochron, plagioclase), the last datum may reflect the superposed metamorphism time. Along the western coast of the South China Sea, nearly horizontal quartzite is present east of the Mekong River, central-eastern Cambodia. The quartzite bears fossils of Middle Cambrian and overlapped over Kon Tum Complex of the Indosinian block. Therefore, the Indosinian and South China Sea blocks were united together before the Red River—Yinggehai fault depression and may be called Indosinian—South China Sea block which extended northwestwards to connect the southern Qiangtang, Lhasa-Bomi and Himalayan blocks. All these blocks were crustal units accreted on the Indian shield margin during the Late Neoproterozoic to the earlier Paleozoic.

Lancangjiang fold belt

It is a narrow, elongated fold belt near the Lancangjiang suture. As above described, the Early Paleozoic ocean apparently was closed in Devonian and reactivated in the Late Paleozoic along the southwestern margin of the accreted terranes to the western margin of the Yangtze Platform. Sporadic ultramafic rocks with basic lavas and siliceous rocks of Carboniferous—Early Permian age are distributed along the northern Southern Qiangtang, this rock complex is overlapped with unconformity by shallow facies strata of Late Permian—Early Triassic, implying the closing of the paleoocean. The closing was responsible for the long, narrow fold belt, collisional granites, UP and UHP zones and continental margin volcanic belt which formed relatively later up to Early—Middle Triassic and folded together with the simultaneous active deposits, forming part of the elongated Lancangjiang fold belt.

There is also a long and narrow Yushu-Jinshajiang fold belt at northern part of Northern Qiangtang block. Here the opening and closing of the Carboniferous—Early Permian ocean was similar to that of the Lancangjiang. It is quite possible that the Yushu-Jinshajiang belt is a northern branch of the Lancangjiang belt, between which the (central part of?) Northern Qiangtang—Changdu block was situated. Considering the distribution of the collisional granites and metamorphic rocks at two sides, we can deduce that the oceans were closed and collided towards the central block from both sides.

As closing and colliding of the two oceans took place, in the northern and eastern parts the extensional activity occurred along the Bayan Har—Songpan-Garzк belt, being accompanied with abundant basalt eruption in the Late Permian. Flysch of huge thickness formed during the Early and Middle Triassic; some small oceanic basins were closed in the Late Triassic and the large flysch basins folded and uplifted. The Bayan Har—Songpan-Garzк structural belt was actually consolidated and formed up to the Middle Triassic.

Pre-Cretaceous structures associated with consolidation of the crust

Yarlung Zangbo fold belt and Nagqu fold belt

The two belts are mainly situated over the Himalayan and Lhasa-Bomi blocks in southern Dianzangchuan orogenic belt (Fig. 13). The two fold belts and blocks can be subdivided into following subunits and are described emphatically (Fig. 14).

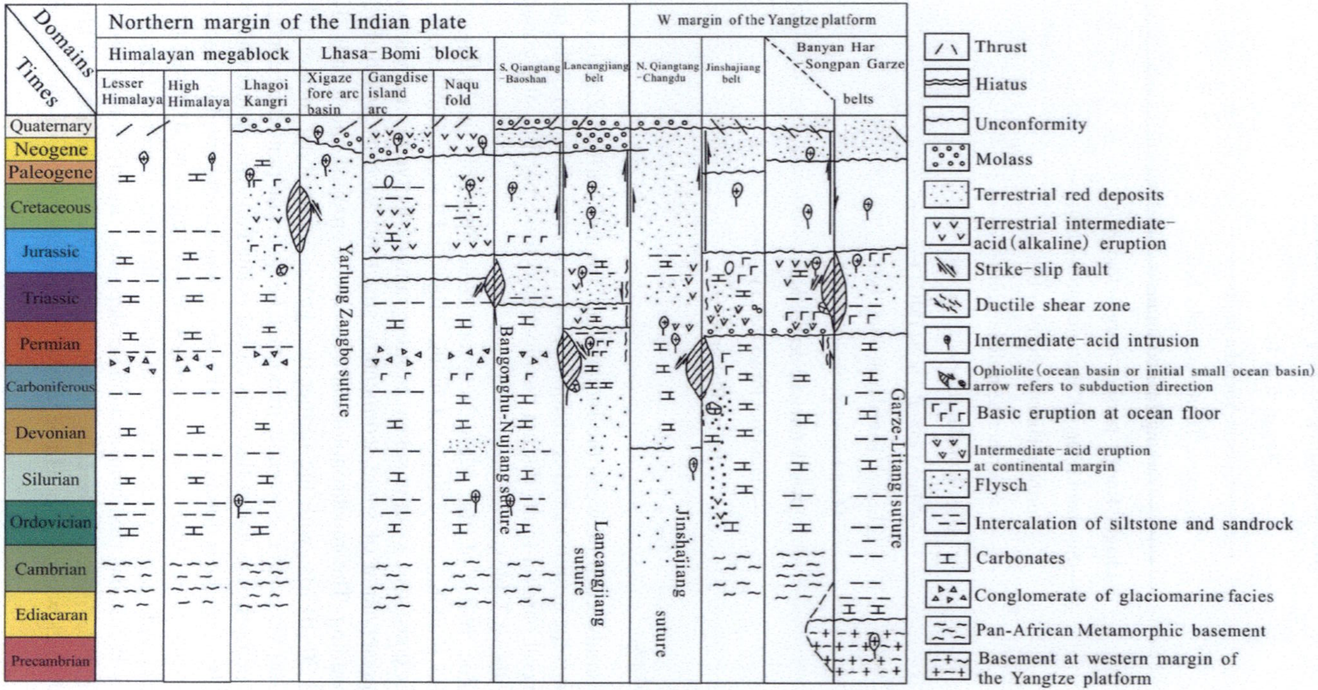

Fig. 14 Correlation of sedimentary sequence—magmatism and structural movements in the the Chuan-Dian-Zang orogenic belt

Southern Qiangtang block

----------Bangong-Nujiang suture---------

Nagqu fold belt

----------Cuoqin-Shenzha fault---------

Gangdise Island arc fold belt

----------Yarlung Zangbo suture---------

Lhagoi Kangri passive margin fold belt

-------Southern Lhagoi Kangri marginal fault------

Northern Himalayan cover belt

------------Main northern slope fault-----------

High Himalayan crystalline belt

----------Main Central Thrust (MCT)---------

Lesser Himalayan thrust belt

----------Main boundary thrust (MBT)---------

Siwalik foreland depression

----------Main front thrust (MFT)---------

Indian Platform

Himalayan block

The block is sandwiched between the Indian Platform and Yarlung Zangbo suture and actually it is the accretion and collision belt on the northern margin of the Indian Platform. Several parallel faults disrupt the block into some parallel strip zones (as shown in the above Table).

Metamorphic rocks of the Meso-Neoproterozoic (possibly including some Cambrian) are predominantly exposed in the High and Lesser Himalayas. The metamorphics consist of kyanite schist, biotite granulite-leptynite, metacarbonate rocks, with preservation of some sedimentary structures such as flysch rhythms and graded bedding. The geochronological data can be grouped into 3 parts, 1250, 600–500 and 20–10 Ma, with the former two groups of regional metamorphism events. 600–500 Ma event is especially important, being accompanied by contemporaneous granitic intrusions and collectively called as late Pan-African tectonic event, while the 20–10 Ma event is dominated by thermo-dynamic metamorphism, anatexis, migmatization and rather scale of leucogranites. Minor Paleozoic strata are present in the two belts and bear abundant fossils of stable platform sequence feature. The Carboniferous-Permian is Gondwanaland facies deposits. The structure of the Lesser Himalaya is a series of thrust structures and accompanying overturned fold klippes and structural windows, while the Higher Himalaya is a series of north-dipped normal shear zones.

All the Main Central Thrust, nappe structures and Main Northern Slope Fault occurred in Miocene (20–10 Ma) when migmatite and great amount of leucogranite related to anatexis also took place. In the northern Himalaya mainly the stable sequence of platform, from the Upper Cambrian or the Lower Ordovician up to the Eocene is widespread. These sediments are shallow sea facies clastics and carbonate deposits, among them the Carboniferous-Permian also bear the features of the Gondwanaland facies deposits. The whole sequence is successive and has the total thickness of nearly 10,000 m. There, magmatism, metamorphism and structure are rather faint. In the Lhagoi Kangri zone, north of the Himaayas, the Precambrian basement rocks and Paleozoic cover sequence are basically exposed near the cores of the domes distributed along the EW zone. The sedimentary feature of the zone is similar to that of the Lesser, Higher and Northern Himalayas, but the Mesozoic beginning from Triassic demonstrates the continental shelf facies of shallow sea transitional to sub-abyssal sea facies of flysch sediments, especially at the northern margin near the Yarlung Zangbo suture where the Upper Jurassic and Lower Cretaceous include the spilites, basalt, radiolarian-bearing siliceous rock, mélange, mafic and ultramafic rocks of the ophiolite. The subduction and closing of the Yarlung Zangbo Paleoocean mainly occurred during the latest Late Cretaceous, while the totally folding and mountain building took place after the Late Eocene, forming the Gangdise molasse and Liuqu conglomerate along two sides of the suture, i.e., the so-called double-molasse.

The southern Himalayan block corresponds to the Siwalik foreland depression which is bounded at the northern margin with the Main Boundary Thrust (MBT) to the Lesser Himalaya and the southern margin with the Main Front Thrust (MFT) to the Indian Platform. The most typical strata in the depression are the Siwalik Group and Lower Pleistocene of molasse formation of coarse clastics and pebbles of total thickness over 5000 m. The molasse layers are rather complicated and three obvious unconformities exist, i.e., the unconformity between the Pliocene and Lower Pleistocene, intra-Pleistocene and the Pleistocene and Holocene. The structure is characterized by the north-dipped thrusts and southwards overturned folds. The deformation of the MBT and molasse occurred during the Late Oligocene. The relatively late MFT formed after the Pleistocene. Most segments of the MFT are buried due to the covering of the Holocene alluvials of the Ganges plain or no obvious activity.

The Himalayan block extends westwards to the southern margin of the Pamirs Plateau and strongly is compressed in the NS direction since Miocene, forming the distinguished Pamir structure syntax (see chapter about the Pamirs). The eastwards extension of the block to Zayu, Tibet forms the abrupt strike change from EW to NS direction and enters into the Arakan Mountain Range belt, Burma.

Lhasa-Bomi (Gangdise) block

The block lies between the Yarlung Zangbo and Bangong-Nujiang sutures. The Precambrian metamorphic rocks and Paleozoic cover strata are similar to that of the Himalayan block to the south. Since the Mesozoic, especially the Late Mesozoic, the block was divided into the southern and northern parts along the Cuoqin-Shenzha fault. The southern part (Gangdise arc folds) consists of the front arc Xigazк basin with Late Cretaceous to Eocene flysch deposits, Gangdise island arc intrusions, andesite volcanic arc and the Precambrian metamorphic rocks in the Nyainqкntanglha Range fault block. The southern part was actually the active margin of the Yarlung Zangbo Paleoocean. Facing the Bangongcuo-Nujiang Paleoocean to the north, the northern part (Nagqu folds) consists of Upper Triassic to Middle Jurassic flysch deposits with intercalations of siliceous rocks, intermediate-basic volcanic rocks and mélange. This part was a passive margin of the southern Bangong-Nujiang paleoocean which closed since the Late Jurassic, and later till the latest Late Cretaceous near Lake Bangongcuo at the western segment. After ocean closing, the suture was strongly deformed and a series of imbricate structures, NE strike slip faults and accompanying Late Cretaceous granite and intrusions of Miocene leucogranites occurred.

The eastwards extension of the Lhasa-Bomi block comes into Tengchong, Yunnan Province, and northwestern Plateau of Burma, while the westwards extension was cut by the Karakorum strike-slip fault in the north, the western end may correspond to part of the Karakorum orogenic belt and its possible further westwards extension to the Central and Southern Pamirs.

Southern Qiangtang-Baoshan (Sino-Burma) block

The block is situated between the Bangong-Nujiang and Lancangjiang sutures. There, both the Precambrian metamorphic basement and Paleozoic cover sequence are similar to that of the Himalayan and Lhasa-Bomi blocks. So, the Himalayan, Lhasa-Bomi and southern Qiangtang-Baoshan blocks correspond to the northern margin of the Indian Platform and components of the Gondwanaland. The western end of the southern Qiangtang block was cut by the Karakorum fault, and the western extension constitutes part of the Karakorum orogenic belt.

2.2 Young Plate (Platform) and Depression Basins

Mesozoic-Cenozoic basins superimposed on the Yangtze Platforms

Sichuan, Chuxiong and Jianghan basins are the major basins

Sichuan Basin is situated in the western Yangtze Platform (Upper Yangtze block) and developed initially in the Late Triassic and its attribute similar to the Ordos basin. The tectonic movement at the end of Middle Triassic made the overall uplift of the Yangtze Platform and formed gentle NE strike anticline structures inside the Sichuan basin. The arc cover folds formed during the Paleozoic are situated in the eastern and southern margins. Along the Kangdian uplift area in the western basin, Late Triassic rifting and volcanic activity occurred resulted in the deposition of Upper Triassic clastic rocks and pyroclastic deposits, which in turn were overlain with unconformity by Jurassic sediments, meanwhile the Jurassic basins migrated eastwards and overlapped the whole Sichuan basin. The tectonic movement during the Late Jurassic was responsible for the arc fold uplifting in eastern Sichuan Province and the westwards migration of the inland lake basins to the Chengdu plain during the Cretaceous—Early Paleogene. The tectonic movement during the Late Cenozoic greatly affected the Sichuan Basin and was responsible for the large scale thrusting from the Longmenshan zone in the western margin to interior basin. Some klippes and tectonic windows are also present. Folding and thrusting took place at the southern margin of the basin. At the Dabashan Mts. in the north, the inwards thrusting occurred and resulted in the nowadays Sichuan structural basin.

Chuxiong Basin is located in the southwestern margin of the Yangtze Platform and southwest of the Kangdian uplift. Large scale basalt eruption from marine to continental facies took place inside the basin during the Middle-Late Permian, while the basin was faulted and uplifted at the end of the Late Permian, with hiatus of Early-Middle Triassic sediments. Some alluvial coarse clastic and coal-bearing clastic sediments of fluvial facies were deposited during the early Late Triassic. The basin subsided over large area at the end of the Late Triassic and individual small fault depressions converted to large depression and connected the Sichuan Basin. Red clastic sediments of fluvio-lacustrine facies deposited during the Jurassic—Early Cretaceous. While in Late Cretaceous—Paleocene, the basin turned into red gypsum-bearing formation of fluvio-lacustrine facies. ***Jianghan basin*** is located east of the Sichuan Basin and includes the region of northern Hunan and Jiangxi Provinces and large area of Hubei Province. The basin deposits of the Late Triassic—Jurassic of the area can be correlated with that of the Sichuan Basin. The difference is that the Jianghan basin was relatively uplifted in Paleozoic and the Precambrian crystalline basement was shallowly buried and the faults developed indicate the block uplifting feature. The Paleozoic cover is rather thin, even absent in places. Some basin sequence of Cretaceous-Paleogene directly overlies the Precambrian metamorphic basement and the structural development of the basin in the period of Cretaceous to Early Paleogene has evolved from small fault depressions to large depression.

East of the Jianghan Basin, there are the Mesozoic ***Luzong, Ningwu volcanic basins*** along the Yangtze River. Further eastwards, there is the ***Subei basin*** as the fault depression basin of Cretaceous-Paleogene, and the southern Yellow Sea to the east is in fact the extension of the Subei basin to the sea. East of the Jianghan basin, including the volcanic basins along the Yangtze River, Subei basin and southern Yellow Sea basin, all are fault depressions overlying the fragments of the Yangtze Platform in the Lower Yangtze sector.

Fault depression basins overlying the fold belts

Basins overlying the Southern China orogenic belt

In the Southeastern Coastal fold belt, the volcanic zone composed of large basin group of intermediateacid volcanic-intrusive rocks developed during the Cretaceous after the enlargement of the Late Triassic—Jurassic coal-bearing fault depression basins. The basin group was located along the NNE to NE strike. This distinguished volcanic zone may stride over the eastern margin of the Southern China fold to the west. The Early Cretaceous volcanic rocks over the coal-bearing deposits of Late Triassic—Jurassic in the large part of the Southeastern Coastal fold belt are rarely developed. But intrusive rocks of the same period were well developed, while the intermediateacid eruptions from the Late Cretaceous to the Paleogene were severely decreased and substituted by groups of small fault depression basins which were filled with red sandstone, conglomerate and mudstone with usual presence of calcium, such as gypsum and glauberite. In some basins, even oil shale and lignite layers occur. During the Quaternary, some basins along the southeastern coast of China could be filled with basalts. It can be deduced that the East China Sea, Taiwan Strait (including zones west of Taiwan Island) and basins in the Southeastern Coastal fold belt all are of the same tectonic setting and basin feature and belong to the same tectonic integrity.

Basins overlying the Dianzangchuan orogenic belt

These basins include the following types.

(1) In the Bayan Har—Songpan-Garzк orogenic belt, Jurassic-Paleogene terrestrial red basins are developed along the strike-slip faults in the folds of flysch in the Early-Middle Triassic.
(2) Basins due to block subsidence between orogenic belts, such as the Changdu and Simao basins, are the depression basins, and the basin sequences are successive to the underlying stable cover deposits. The basins were Jurassic-Cretaceous in time and of great thickness in sediments which are dominated by red-colored layers. Sometimes small Cenozoic basins may be superimposed and were filled with thick Late Paleogene-Quaternary basalts.
(3) Cenozoic fault depression basins isolated along some compression faults, such as the Lunpola, Gaize and Bange basins, are filled with clastic sediments of fluvial and lacustrine facies.
(4) Foreland depression basins are mainly developed in forelands of the Himalayas, filled with molasse of Pliocene epoch—Pleistocene epoch the Siwalik pre-Himalaya foreland depression, and the molasse of Late Eocene epoch—Oligocene epoch in the Gangdise belt in the northern margin.
(5) In addition, it is noteworthy that a series of basins distributed along the NS strike and NE en echelon pattern are controlled by the extensional faults in the Dianzangchuan orogenic belt.

Offshore basins

(1) ***Bohai Sea and Yellow Sea basins*** are overlying the Sino-Korean Platform and are filled with basalts and gypsum-bearing red clastics of the Late Cretaceous-Paleogene.
(2) The southern part of ***Yellow Sea basin*** may be correlated with the Subei basin and is the extension of the Yangtze Platform in the Lower Yangtze reaches into the sea. From north to south, the southern Yellow Sea can be subdivided into 5 secondary units, i.e., the northern Qianliyan uplift; northern depression; central uplift; southern depression and Wunansha uplift (3 uplifts and 2 depressions). The southern Yellow Sea depression began to form in the Late Cretaceous. There, Lower Paleogene sediments are probably absent in some uplifts while the Upper Paleogene and Quaternary are well developed.

(3) ***East China Sea basin*** on the whole can be subdivided into 3 structure subunits, i.e., the western depression zone, the central uplift (folding) zone and the eastern depression zone. The western depression zone is dominated by the Paleogene deposits with the thickness of over 5000 m. While the Upper Paleogene sequence can be connected with the oiland gas-bearing basin deposits of Upper Paleogene in western Taiwan.

The central uplift (folding) zone, through the Diaoyudao Island to the south, can be connected with the Suao zone of Lower Paleogene in Taiwan. The folded sequence is composed of Early Paleogene and older strata and the folding occurred mainly during the Early Miocene, called Puli movement in Taiwan. The eastern depression zone is bounded by the Okinawa trough in the east and the central uplift zone in the west. The crust is only 21 km in thickness. The huge thickness of Pliocene to Quaternary deposits covers the folded Miocene and older sequence. From the volcanic activity, earthquake, folding, fault and high heat flow, it can be deduced that the eastern depression zone is at the transitional area from continental to ocean crust.

(4) ***South China Sea basin***

Along the northern margin of the basin and near the coast of the main China continent, the Beibuwan basin of Late Mesozoic and Zhujiangkou basin of Early Cenozoic are distributed. The deposits and structure feature of the basins can be correlated with the Southeast Coastal fold belt of the continent, i.e., extension of the continent into the sea.

The main part of the South China Sea and various basins were formed through faulting and cracking of the not too rigid crust (Indosinian—South China Sea block).

The central oceanic basin formed as a result of several events of the spreading from the Late Cretaceous to the Early Miocene on the oceanic crust occurred in the central sea basin and covered by recent deposits of only 500 m thick. A great amount of basalt eruptions formed many seamounts. The crust is only 6–9 km thick and the sea depth is 3000 m. West of the central ocean basin and east of the Yinggehai fault depression there is a series of basins, such as the southeastern Hainan basin, southwestern Taiwan basin, Wan'an basin that bear residuals of the continental crust. According to drilling data of Well 1 of Xiyong, the basement is overlain by 28 m-thick weathered crust, covered with the 1300 m-thick reef limestone and chalk of Late Paleogene.

The Yinggehai basin is situated along the Yinggehai fault depression in which huge amount of Paleogene terrestrial clastics of littoral to shallow sea facies deposited. The depression was extended southwards and separated the Wan'an basin into two basins, i.e., the northern Wan'an basin in the east and western Wan'an basin in the west; the latter has the sediments approximately similar to those of the Yinggehai basin. But basic volcanic rocks of oceanic feature can be found along the NE strike fault in central part of the western Wan'an basin.

Cenozoic basins are situated in the southeast of the central ocean basin and distributed around the Nansha archipelago and adjacent areas, including the Liyue basin, Zhenghe basin, Zengmu basin and Palawan basin. The basement of the Nansha Archipelago may be composed of the Precambrian sequence. According to drilling data from the Liyue beach, the drilling down to 4000 m has not penetrated the Lower Paleogene. In the continental shelf area, northwest of the Palawan, there is an unconformity between the Miocene and Pliocene. Above the unconformity, the successive sediments from Pliocene to Pleistocene are accumulated, while the folded Miocene is below the unconformity.

2.3 Main Stages of Tectonic Evolution

According to the above descriptions on the tectonic skeleton of South China, we can give the following summary to the structural evolution of the region.

(1) The Yangtze Platform formed during the Late Neoproterozoic, experiencing the initial consolidation at 800 Ma and complete consolidation at 700 Ma to platform stage.
(2) Paleozoic accretion to the southern margin of the Yangtze Platform took place in two stages: one to the southwestern margin during the Silurian (the Early Paleozoic folding in northern Dianzangchuan orogenic belt) and another to the southeastern margin (the Early Paleozoic folding).
(3) The subsequent accretion occurred in the nowadays southeastern coastal region, where the pre-Upper Carboniferous of the Late Paleozoic and older sequence were folded and consolidated, while in the southwestern China the opening and closing evolution of the Tethys began.
(4) In southwestern China, the development of Paleotethys began in the Carboniferous or even earlier and was closed at the end of the Middle Permian when the evolution of the middle Tethys (or Mesotethys) began. In eastern China, the formation and evolution of the Pacific realm tectonics continued and strong tectono-magmatism took place.
(5) Formation and spreading of the Indian Ocean in Cretaceous caused the closing of the middle Tethys (Mesotethys) at the end of the Cretaceous. The continuous spreading of the Indian Ocean resulted in northward movement of the Indian Plate with subsequent intensive displacement of and compressing effect in the Dianzangchuan orogenic belt, which were responsible for the crustal shortening, thickening and intensive uplifting of the Qinghai-Tibetan Plateau.

The uplifting of the Plateau can be subdivided into 4 stages (Li et al. 2010). The first stage, collision and subsequent local uplifting in K_2–E_2, caused the formation of the Yarlung Zangbo suture and local uplift of the Gangdise arc. The second stage resulted in slow and differential uplifting of large area in E_3–N_1. The third stage was reflected in severe intracontinental compressing and rapid uplifting in N_2–Q_1. The fourth stage caused the geothermal static uplifting since Q_2. But in eastern China, the tectonic regime was expressed in the trench-arc system of western Pacific domain from the end of Mesozoic to Cenozoic, forming the epicontinental and marginal seas through extension and crust thinning.

3 Tectonics of Korean Peninsula

3.1 Main Tectonic Structures of Korean Peninsula

The Korean Peninsula, lying in the continental margin of the eastern Eurasian Continent, represents an important tectonic link between continental blocks of North and South China and the island arcs of Japan. In general, the tectonic provinces of the peninsula can be divided into nine lithotectonic units (Fig. 15): (1) Hambuk Fold Belt; (2) Nangnim Massif (Pyeongbuk Massif); (3) Pyeongnam Basin; (4) Imjingang Belt; (5) Gyeonggi Massif; (6) Okcheon Belt; (7) Yeongnam Massif; (8) Gyeongsang Basin; and (9) Circum-Pacific Volcanic Belt. Each of them has distinct tectonic structure and evolution history in the build-up of the present-day Korean Peninsula. The birth of the Peninsula's rocks can be dated back to as early as the Paleoarchean (~3.8 Ga), as evidenced by rare occurrence of very old zircon grains in the Precambrian rocks. The Peninsula is a composite landmass consisting of three basement blocks that were created largely during Neoarchean to Paleoproterozoic. They are called, from north to south, the Nangnim, Gyeonggi and Yeongnam massifs. Each massif is bordered by two intervening fold-and-thrust belts, the Imjingang and Okcheon (Fig. 15). Despite of much uncertainty on the architecture of their boundaries, the current shape of the Korean Peninsula were largely accomplished by continental collision among these Precambrian massifs during the Late Paleozoic to Middle Mesozoic. Before their final amalgamation, the Early Paleozoic intracratonic Pyeongnam Basin in the Nangnim Massif and Taebaeksan Basin in the Yeongnam Massif had been developed, whereas the Late Proterozoic to Paleozoic sedimentary successions in the northern (e.g., Imjingang Basin) and southern (e.g., Okcheon Basin) margins of the Gyeonggi Massif continued to grow in thickness. After completion of continental amalgamation during the Early to Middle Mesozoic, tectonic regime in the Korean peninsula changed from continental collision system to continental magmatic arc system. The change in tectonic regime has caused not only widespread arc-related magmatism (so-called Songlim-Daebo granitoids) in the entire peninsula, but also the development of the Cretaceous non-marine Gyeongsang Basin in the southeastern part of the Peninsula. Sustained arc system caused back-arc opening in the East Sea and resulted in pull-apart basins along the easternmost margin of the Peninsula (Fig. 15).

- Thus, in limits of the Korean Peninsula there are the following main tectonic structures with consolidated crust, formed:
- by the beginning of the Mesoproterozoic (Sino-Korean Platform);
- by the end of the Neoproterozoic (Yangtze Platform);
- by the beginning of the Cretaceous (Sikhote-Alin area of the Pacific mobile belt).

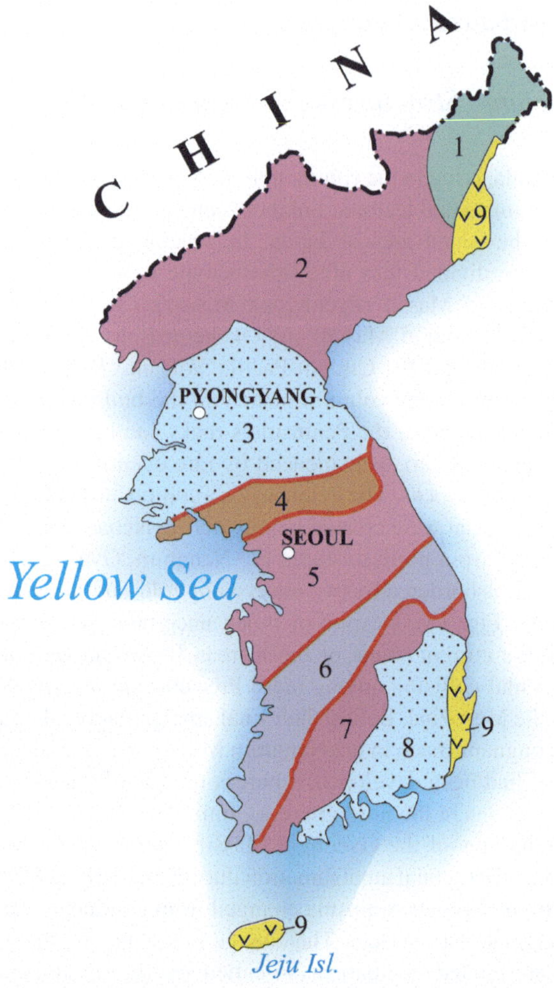

Fig. 15 Tectonic provinces of the Korean Peninsula. Inserted numbers are for: (1) Hambuk Fold Belt; (2) Nangnim Massif; (3) Pyeongnam Basin; (4) Imjingang Belt; (5) Gyeonggi Massif; (6) Okcheon Belt; (7) Yeongnam Massif; (8) Gyeongsang Basin; and (9) Circum-Pacific Volcanic Belt

3.2 Sino-Korean Platform

Tectonic framework

The platform occupies southeastern portion of the Eurasian Continent including the Huangho Basin, southern Manchuria and approximately entire Korean Peninsula. It is separated from the Siberian Platform by the Mongol-Okhotsk Fold Belt. The platform is composed of extensive Precambrian crystalline basement rocks which are overlain by the Phanerozoic strata. The East–West trending ***Imjingang Fold Belt*** (Fig. 15—4) divides the ***Nangnim Massif*** (Fig. 15—2) from the ***Gyeonggi Massif*** (Fig. 15—5) to the south, which comprises the boundary of the Sino-Korean (North China) Platform and Yangtze Platform in the Korean Peninsula. The Nangnim Massif is the Precambrian terrane in northern Korea consisting of high-grade schists and gneisses, overlain with unconformity by the Mesoproterozoic to Phanerozoic supracrustal rocks (Wu et al. 2007). Geosynclinal sedimentation, volcanism, magmatism and folding are crustal characteristics of the Korean Peninsula during the late Paleoproterozoic. After the Machollyong tectonic events (Late Paleoproterozoic tectonic movements) in northern Korea, Sino-Korean Platform became stable and began to undergo transgression (Paek 1996). Thick accumulation of a sedimentary sequence is produced by the transgression. As the Gyeonggi Massif, a tectonic segment of the middle of the Korean Peninsula, crustal materials were extracted from the depleted adjoining stable craton from Neoarchean to Paleoproterozoic time. A marginal deep basin formed along the middlewestern margin of the platform in Korea.

Yeongnam Massif (Fig. 15—7) is another major Precambrian terrane in Korea including the Sino-Korean Platform, which is located in the southern ***Okcheon Fold Belt*** (Fig. 15—6). The massif is composed of high-grade schists and gneisses, unconformably overlain by the Phanerozoic supracrustal rocks. The fossil fauna occurrence in the CambrianOrdovician Taebaeksan Basin, the Taebaek Group, correlates to the North China faunal province (Kobayashi 1953). Recent analyses on trilobite assemblages (Choi et al. 2003) and Archaeoscyphia and Calathium association (Kwon et al. 2003a) clearly demonstrate that the Taebaeksan Basin was connected to the carbonate platform of North China (Kwon et al. 2006). And some paleomagnetic data indicate a clockwise rotation of the Taebaek area. Those evidences reveal that a half of the southern Korean peninsula presumably belongs to a part of the Sino-Korean Platform.

Main Fracture System

Complex fracture systems are developed in the Nangnim Massif and neighboring area. The nature and tectonic characteristics of the fracture systems, however, are not established, so no available literature related to the fracture systems in North Korea region are present. Many of fractures are reactivated after they were firstly formed through the multiple geologic events. The major trend of the large scale normal faults is NNE–NW. These faults are developed along the large water systems and have strike-slip component which is mostly dextral shear sense. However, strike-slip fault developed along the Amnok River shows sinistral shear sense. Tertiary and Quaternary volcanic and intrusive rocks occur along the fault systems (Gilju–Myeongcheon, Chugaryeong Fault Zone, Goksan Fault Zone). It was suggested that these faults might be primarily formed in the Precambrian, and reactivated during Late Mesozoic to Cenozoic time.

Tectonic evolution

The basement rocks of the Nangnim Massif are mainly formed (2538–2636 Ma) and metamorphosed (2462–2433 Ma) in the Neoarchean (Zhao et al. 2006). The Mesoarchean grey gneiss, however, rarely occurs in the Nangnim Massif (3.1–3.4 Ga) (Zhao et al. 2006). The presence of the Mesoarchean crust is partly supported by zircon Hf model age of 4.0 Ga (Wu et al. 2007). All these facts suggest that the igneous and sedimentary rocks of the Nangnim Massif were emplaced and deposited on the Archean basement. Paleoproterozoic orogeny (2.1–2.0 Ga) deformed the Archean strata (e.g., Gimchaek metamorphic complex) which were intruded by ultramafic to mafic rocks. Paleoproterozoic orogeny comprised the emplacement of granitoids (ca. 1.9 Ga) and the metamorphism of the Jungsan and Macheolnyeong groups which resulted in the formation of garnet/sillimanite-bearing granite (1.9–1.1 Ga). Post-tectonic porphyritic monzogranites, similar to rapakivi granite, intruded the basement complex, and the Nangnim and Jungsan groups at 1865–1843 Ma (Zhao et al. 2006). Recently, (Wu et al. 2007) reported zircon U–Pb ages of ca. 2.1 and 1.9–1.8 Ga as the periods of dominant granite magmatism in the Nangnim Massif, and emphasized these periods as the timing of major cratonization of the Nangnim Massif.

Presence of the Archean crust in the Yeongnam Massif is not evident. Recent geochronologic studies, however, reported Neoarchean age at the inherited core of zircons separated from the Paleoproterozoic biotite gneiss (Lee et al. 2002). The granitic gneisses and amphibolites of the Yeongnam Massif were emplaced at ca. 1.9 Ga through the reworking of the Neoarchean basement rocks. These Paleoproterozoic rocks underwent high temperature regional metamorphism during 1.8–1.7 Ga and 1.6–1.4 Ga over the large area of the massif (Park et al. 2000; Lee et al. 2002; Kim and Cho 2003; Sagong et al. 2003; Lee et al. 2005).

At the beginning of the Paleoproterozoic, the Korean Peninsula territory was probably uplifted to the land surface except the northeastern part. It is uncertain whether this part subsided to form a geosyncline or the geosyncline existed as an intercratonic in the Nangnim Massif. The general trend of the geosyncline is N–NW in the northern extension and SW–NE in the western extension. The former is called the Macheolnyeong Geosyncline (or Hyesan–Iwon Geosyncline in North Korea) and the latter, Amnok Geosyncline. Thick terrestrial and marine sediments were deposited in eugeosynclinal parts of the Macheolnyeong Geosyncline. These sedimentary sequences are known as the Macheolnyeong System which is subdivided into the Songjin, Bukdaecheon and Namdaecheon Series in ascending order. The Macheolnyeong System was then folded and intruded by mafic rocks in the early stage and by granite in the later stage. These igneous rocks are known as the Iwon Group dated as 1370 Ma.

In general, three episodes of metamorphism have been recognized in the Yeongnam Massif: (1) upper amphibolite to granulite facies, (2) amphibolite facies, and (3) epidote–amphibolite to greenschist facies (Lee et al. 1981). The P–T evolutionary paths of the massif, however, are still controversial: (1) clockwise (Oh et al. 2000; Kwon et al. 2003b) or (2) anticlockwise paths (Kim et al. 2002). Furthermore, the timing of each metamorphic event has not been clearly constrained yet.

Phanerozoic cover rock sequences of marine and terrestrial sediments unconformably overlaid the Precambrian basement rocks. The basement rocks and cover rock sequences of the Yeongnam Massif were deformed and metamorphosed during the accretion of the Gyeonggi Massif, whereas those of the Nangnim Massif were relatively deformed. Syn-orogenic terrigenous sedimentary basins were developed along the NNE to NE striking thrusts in the Middle Jurassic (Han et al. 2006; Jeon et al.

2007). Synand post-orogenic Mesozoic granitoids intruded the basement gneisses and cover rock sequences. These magmatic activities accompanied significant juvenile crustal growth related to the subduction of the Pacific Ocean under Eastern Asia. Compressive force prevailed under active continental margin environment with intermittent tensional force in within-plate (Park et al. 2005). Northward subduction of the Izanagi oceanic plate comprised sinistral, brittle shearing in a retro-arc environment. Nonmarine sedimentary sequences and volcanic rocks are mainly present in the southeastern part of the Yeongnam Massif with subordinate exposures along the NE–SW trending transtensional basins (Chough et al. 2000 and others).

3.3 Yangtze Platform

Tectonic framework

The extension of the Yangtze Platform to the Korean Peninsula is still debatable. A prior correlation was the ***Imjingang Fold Belt*** at the middle of the Korean Peninsula to the Dabie-Sulu Belt in China. But non coeval data dissent from the idea. Towards the correlation of the ***Gyeonggi Massif***, central part of the Korean Peninsula bounded with the Imjingang Fold Belt and the ***Okcheon Fold Belt***, it belongs to the Sino-Korean Platform, because the crustal rocks of the Korean Peninsula were consolidated during the Paleoto Mesoproterozoic after initiation of crustal growth during the Paleoarchean.

Nevertheless, many of tectonic models favor the correlation of the Gyeonggi Massif to the South China (or Yangtze) Craton (Yin and Nie 1993; Li 1994; Ree et al. 1996; Lee et al. 1998, 2000; Chough et al. 2000; Uno and Chang 2000). Recent data reveal that the Korean Peninsula is a composite landmass originated from both the North and South China blocks emerged by contrasting fossil fauna occurrences from the Cambrian-Ordovician Taebaeksan and Yeongweol Groups, each of which belongs to the North and South China faunal provinces, respectively (Choi et al. 2003). Recently new evidence has come from the new findings of eclogite remnants in the southwestern part of the Gyeonggi Massif (Oh et al. 2005). The timing of eclogite-facies metamorphism was dated to be ca. 230–220 Ma (Kim et al. 2006), which is clearly coeval with that of the continental collision event between the North and South China cratons. The crustal affinity of the Gyeonggi Massif with the South China (Yangtze) Craton is also supplemented by a notable evidence of Neoproterozoic riftor extension-related magmatic events in the massif: alkaline magmatism [SHRIMP U–Pb zircon age of 742 $\pm$ 13 Ma (Lee et al. 2003a)] in the north-western Gyeonggi Massif; and withinplate mafic magmatism [Sm–Nd whole-rock age of ca. 850 Ma (Lee and Cho 1995)] in the central Gyeonggi Massif. These events are coeval with U–Pb zircon age of 756 $\pm$ 1 Ma, reported from metatrachyte in the Okcheon Fold Belt (Lee et al. 1998), and with SHRIMP U–Pb zircon age of 861 $\pm$ 7 Ma, reported from metabasalt in the Imjingang Fold Belt (Cho et al. 2001). Neoproterozoic intracontinental riftrelated magmatism is well known in the South

China (Yangtze) Block (Ames et al. 1996; Xue et al. 1997; Rowley et al. 1997; Li 1999; Hacker et al. 2000). The similarities in age thus suggest that the Gyeonggi Massif may correlate be a part of the South China (Yangtze) Craton.

The ***Imjingang Fold Belt***, east-trending fold-andthrust belt, is situated in the northern margin of the Gyeonggi Massif. This belt is characterized by the occurrence of the Devonian–Carboniferous sedimentary sequence which is underlain unconformably by the Proterozoic basement rocks. Peak metamorphic P–T conditions are 8.5–11.5 kbar, 660–780 °C, respectively. These P–T estimates and isothermal decompression reported from the metapelitic unit suggest a clockwise P–T path of the Imjingang Belt evolved from eclogite-facies conditions (Cho et al. 2007).

Main Fracture System

The Gyeonggi Massif experienced at least three old events in which the first generation of folds (N–S trending isoclinal folds) was genetically associated with N–S directed ductile shearing (Kim et al. 2000). The present regional structure is dominated by N–S trending open folds of the third generation. These contractional structures are cut by an E–W striking ductile shear zone [Gyeonggi Shear Zone (Kim et al. 2000)] with a top-down-to-the-north (normal) sense of shear. The Gyeonggi Shear Zone, at least 4 km wide, occurs along the boundary between the Gyeonggi Massif and the Imjingang Belt to the north (Ree et al. 1996; Kim et al. 2000). Based on field observations and microstructural analysis, investigations (Kim et al. 2000) suggest that this crustal-scale shear zone evolved from deep crustal ductile regime to shallow crustal brittle regime. Since the Jurassic, Daedong Group (Chun et al. 1988) unconformably overlies the shear zone, and syntectonic muscovites in cyclonites are dated as 226 $\pm$ 1.2 Ma (Rb–Sr age). This age is interpreted as the timing of the extensional ductile shearing in the Late Triassic (Kim et al. 2000). On the other hand, east-trending tight folds verging to the south occur both in the Gyeonggi Shear Zone and Gyeonggi Massif. The compressional structures observed in the Gyeonggi Shear Zone are overprinted by an extensional ductile shearing, which suggests that extensional deformation followed the compressional deformation. Although it is not clear whether the compressive deformation producing E–W trending folds in the northwestern Gyeonggi Massif is correlated with E–W trending folding and thrusting in the Imjingang Belt, Kim with co-authors (2000) suggest that at least some portion of the Precambrian Gyeonggi Massif was involved in the Late Paleozoic to Early Mesozoic collisional process between the North and South China blocks.

The southeastern boundary between the Gyeonggi Massif and the Okcheon Belt is also a ductile shear zone with a top-down-to-the-SSE normal sense of shear (Ree et al. 1995; Sagong and Kwon 1998). Lee S. R. and Cho M. suggest that the Precambrian supracrustal rocks in the northeastern Gyeonggi Massif experienced an upper amphibolite-facies peak metamorphism of medium-pressure type, followed by an isothermal decompression (Lee and Cho 1995). This P–T evolution, together with the extensional ductile shear zones along the northern and southeastern margins of the Gyeonggi Massif, suggests rapid uplift of the massif, probably in the late stage of or posterior to the Permo–Triassic collision between the North and South China blocks. More sophisticated data are required to unravel tectonic evolution of the Gyeonggi Massif, particularly during the Precambrian.

Tectonic evolution

The Gyeonggi Massif experienced polyphase deformations and metamorphisms since its formation. Recently, the SHRIMP U–Pb magmatic zircon age of Neoarchean was determined from tonalite in the western Gyeonggi Massif (Cho et al. 2004a). Thus, it is believed that some granitic intrusion took place at the end of the Archean. The Gyeonggi Massif suffered from high temperature metamorphism reaching granulite facies and related magmatism in the Paleoproterozoic (ca. 1850 Ma) (Sagong et al. 2003; Cho et al. 2004a).

During the Neoproterozoic (ca. 850–742 Ma), bimodal magmatism was dominant within the plate in the Gyeonggi Massif and Okcheon Metamorphic Belt (Lee et al. 1998; Lee et al. 2003a, b; Cho et al. 2004b). The alkaline granitoids intruding along the boundary between northern margin of the Gyeonggi Massif and Imjingang Belt originated from the magma derived from ancient continental crust with addition of juvenile mantle-derived basaltic magma (Lee et al. 2003a, b). The granitoids experienced ductile, extensional deformation, following the Permo–Triassic regional metamorphism (Ree et al. 1996; Kim et al. 2000). The Okcheon Metamorphic Belt experienced regional peak metamorphism during the Late Carboniferous to Early Permian (ca. 285 Ma) (Kim et al. 2007). The Gyeonggi Massif, and Imjingang and Okcheon Metamorphic belts were juxtaposed with the Yeongnam and Nangnim Massif during the Permo–Triassic collisional orogeny (Ree et al. 1996; Chough et al. 2000; Cho et al. 2007).

The final phase of deformation occurred within the massif during the Daebo Orogeny in the Jurassic. The deformation and metamorphism of the Early Jurassic cover rocks which occur in small patches on the massif, is locally strong that the Jurassic cover rocks are folded and thrusted over the older schists and gneisses, but elsewhere they discordantly lie on a folded and metamorphosed basement. The Jurassic Chungnam System develops in the depression zone in the southwestern part of the Gyeonggi massif which is known as the Chungnam Depression Zone, and a few isolated other localities in the western and northwestern parts of Seoul.

The Gyeonggi Massif experienced the Daebo Orogeny and the intrusion of related granitoids during the Jurassic to Early Cretaceous. The orogenic event comprised the depressions in the massif where the Cretaceous Gyeongsang Supergroup was deposited. The Gongju Depression Zone is characterized by volcanic lavas and volcanic debris resting upon the terrestrial deposits.

3.4 *Sikhote-Alin Area of the Pacific Mobile Belt*

Tectonic framework

Southeastern extreme of the Korean Peninsula is a possible extension of the Sikhote-Alin area as a structure with consolidated crust, formed by the beginning of the Cretaceous. At the beginning of Cretaceous of Korea where extensive crustal upheaval and enormous nonmarine sedimentary basins took place. The sequence is composed of fluvio-lacustrine sediments, granitoids and volcanic rocks of the early Hauterivian to the latest Mesozoic. The filling materials of the ***Gyeongsang Basin***, the largest Cretaceous basin in Korea, were derived from adjacent Okcheon Orogenic Belt and uplifted Paleozoic to Early Mesozoic accretional prisms located to the east (Chang 1988). Volcanism and intrusive magmatism in the Gyeongsang Basin began in the mid-Cretaceous.

The ***Duman River Basin*** is located in the northeastern part of the Korean Peninsula extending from the southern side of the Duman River to the east of the ***Gilju–Myeongcheon Rift Basin***. The Neogene sediments are distributed along the lower reaches of the Duman River and in the Gilju–Myeongcheon Rift Basin. Some Tertiary to Quaternary alkali volcanic rocks intruded into and extruded over the sediments. The ***Pohang Basin*** (Yeonil) is situated in a small area in the southeastern maritime region of the east coastal zone of the Gyeongsang Basin. The basin includes the Jeju Island off the south coast and the Ulleung and Dok Islands off the east coast of the Korean Peninsula.

Main Fracture System

The NE–SW trending large fracture zone is developed in the central part of the Gilju–Myeongcheon Basin. The oligoclase sediments and some of the basements are exposed in the western side of the fracture zone, whereas the Pliocene volcanic and intrusive rocks are exposed in the eastern side of the fracture zone.

The north–northeasterly trending fault zones are developed in the middle part (Chugaryeong Fault Zone) and in the western part (Goksan Fault Zone) of the Korean Peninsula. Cenozoic basalt flooded out along these faults. It was suggested that these faults might be primarily formed during the Precambrian, and reactivated during the Late Mesozoic to Cenozoic.

The Pohang Basin is bounded on the west by the Yangsan fault which experienced about 35 km of post-Eocene dextral strike-slip movement (Chang et al. 1990; Chae and Chang 1994; Yoon and Chough 1995). During the Early–Middle Miocene, however, the strike-slip movement was negligible, whereas the western margin of the Pohang Basin was under extensional regime.

Tectonic evolution

The Gyeongsang Basin and scattered small basins were extinguished at the end of the Cretaceous, and the whole of the Korean Peninsula had been uplifted as a land except small Tertiary basins and grabens developed along the eastern coast. The Paleogene sediments were deposited in the Gilju–Myeongcheon Rift Basin and Duman River Basin in the northeastern part of the Korean Peninsula.

There exist unconformities between the Paleogene and Neogene strata, and between the Neogene and Pleistocene strata in the Gilju–Myeongcheon Rift Basin. The Neogene sediments accumulated in the Yeonil Basin where an unconformity exists between the Lower Miocene Janggi Group of terrestrial sediments and the Upper Miocene Yeonil Group of littoral sediments. The unconformity was caused by minor movement, so-called Yeonil Disturbance. The disturbance was not intense and just comprised minor warping of the sediments. Mafic to intermediate volcanic activities took place in the Gilju-Myeongcheon and Chugaryeong Rift Zones. Alkali volcanic eruption took place in these north–northeast trending rift zones, the Baekdu Mountain, Gaema Plateau, and Jeju, Ulleung and Dok Islands during the Plio–Pleistocene.

The age of the volcanic rocks in the Baekdu Mountain and Gaema Plateau area is Plio–Pleistocene, whereas that of Jeju, Ulleung and Dok Islands is slightly younger.

References

Ames L, Zhou G, Xiong B (1996) Geochronology and geochemistry of ultrahigh-pressure metamorphism with implications for collision of the Sino-Korean and Yangtze cratons, Central China. Tectonics 15:472–489

Chae BG, Chang TW (1994) Movement history of the Yangsan Fault and its related fractures in the Chongha–Yongdok Area, Korea. J Geol Soc Korea 30:379–394 (in Korean with English abstract)

Chang KH (1988) Cretaceous Strata. In: Lee DS (ed) Geology of Korea. Kyohak-Sa, pp 175–194 (in Korean)

Chang KH, Woo BG, Lee JH, Park SO, Yao A (1990) Cretaceous and early Cenozoic stratigraphy and history of the eastern Kyongsang Basin, S. Korea. J Geol Soc Korea 26:471–487

Chekhovich VD (1993) Tectonics and geodynamics of folded framework of minor Oceanic basins. Nauka, Moscow, 272 p. (in Russian)

Cheng Y et al (eds) (1994) Introduction to geology of China. Geological Publishing House, Beijing, 480 p

Cho DL, Kwon ST, Jeon EY, Armstrong R (2001) SHRIMP U-Pb zircon geochronology of an amphibolite and a paragneiss from the Samgot Unit, Yeoncheon Complex in the Imjingang Belt, Korea: Tectonic implication. Geol Soc Korea, Abstract volume, 89

Cho M, Kim H, Kim J, Wan Y, Liu D (2004a) Geochronologic correlation between the Gyeonggi Massif, Korea, and the South China Craton: evidence from the SHRIMP UPb Zircon Ages. The 1st AOGS Meeting Abstracts, Singapore, p 78

Cho M, Kim T, Kim H (2004b) SHRIMP U–Pb zircon age of a felsic metatuff in the Ogcheon metamorphic belt, Korea: Neoproterozoic (ca. 750 Ma) volcanism. J Petrol Soc Korea 13:119–125

Cho M, Kim Y, Ahn J (2007) Metamorphic evolution of the Imjingang Belt, Korea: implications for Permo-Triassic collisional Orogeny. Int Geol Rev 49:30–51

Choi DK, Kim DH, Sohn JW, Lee SB (2003) Trilobite faunal successions across the Cambrian—Ordovician boundary intervals in Korea and their correlation with China and Australia. J Asian Earth Sci 21:781–793

Chough SK, Kwon ST, Ree JH, Choi DK (2000) Tectonic and sedimentary evolution of the Korean Peninsula: a review and new view. Earth Sci Rev 52:175–232

Chun HY, Bong PY, Lee HY, Choi SJ, Kim CB, Kwon YI (1988) Paleontology and Stratigraphy of the Kyunggi Coalfield. Research Report KR-88-1A, Korea Institute of Energy and Resources, Seoul, 41 p. (in Korean with English abstract)

Gan X, Zhao F, Jin W, Sun D (1996) The U-Pb ages of early Proterozoic-Archean zircons captured by igneous rocks in southern China. Geochimica 25(2):112–120

Ganelin AV (2011) Geochemistry and geodynamic significance of the Dike series of the Aluchin Ophiolite Complex, Verkhoyansk–Chukotka Fold Zone, Northeast Russia. Geochem Int 7:654–675 (in Russian)

Gao S, Zhang B (1990) The discovery of Archean TTG gneisses in the northern Yangtze Platform and their implications. Earth Sci J China Univ Geosci 25(6):675–679

Grantz A, Hart PE, Childers VA (2011) Geology and tectonic development of the Amerasian and Canadian basins, Arctic Ocean. In: Spencer AM, Embry AF, Gautier DL, Stoupakova AV, Sørensen K (eds) Arctic petroleum geology. Geological Society, London, Memoirs, vol 35, pp 771–800

Hacker BR, Ratschbacher L, Webb L, McWilliams MO, Ireland T, Calvert A, Dong S, Wenk HR, Chateigner D (2000) Exhumation of ultra high pressure continental crust in east central China: late Triassic-Early Jurassic tectonic unroofing. J Geophys Res 105:13339–13364

Han R, Ree JH, Cho DL, Kwon ST, Armstrong R (2006) SHRIMP U–Pb zircon ages of pyroclastic rocks in the Bansong Group, Taebaeksan Basin, South Korea and their implication for the Mesozoic tectonics. Gondwana Res 9:106–117

Jeon H, Cho M, Kim H, Horie K, Hidaka H (2007) Early Archean to Middle Jurassic evolution of the Korean Peninsula and its correlation with Chinese Cratons: SHRIMP U–Pb Zircon age constraints. J Geol 115:525–539

Khanchuk AI, Golozubov VV, Panchenko IV, Ignatiev AV, Chudaev OV (1992) Ganychalan terrane of the Koryak Upland. Pac Geol 4:84–93

Kim J, Cho M (2003) Low-pressure metamorphism and leucogranite magmatism, northeastern Yeongnam Massif, Korea: implication for Paleoproterozoic crustal evolution. Precambrian Res 122:235–251

Kim JN, Ree JH, Kwon ST, Park Y, Choi SJ, Cheong CS (2000) The Kyonggi shear zone of the central Korean Peninsula: Late orogenic imprint of the North and South China collision. J Geol 108:469–478

Kim DY, Song YS, Park KH (2002) Petrogenesis and metamorphism of charnockite of eastern Jirisan area. J Petrol Soc Korea 11:138–156 (in Korean with English abstract)

Kim SW, Oh CW, Williams IS, Rubatto D, Ryu I-C, Rajesh VJ, Kim C-B, Guo J, Zhai M (2006) Phanerozoic high-pressure eclogites and intermediate-pressure granulite facies metamorphism in the Gyeonggi massif, South Korea: Implications for the eastward extension of the Dabie-Sulu continental collision zone. Lithos 92:357–377

Kim H, Cho M, Cheong CS (2007) The effect of allanite inclusions on U–Pb step-leaching ages and Sm–Nd isotope systematics of garnet from the Ogcheon metamorphic belt, South Korea. Chemical Geol 236:27–41

Kobayashi T (1953) Geology of South Korea with special reference to the limestone plateau of Kogendo. J Faculty Sci Univ Tokyo, Sec. II 8:145–293

Kovalenko DV (2003) Geological interpretation of the paleomagnetic data from Cretaceous—Paleogene rocks of the southern Koryak Highland, Doctoral Thesis, Institute of Lithosphere, Ac Sci Rus, 25 p. (in Russian)

Kwon YK, Lee DJ, Choi DK, Chough SK (2003a) Lower Ordovician sponge bioherms in the Makkol Formation, Taebaeksan Basin, Mideast Korea. Facies 48:79–90

Kwon YW, Oh CW, Kim HS (2003b) Granulitefaceis metamorphism in the Punggi area, northeastern Yeongnam Massif, Korea and its tectonic implications for East Asia. Precambrian Res 122:253–273

Kwon YK, Chough SK, Choi DK, Lee DJ (2006) Sequence stratigraphy of the Taebaek Group (Cambrian–Ordovician), Mideast Korea. Sed Geol 192:19–55

Ledneva GV, Pease VL, Sokolov SD (2011) PermoTriassic hypabyssal mafic intrusions and associated tholeiitic basalts of the Kolyuchinskaya Bay, Chukotka (NE Russia): Links to the Siberian LIP. J Asian Earth Sci 40:737–745
Lee SR, Cho M (1995) Tectonometamorphic evolution of the Chuncheon amphibolite, central Gyeonggi Massif, South Korea. J Metamorph Geol 13:315–328
Lee SM, Na KC, Lee SH, Park BY, Lee SW (1981) Regional metamorphism of the Metamorphic Rock Complex in the Southeastern Region of the Sobaegsan Massif. J Geol Soc Korea 17:69–188 (in Korean with English abstract)
Lee KS, Chang HW, Park KH (1998) Neoproterozoic bimodal volcanism in the central Ogcheon Belt, Korea: age and tectonic implication. Precambrian Res 89:47–57
Lee SR, Cho M, Yi KW, Stern R (2000) Early Proterozoic granulite in central Korea: tectonic correlation with Chinese cratons. J Geol 108:729–738
Lee HS, Song YS, Park KH (2002) Zircon chemical age of the Precambrian gneisses from Gimcheon area in the central Yeongnam massif, Korea. J Petrol Soc Korea 11:157–168 (in Korean with English abstract)
Lee SR, Cho M, Cheong C-S, Kim H, Wingate MTD (2003a) Age, geochemistry and tectonic significance of Neoproterozoic alkaline granitoids in the northwestern margin of the Gyeonggi Massif, South Korea. Precambrian Res 122:297–310
Lee SR, Cho M, Hwang JH, Lee B, Kim Y, Kim JC (2003b) Crustal evolution of the Gyeonggi Massif, South Korea: Nd isotopic evidence and implications for continental growths of East Asia. Precambrian Res 121:25–34
Lee SG, Kim Y, Kim KH (2005) Geochemistry and Sm–Nd isotope systematics of Precambrian granitic gneiss and amphibolite core at the Muju area, middle Yeongnam Massif. J Petrol Soc Korea 14:127–140 (in Korean with English abstract)
Li ZX (1994) Collision between North and South China blocks: a crustal-detachment model for suturing in the region east of the Tanlu Fault. Geology 22:739–742
Li XH (1999) U-Pb zircon ages of granites from the southern margin of the Yangtze Block: timing of Neoproterozoic Jinning Orogeny in SE China and implications for Rodinia Assembly. Precambrian Res 97:43–57
Li J (2001) Collision time and style of the Sino-Korean and Yangtze blocks—the evolution of the sedimentary setting in the Sinian to Jurassic, middle and lower reaches of the Yangtze River. Acta Geol Sin 75(1):25–34
Li J (2004) Structure characteristics of crustal "mosaic structure and superimposition" of the continental crust of China and its evolution. Geol Bull China 23(9):986–1004
Li C, Zhai G, Yin F, Mao X (2009) An important window for understanding the Qinghai-Tibetan Plateau—a review on research progress in recent years of Qiangtang area, Tibet, China. Geol Bull China 28(9):1169–1177
Li T, Chen B, Dai W (2010) Geological map series of the Qinghai-Tibet Plateau. Guangdong Science and Technology Press
Oh CW, Jeon EY, Park BY, Ahn KS, Lee JH (2000) Metamorphic evolution of granitic and porphyroblastic gneisses in the Seungju–Suncheon area, the southwestern part of the Sobaeksan Massif. J Petrol Soc Korea 9:121–141 (in Korean with English abstract)
Oh CW, Kim SW, Choi SG, Zhai M, Guo J, Sajeev K (2005) First finding of eclogites facies metamorphic event in South Korea and its correlation with the Dabie-Sulu collision belt in China. J Geol 113:226–32
Paek RJ (1996) Crust development of the Korean Peninsula. In: Paek RJ, Kang HG, Jon GP (eds) Geology of Korea. Institute of Geology, State Academy of Sciences DPR of Korea, pp 508–511
Parfenov LM, Natapov LM, Sokolov SD, Tsukanov NV (1993) Terranes analysis and accretion in northeast Asia. The Islands Arc 2:35–54
Park KH, Song YS, Park ME, Lee SG, Ryu HJ (2000) Petrological, geochemical and geochronological studies of Precambrian basement in Northeast Asia region: 1. Age of the metamorphism of Jirisan area. J Petrol Soc Korea 9:29–39 (in Korean with English abstract)

Park KH, Lee HS, Cheong CS (2005) Sphene U–Pb ages of the granodiorites from Gimcheon, Seogju and Anui areas of the middle Yeongnam Massif. J Petrol Soc Korea 14:1–11 (in Korean with English abstract)
Ree JH, Kwon ST, Cho M (1995) Extensional ductile shear zones at northern and southeastern margins of the Gyeonggi Massif, Korea: a rapid uplift (?). Chin Sci Bull Suppl 40:134
Ree JH, Cho M, Kwon ST, Nakamura E (1996) Possible eastward extension of Chinese collision belt in South Korea: the Imjingang Belt. Geology 240:1071–1074
Ren J, Jiang C, Zhang Z, Qin D (1980) Geotectonic evolution of China—Brief notes to the Tectonic map of China. Science Publishing House, Beijing, 124 p
Ren J et al (1999) The tectonics of China from a global view—a guide to the tectonic map of China and adjacent regions. Geological Publishing House, Beijing, 46 p
Rowley DB, Xue F, Tucker RD, Peng ZX, Baker J, Davis A (1997) Ages of ultrahigh-pressure metamorphism and protolith orthogneisses from the eastern Dabie Shan: U/Pb zircon geochronology. Earth Planet Sci Lett 151:191–203
Sagong H, Kwon ST (1998) Pb–Pb age and uplift history of the Busan gneiss complex in the Okchon Belt, Korea: a comparison with the Bagdalryeong gneiss complex in the Kyongki Massif. Geosciences J 2:99–106
Sagong H, Cheong CS, Kwon ST (2003) Paleoproterozoic orogeny in South Korea: evidence from Sm-Nd and Pb step-leaching garnet ages of Precambrian basement rocks. Precambrian Res 122:275–295
Sokolov SD (1992) Accretionary tectonics of the Koryak—Chukotka segment of the Pacific belt. Nauka, Moscow (Trans. of Geological Institute, Russian Academy of Sciences, 479), 182 p. (in Russian)
Sokolov SD, Grigoriev VN, Peyve AA, Batanova VG, Krylov KA, Luchitskaya MV, Aleksutin MV (1996) Structural and compositional patterns in serpentinite mélanges. Geotectonics 1:3–16 (in Russian)
Sokolov SD, Bondarenko G Ye, Layer PW, Kravchenko-Berezhnoy IR (2009a) South Anyui suture: tectono-stratigraphy, deformations, and principal tectonic events. Stephan Mueller Spec Publ Ser 4:201–221
Sokolov SD, Bondarenko G Ye, Khudoley AK, Morozov OL, Luchitskaya MV, Tuchkova MI, Layer PW (2009b) Tectonic reconstruction of the Uda-Murgal arc and the Late Jurassic and Early Cretaceous convergent margin of Northeast Asia—Northwest Pacific. Stephan Mueller Spec Publ Ser 4:273–288
Solov'ev AV (2008) Study of tectonic processes in regions of convergence of lithospheric plates. Methods of fission track dating and structural analysis. Nauka, Moscow, 315 p. (in Russian).
Tuchkova MI (2011) Lithology of terrigenous rocks from Mesozoic fold belt of the continental margin (Greater Caucasus, North-East Asia). LAP, Moscow (Trans. of Geological Institute, Russian Academy of Sciences, 600), 334 p. (in Russian)
Uno K, Chang KH (2000) Paleomagnetic results from the lower Mesozoic Daedong Supergroup in the Gyeonggi Block, Korean peninsula: an eastern extension of the South China Block. Earth Planet Sci Lett 182:49–59
Wu FY, Yang JH, Wilde SA, Liu XM, Guo JH, Zhai MG (2007) Detrital zircon U-Pb and Hf isotopic constraints on the crustal evolution of North Korea. Precambrian Res 159:155–177
Xing F, Xu X, Li Z (1993) The discovery and significance of the Early Proterozoic basement from the lower and middle reaches of the Changjiang River. Chin Sci Bull 38(20):1883–1886 (in Chinese)
Xue F, Rowley DB, Tucker RD, Peng ZX (1997) U-Pb zircon ages of granitoid rocks in the north Dabie complex, eastern Dabie Shan, China. J Geol 105:744–753
Yin A, Nie S (1993) An indentation model for the North and South China collision and the development of the Tan-Lu and Honam fault systems, Eastern Asia. Tectonics 12:801–813
Yoon SH, Chough SK (1995) Regional strike-slip in the eastern continental margin of Korea and its tectonic implications for the evolution of Ulleung Basin, East Sea (Sea of Japan). Geol Soc America Bull 107:83–97

Ziye Y (1989) The determination of the Early Permian glaciomarine deposit in Hainan Island and its tectonic significance. J Nanjing Univ (Natural Science edition) 25(1):108–121

Zhao G, Cao L, Wilde SA, Sun M, Choe WJ, Li S (2006) Implications based on the first SHRIMP U–Pb zircon dating on Precambrian granitoid rocks in North Korea. Earth and Planetary Sci Lett 251:365–379

Zuyi Z (1992) Geologic-geophysical features of China Seas and adjacent regions. In: Liu G (ed). Science Press, Beijing, 424 p

Conclusion

The monograph presents results of tectonic, geological and geophysical studies of Northern, Central and Eastern Asia obtained over last 20 years by specialists from geological surveys and national academies of sciences of Russia, China, Mongolia, Kazakhstan and Korea. All the collected materials provided a basis for the *Tectonic Map of Northern-Central-Eastern Asia and Adjacent Areas at 1:2,500,000 scale* compiled under the aegis of the Commission for the Geological Map of the World in 2012. This map demonstrates the long way the tectonic mapping has done since the compilation of the very first tectonic map for this territory in 1922. Since then, mapping concepts have changed dramatically as well as basic paradigm of geotectonics reflected in latest tectonic maps. CGMW has included the Tectonic Map into its catalogue and today it is available on its official website.

The map confirms the principal concept of tectonic mapping based on the time of consolidated crust formation (oceanic, transitional, continental), which was first used in the compilation of the *International Tectonic Map of Europe, scale 1:5,000,000* (1996). However, in the Tectonic Map of Northern-Central-Eastern Asia and Adjacent Areas at 1:2,500,000 scale and in the *Tectonic Map of Central Asia and Adjacent Areas, scale 1:2,500,000* (2008), special attention was paid not only to the solid Earth formation, but also to its destruction processes. Destruction of the consolidated crust was shown in the form of continental rifting, intraisland-arc rifting and spreading while thermal-magmatic reprocessing of continental crust was shown by large igneous provinces (LIPs). This fact was considered while compiling the *Metallogenic Map of Northern, Central and Eastern Asia*, which was based on the Tectonic Map for the same region, since a certain group of ore minerals is associated with destruction processes.

Both the monograph and the Tectonic map of Northern, Central and Eastern Asia confirm the modern geotectonic paradigm of the existence of two main types of crust (oceanic and continental) and transformation during the evolution of oceanic crust into continental crust. Both the map and the monograph reflect the priority of horizontal movements in the formation of fold and fold-thrust belts, zones, regions, and systems. This evidences the multilayer character and tectonic layering of the lithosphere confirmed by geophysical data that are becoming a necessary attribute

O. V. Petrov and S. Dong (eds.), *Tectonics of Asia (Northern, Central and Eastern Asia)*, Springer Geology, https://doi.org/10.1007/978-3-030-62001-1

of tectonic maps and explanatory notes. Finally, the published map and monograph reflect the replacement of accretion (Pacific) style of crustal evolution by collision (Indo-Atlantic) style. The Central Asian mobile belt is shown as a typical ancient (Neoproterozoic–Middle Paleozoic) accretion fold-thrust structure, which later got overlain by younger (Late Paleozoic–Mesozoic) collision belts (Beishan, Solonker, Mongol-Okhotsk and others).

It should be noted that the compilation of the map and monograph required generalizing a huge amount of factual material on stratigraphy, magmatism, metamorphism, structural geology, absolute age and deep geophysics. Moreover, the implementation of such a large and challenging international project called for the correlation of a significant amount of geological data especially in cross-border areas. One of important achievements of the project is agreement of geologists from all participating countries on the main problems of theoretical geotectonics and tectonic interpretation of geological data. Many issues were worked out in course of numerous international working meetings and field trips.

The monograph and the Tectonic Map of Northern-Central-Eastern Asia would not be possible without utilization of materials on potential fields and deep structure of the region. Analysis of geophysical data made it possible to clarify boundaries of main tectonic structures, get a better understanding of sedimentary basins' structure, and identify the Earth's crustal types within the mapped tectonic structures and geodynamic settings of their formation, especially in transitional zones continent-ocean. The tectonic map is supplemented by the magnetic and gravimetric anomaly maps, the maps of the Earth's crust thickness and the sedimentary cover thickness, the map of crustal types and the geotransect crossing main structures of Northern, Central and Eastern Asia.

The *Tectonic Map of Northern-Central-Eastern Asia and Adjacent Areas* was displayed at the 35th session of the International Geological Congress (South Africa, Cape Town 2016) and attracted attention of the international geological community.

The *Tectonic Map of Northern, Central and Eastern Asia and Adjacent Areas* is the very first map of the 1:2,500,000 scale compiled for such a vast area; it was compiled by a big group of specialists from geological surveys and national academies of sciences from different countries. This comprehensive scientific work demonstrates how well specialists can work together enriching our knowledge of the geological structure and tectonic development of this vast region.

The manufacturer's authorised representative in the EU is Springer Nature Customer Service Centre GmbH, Europaplatz 3, 69115 Heidelberg, Germany. If you have any concerns regarding our products, please contact ProductSafety@springernature.com

Printed and bound by CPI Group (UK) Ltd, Croydon, CR0 4YY
30/09/2026
02225699-0004